Communications in Computer and Information Science 910

Commenced Publication in 2007
Founding and Former Series Editors:
Phoebe Chen, Alfredo Cuzzocrea, Xiaoyong Du, Orhun Kara, Ting Liu,
Dominik Ślęzak, and Xiaokang Yang

More information about this series at http://www.springer.com/series/7899

Leonid Sokolinsky · Mikhail Zymbler (Eds.)

Parallel Computational Technologies

12th International Conference, PCT 2018
Rostov-on-Don, Russia, April 2–6, 2018
Revised Selected Papers

 Springer

Editors
Leonid Sokolinsky
South Ural State University
Chelyabinsk
Russia

Mikhail Zymbler
South Ural State University
Chelyabinsk
Russia

ISSN 1865-0929 ISSN 1865-0937 (electronic)
Communications in Computer and Information Science
ISBN 978-3-319-99672-1 ISBN 978-3-319-99673-8 (eBook)
https://doi.org/10.1007/978-3-319-99673-8

Library of Congress Control Number: 2018952058

This Springer imprint is published by the registered company Springer Nature Switzerland AG
The registered company address is: Gewerbestrasse 11, 6330 Cham, Switzerland

Preface

This volume contains a selection of the papers presented at the 12th International Scientific Conference on Parallel Computational Technologies, PCT 2018, held during April 2–6, 2018, in Rostov-on-Don, Russia.

The PCT series of conferences aims at providing an opportunity to discuss the future of parallel computing, as well as to report the results achieved by leading research groups in solving both scientific and practical issues using supercomputer technologies. The scope of the PCT series of conferences includes all aspects of high performance computing in science and technology such as applications, hardware and software, specialized languages, and packages.

The PCT series is organized by the Supercomputing Consortium of Russian Universities and the Federal Agency for Scientific Organizations. Originated in 2007 at the South Ural State University (Chelyabinsk, Russia), the PCT series of conferences has now become one of the most prestigious Russian scientific meetings on parallel programming and high-performance computing. PCT 2018 in Rostov-on-Don continued the series after Chelyabinsk (2007), St. Petersburg (2008), Nizhny Novgorod (2009), Ufa (2010), Moscow (2011), Novosibirsk (2012), Chelyabinsk (2013), Rostov-on-Don (2014), Ekaterinburg (2015), Arkhangelsk (2016), and Kazan (2017).

All papers submitted to the conference were scrupulously evaluated by three reviewers on the relevance to the conference topics, scientific and practical contribution, experimental evaluation of the results, and presentation quality. PCT's Program Committee selected the 24 best papers to be included in this CCIS proceedings volume.

We would like to thank the Russian Foundation for Basic Research for their continued financial support of the PCT series of conferences, as well as the respected PCT 2018 sponsors, namely platinum sponsors, RSC Group and Intel, gold sponsors, NVIDIA and Hewlett Packard Enterprise, and silver sponsor AMD.

We would like to express our gratitude to every individual who contributed to the success of PCT 2018. Special thanks go to the Program Committee members and the external reviewers for evaluating papers submitted to the conference. Thanks also go to the Organizing Committee members and all the colleagues involved in the conference organization from Don State Technical University, the South Ural State University, and Moscow State University. We thank the participants of PCT 2018 for sharing their research and presenting their achievements as well.

Finally, we thank Springer for publishing the proceedings of PCT 2018 in the Communications in Computer and Information Science series.

May 2018

Leonid Sokolinsky
Mikhail Zymbler

Organization

The 12th International Scientific Conference on Parallel Computational Technologies, PCT 2018, was organized by the Supercomputing Consortium of Russian Universities and the Federal Agency for Scientific Organizations, Russia.

Steering Committee

Berdyshev, V. I.	Krasovskii Institute of Mathematics and Mechanics, Yekaterinburg, Russia
Ershov, Yu. L.	United Scientific Council on Mathematics and Informatics, Novosibirsk, Russia
Minkin, V. I.	South Federal University, Rostov-on-Don, Russia
Moiseev, E. I.	Moscow State University, Russia
Savin, G. I.	Joint Supercomputer Center, Russian Academy of Sciences, Moscow, Russia
Sadovnichiy, V. A.	Moscow State University, Russia
Chetverushkin, B. N.	Keldysh Institute of Applied Mathematics, Russian Academy of Sciences, Moscow, Russia
Shokin, Yu. I.	Institute of Computational Technologies, Russian Academy of Sciences, Novosibirsk, Russia

Program Committee

Sadovnichiy, V. A. (Chair)	Moscow State University, Russia
Dongarra, J. (Co-chair)	University of Tennessee, USA
Sokolinsky, L. B. (Co-chair)	South Ural State University, Russia
Voevodin, Vl. V. (Co-chair)	Moscow State University, Russia
Zymbler, M. L. (Academic Secretary)	South Ural State University, Russia
Ablameyko, S. V.	Belarusian State University, Belarus
Afanasiev, A. P.	Institute for Systems Analysis, Russian Academy of Sciences, Russia
Akimova, E. N.	Krasovskii Institute of Mathematics and Mechanics, UrB, Russian Academy of Sciences, Russia
Andrzejak, A.	Heidelberg University, Germany
Balaji, P.	Argonne National Laboratory, USA
Boldyrev, Y. Ya.	Saint-Petersburg Polytechnic University, Russia
Carretero, J.	Carlos III University of Madrid, Spain
Gazizov, R. K.	Ufa State Aviation Technical University, Russia
Gergel, V. P.	Lobachevsky State University of Nizhny Novgorod, Russia

Glinsky, B. M.   Institute of Computational Mathematics and
                                 Mathematical Geophysics SB, Russian Academy
                                 of Sciences, Russia
Goryachev, V. D.   Tver State Technical University, Russia
Il'in, V. P.   Institute of Computational Mathematics and
                                 Mathematical Geophysics SB, Russian Academy
                                 of Sciences, Russia
Kobayashi, H.   Tohoku University, Japan
Kunkel, J.   University of Hamburg, Germany
Labarta, J.   Barcelona Supercomputing Center, Spain
Lastovetsky, A.   University College Dublin, Ireland
Ludwig, T.   German Climate Computing Center, Germany
Lykosov, V. N.   Institute of Numerical Mathematics, Russian Academy
                                 of Sciences, Russia
Mallmann, D.   Julich Supercomputing Centre, Germany
Michalewicz, M.   A*STAR Computational Resource Centre, Singapore
Malyshkin, V. E.   Institute of Computational Mathematics and
                                 Mathematical Geophysics SB, Russian Academy
                                 of Sciences, Russia
Modorsky, V. Ya.   Perm Polytechnic University, Russia
Shamakina, A. V.   High Performance Computing Center in Stuttgart,
                                 Germany
Shumyatsky, P.   University of Brasilia, Brazil
Sithole, H.   Centre for High Performance Computing, South Africa
Starchenko, A. V.   Tomsk State University, Russia
Sterling, T.   Indiana University, USA
Taufer, M.   University of Delaware, USA
Turlapov, V. E.   Lobachevsky State University of Nizhny Novgorod,
                                 Russia
Wyrzykowski, R.   Czestochowa University of Technology, Poland
Yakobovskiy, M. V.   Keldysh Institute of Applied Mathematics,
                                 Russian Academy of Sciences, Russia
Yamazaki, Y.   Federal University of Pelotas, Brazil

Organizing Committee

Sukhinov, A. I. (Chair)   Don State Technical University, Russia
Chistyakov, A. E.   Don State Technical University, Russia
  (Co-chair)
Sidoryakina, V. V.   Chekhov Taganrog Institute, Russia
  (Secretary)
Antonov, A. S.   Moscow State University, Russia
Antonova, A. P.   Moscow State University, Russia
Bardina, M. G.   South Ural State University, Russia
Bednaya, T. A.   Don State Technical University, Russia
Belova, Yu. V.   Don State Technical University, Russia

Contents

High Performance Architectures, Tools and Technologies

Hierarchical Domain Representation in the AlgoWiki Encyclopedia:
From Problems to Implementations 3
   Alexander Antonov, Alexey Frolov, Igor Konshin,
   and Vladimir Voevodin

A Toolkit for the Development of Data-Driven Functional
Parallel Programmes 16
   Alexander I. Legalov, Vladimir S. Vasilyev, Ivan V. Matkovskii,
   and Mariya S. Ushakova

Machine Learning Techniques for Detecting Supercomputer Applications
with Abnormal Behavior 31
   Alexander Bezrukov, Mikhail Kokarev, Denis Shaykhislamov,
   Vadim Voevodin, and Sergey Zhumatiy

Role-Dependent Resource Utilization Analysis for Large HPC Centers 47
   Dmitry Nikitenko, Pavel Shvets, Vadim Voevodin, and Sergey Zhumatiy

High-Performance Reconfigurable Computer Systems with
Immersion Cooling 62
   Ilya Levin, Alexey Dordopulo, Alexander Fedorov,
   and Yuriy Doronchenko

Hybrid Supercomputer Desmos with Torus Angara Interconnect:
Efficiency Analysis and Optimization 77
   Nikolay Kondratyuk, Grigory Smirnov, Ekaterina Dlinnova,
   Sergey Biryukov, and Vladimir Stegailov

Performance of Elbrus Processors for Computational Materials Science
Codes and Fast Fourier Transform. 92
   Vladimir Stegailov, Alexey Timofeev, and Denis Dergunov

Performance and Energy Analysis of Nighttime Satellite Image Archive
Processing Module 104
   Ekaterina Tyutlyaeva, Sergey Konyukhov, Igor Odintsov,
   Alexander Moskovsky, and Mikhail Zhizhin

Parallel Numerical Algorithms

Fully Homomorphic Encryption for Parallel Implementation
of Approximate Methods for Solving Differential Equations   119
   Artem K. Vishnevsky and Sergey F. Krendelev

Static Balancing Methods in Projection-Based Mesh
Generation Algorithm   135
   Sergej K. Grigorjev and Mikhail V. Yakobovskiy

Fine-Grained Parallel Algorithms in TIM-3D Code   147
   Andrey Alexandrovich Voropinov and Ivan Gennadievich Novikov

Modified Componentwise Gradient Method for Solving Structural
Magnetic Inverse Problem   162
   Elena N. Akimova, Vladimir E. Misilov, and Andrey I. Tretyakov

Parallel Multipoint Approximation Method for Large-Scale
Optimization Problems.   174
   *Victor P. Gergel, Konstantin A. Barkalov, Evgeny A. Kozinov,
   and Vassili V. Toropov*

High-Performance Computation of Initial Boundary Value Problems   186
   Valery Il'in

A Study of Euclidean Distance Matrix Computation on Intel
Many-Core Processors.   200
   Timofey Rechkalov and Mikhail Zymbler

Parallel Method of Pseudoprojection for Linear Inequalities   216
   Irina Sokolinskaya

Supercomputer Simulation

GPU Acceleration of Bubble-Particle Dynamics Simulation   235
   *Ilnur A. Zarafutdinov, Yulia A. Pityuk, Azamat R. Gainetdinov,
   Nail A. Gumerov, Olga A. Abramova, and Iskander Sh. Akhatov*

VM2D: Open Source Code for 2D Incompressible Flow Simulation
by Using Vortex Methods   251
   Kseniia Kuzmina, Ilia Marchevsky, and Evgeniya Ryatina

Modeling of Nonstationary Two-Phase Flows in Channels Using
Parallel Technologies.   266
   *Yury Perepechko, Sergey Kireev, Konstantin Sorokin,
   and Sherzad Imomnazarov*

Supercomputer Simulation of Cathodoluminescence Transients
in the Vicinity of Threading Dislocations   280
   Karl K. Sabelfeld and Anastasiya Kireeva

Supercomputer Simulation of Promising Nanocomposite Anode Materials
for Lithium-Ion Batteries: New Results   294
   Vadim M. Volokhov, Dmitry A. Varlamov, Tatyana S. Zyubina,
   Alexander S. Zyubin, Alexander V. Volokhov, Elena S. Amosova,
   and Gennady A. Pokatovich

Parallel Solution of Sediment and Suspension Transportation Problems
on the Basis of Explicit Schemes   306
   Alexander I. Sukhinov, Alexander E. Chistyakov,
   and Valentina V. Sidoryakina

Three-Dimensional Mathematical Model of Wave Propagation Towards
the Shore   322
   Alexander Sukhinov, Alexander Chistyakov, and Sophia Protsenko

Supercomputer Modeling of Hydrochemical Condition of Shallow Waters
in Summer Taking into Account the Influence of the Environment   336
   Alexander I. Sukhinov, Alexander E. Chistyakov, Alla V. Nikitina,
   Yulia V. Belova, Vladimir V. Sumbaev, and Alena A. Semenyakina

Author Index   353

High Performance Architectures, Tools and Technologies

Hierarchical Domain Representation in the AlgoWiki Encyclopedia: From Problems to Implementations

Alexander Antonov[1]([✉]), Alexey Frolov[2], Igor Konshin[2],
and Vladimir Voevodin[1]

[1] Lomonosov Moscow State University, Moscow, Russia
{asa,voevodin}@parallel.ru
[2] Institute of Numerical Mathematics of the Russian Academy of Sciences,
Moscow, Russia
frolov@inm.ras.ru, igor.konshin@gmail.com

Abstract. Algorithm description is the basic unit in the AlgoWiki Open Encyclopedia of Algorithmic Features. However, computational algorithms are not objectives in and of themselves: they are needed to address problems encountered in various fields of science and industry. On the other hand, there are many practical problems that can be tackled using various methods. This warrants the introduction of another basic term that fits between the concepts of a problem and an algorithm. Also, any algorithm can have different implementations, whether related to a single computing platform or to different platforms. The "problem–method–algorithm–implementation" chain is the basis for describing any subject area in AlgoWiki. This paper describes the permitted freedom in describing such chains, which arises when studying the approaches to address various practical problems.

Keywords: AlgoWiki · Problem · Method · Algorithm
Implementation · Parallelism resource · Parallel computing
Supercomputers

1 Introduction

The issues of efficiency and parallelism support throughout the entire supercomputer software stack are central to all global supercomputing forums today [1–3]. Judging by current trends, the degree of computing system parallelism will grow by an order of magnitude every several years [4]. This illustrates the relevance

The results described in Sects. 2–5 were obtained at Lomonosov Moscow University with the financial support of the Russian Science Foundation (agreement № 14-11-00190). The research was carried out using the equipment of the shared research facilities of HPC computing resources at Lomonosov Moscow State University.

© Springer Nature Switzerland AG 2018
L. Sokolinsky and M. Zymbler (Eds.): PCT 2018, CCIS 910, pp. 3–15, 2018.
https://doi.org/10.1007/978-3-319-99673-8_1

and importance of addressing this issue throughout the entire range of computing devices, from mobile platforms to exascale supercomputers. Scientists are currently examining architectures that would potentially form the foundation for new generations of computers. Light and/or heavy computing cores, accelerators, SIMD and data-flow processing concepts can be used. Vector processing seems to be getting new life today. This is true for the new-generation Intel Xeon and Intel Xeon Phi processors. Scalable Vector Extensions (SVE) to ARM processor instruction set [5] were announced in August 2016. These extensions were developed together with Fujitsu, which is planning to use it in a new generation of CPU for creating the Post-K supercomputer [6]. Intel has started working on integrating classical multi-core CPUs and ARM processors with FPGA segments on the same chip [7,8].

The Sunway TaihuLight supercomputer [9], the current leader on the TOP500 list, consists of more than 10 million cores, an unprecedented degree of parallelism. At the same time, its architecture has a few features that dictate a need to follow a certain style for writing efficient programs [10], namely to maintain a very high ratio of arithmetic operations execution speed to data transmission speed from RAM (flops/bytes)... These facts clearly show how important it is to know the properties and features of parallel algorithm structures today and in the future. Obviously, the full utilization of future computer capacity would require redesigning and rewriting source code; the important thing, though, is that the basic properties of the algorithms will remain the same.

2   About the AlgoWiki Project

The AlgoWiki Open Encyclopedia of Parallel Algorithmic Features [11] project has been running at Lomonosov Moscow State University since 2014. During this time, the project has drawn attention from the computing community [12, 15]. All computational algorithms are described in AlgoWiki using the same universal structure, with special emphasis on properties related to parallelism. The description consists of two primary parts. The first describes the machine-independent properties of an algorithm. The second part describes the properties of its specific implementations [13]. As the number of algorithms described in the AlgoWiki Encyclopedia rapidly grew, the issue of classification and distribution by thematic categories was raised shortly after the project was launched.

A logical extension to the AlgoWiki project was the specific algorithm analysis [14] with expert-quality reviews of various approaches to addressing individual applied problems. Each problem can usually be addressed with several different algorithmic approaches or methods. Each method has its own features, and those algorithms that fit well a specific class of computers may not always be suitable for another class. The AlgoWiki project is acquiring new dimensions that help researchers move from analyzing individual algorithms to analyzing various algorithmic approaches to tackling problems.

3 The Basic Concepts Are: Problem, Method, Algorithm and Implementation

Algorithm description the basic unit in the AlgoWiki Open Encyclopedia of Algorithmic Features. The notion of an algorithm has been in existence for a long time, with the actual word originating from the name of al-Khwarizmi, an Arab scientist who lived between the 8th and 9th centuries [16]. Many definitions of the term "algorithm" exist, probably the most famous one being attributed to Knuth [17]: an algorithm is "a finite set of rules that gives a sequence of operations for solving a specific type of problem," and it "has five important features:

– Finiteness. . .
– Definiteness. . .
– Input. . .
– Output. . .
– Effectiveness."

A common universal structure was proposed for describing in a standard form these and other important properties of various computational algorithms in the AlgoWiki project. According to D. Knuth, "[a]lgorithms are the threads that tie together most of the subfields of computer science." In the AlgoWiki Encyclopedia, algorithm descriptions also act as the central link in the chain connecting applied scientific problems with the results of supercomputer experiments.

At its initial stage of development, AlgoWiki was a simple list of algorithm descriptions. As the project expanded, the algorithms were divided into thematic subcategories, eventually developing into a form of algorithm classification. However, computational algorithms are not objectives in and of themselves: they are needed to solve *problems* faced in various areas of science and industry. For this reason, descriptions of the practical problems to be addressed were added to the AlgoWiki Encyclopedia. The problems can be described at various levels: from a specific practical problem being addressed (for example, "a general atmospheric circulation model") to mathematical formulations ("solving elliptical equations").

Many practical problems can be addressed using various *methods*: this warrants the introduction of another basic term that fits between the concepts of a problem and an algorithm. A problem can potentially be addressed via different methods (for example, elliptical equations can be solved using the direct method, Fourier transforms or iterative methods [18–20]). Each method has its advantages and may be preferred over others under certain conditions. These conditions can be dictated by the target software and hardware environment (e.g., iterative methods are preferred when solving elliptical equations on parallel computers with distributed memory). Some methods can also use other methods in sequence, which requires AlgoWiki to have several method levels.

Also, any algorithm can have different *implementations*, whether related to a single computing platform or to different platforms. For example, the following parallel implementations are known for three-dimensional Fast Fourier Transform: FFTW (using MPI+OpenMP technologies) [21], MKL FFT (MPI) [22], AccFFT (MPI, CUDA) [23], etc. Within the algorithm description, AlgoWiki allows the properties of each specific implementation of the algorithm to be described, with visualization of the results obtained on various software and hardware platforms.

Thus, a *"problem–method–algorithm–implementation" chain* is being built as part of the AlgoWiki Open Encyclopedia of Algorithmic Features, which forms the basis for describing any subject area and follows the concept of linked representation of various algorithmic approaches to addressing one and the same problem. Effectively, the AlgoWiki project is acquiring new dimensions that help researchers move from analyzing individual algorithms to analyzing various algorithmic methods for addressing problems. Whether evaluating a definite integral, finding characteristic vectors, or searching for the minimum spanning tree of a graph, multiple algorithms can be offered to address the problem at hand, each having its own properties that can be key to efficiently implementing it on a specific computing system.

The classification of algorithms in the AlgoWiki project by their compatibility with specific supercomputer infrastructure will become the basis for comparing various algorithms to each other, which is needed for switching from analyzing individual algorithms to analyzing algorithmic methods for addressing problems. This markup makes it possible to compare the compliance of algorithms to the properties of a specific computer architecture, understand the advantages of each specific approach compared to others, compare the theoretical potential of various algorithmic approaches to the same problem, and draw a variety of other conclusions.

4   Interrelationships Between Basic Concepts

4.1   From Problem to Method

The Problem (P) level is the most general of the basic concepts considered in AlgoWiki. Algorithms addressing problems from various areas of study can be described in the AlgoWiki Encyclopedia. From various areas in mathematics and more specific problems in computational physics, quantum chemistry and biomathematics, to practical issues in construction, design and all areas of science where computer simulation can be applied. More complex problems may require addressing additional problems (P–P link), for some of which solutions are already available (see Fig. 1a).

The way a problem is addressed can be called a Method (M); the resulting link is denoted as P–M (Fig. 1a). The description of a method usually contains a mathematical description of the way it addresses the problem, an explanation of its precision level, some considerations on the solution accuracy, the way to arrive at a solution and a justification for this being the only solution. Let us assume that a method is a mathematically justified sequence of actions by which one can follow to arrive at a solution to the original problem.

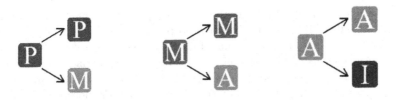

a) Possible connections
from P level

b) Possible connections
from M level

c) Possible connections
from A level

Fig. 1. Possible connections between nodes and links in the "problem–method–algorithm–implementation" chain. We denote with "P" a problem, with "M" a method, with "A" an algorithm, and with "I" an implementation

As an example of the sub-chain being considered, we can describe a solution to a specific problem (P level), for instance, the analysis of the flow around an aircraft. The solution to gas dynamic equations in a complex-shaped area requires solving the subproblem (P) of construction of a computational grid in the given region. The latter problem is usually reduced to solving a set of linear elliptical equations (P), which requires performing the decomposition (P) of a coefficient matrix into triangular factors (P). Matrix decomposition can be done using the Cholesky factorization method (M level). Thus, the sequence considered above can be presented as follows: P–P–P–P–P–M. It should be noted that the last three nodes in this sub-chain are already described in AlgoWiki.

4.2 From Method to Algorithm

The method level (M) stands for a known solution to a certain (most commonly mathematical) problem. Methods usually operate with mathematical objects of a general nature (matrices, vectors, graphs, arrays, etc.). Very complex methods can include solving certain secondary subproblems using other known methods (M–M links; Fig. 1b). These contain a description of the general method for tackling a problem, frequently providing an opportunity to detail the method further.

The method is detailed by recording a fixed sequence of computations that result in the concept of an algorithm (A) and, consequently, an M–A link (Fig. 1b). One method can lead to a variety of algorithms with different computational properties. An algorithm implies a precise indication of the entire set of operations needed to implement the method and their sequence of operations. Even though a Gaussian elimination method for a linear system is not sensitive to the order of calculation of the partial sums in each line of the matrix, this order needs to be defined in the algorithm. At the same time, it should be noted that two algorithms related to the same Gaussian method but with a different order of calculation of partial sums (ascending or descending order), besides difference in accuracy due to round-off errors, would have completely different properties: one is strictly serial, while the other has a good parallel implementation.

As an example, we can quote a sub-chain fully described in AlgoWiki. If matrix factoring is performed using Gaussian elimination by finding a triangular factorization (M), and its option LU-decomposition [24] without using transposition (M), with the compact Gaussian elimination method (M), or more specifically, the compact Gaussian elimination for tridiagonal matrices (M), then, for example, serial algorithm (A) for the compact factoring scheme described above can be considered. The chain above can be represented as follows: M–M–M–M–A.

4.3   From Algorithm to Implementation

The algorithm level (A) is the most branched structure in AlgoWiki, owing to the use of various clarifications of a specific method, or to the application of certain tricks to solve the problem faster. Some algorithms can engage many different operations, each being an independent algorithm (A–A link; Fig. 1c). For example, an algorithm (A) for solving a system of linear equations by the conjugate gradients method [20], in which every iteration uses auxiliary algorithms, is also described in AlgoWiki Encyclopedia: multiplication of a densely populated matrix by a vector (A), scalar multiplication of vectors (A), finding the vector norm (A), and vector operations (A) such as AXPY. This can result in a long chain of algorithms or trees of algorithms.

Once an algorithm is selected for solving a problem, a data structure should be developed. After the data structure is fixed, the most suitable programming language can be chosen and work can start on writing a specific implementation (I) of the chosen algorithm as a computer program (A–I link; Fig. 1c). It should be noted that the preliminary design and data structure development for complex algorithms may take a substantial amount of time, comparable to the time it takes to write the actual computer program.

As a result, each algorithm can be implemented using different programming languages. Many developers publish their implementations with detailed descriptions. This makes it possible to find either the necessary program or the information on how to write the most efficient program. Most pages in the AlgoWiki Encyclopedia, together with algorithm descriptions, contain reviews of existing implementations and direct links to open source codes.

4.4   From Implementation to Computations

When running the computations using the written program or a specific existing implementation, one should pay attention to the correct operation of the program; the most experienced developers design tests to debug the program. An important indicator of the quality of a program is its computational performance. This issue receives much attention in AlgoWiki Encyclopedia articles. The computation locality and data usage locality are analyzed, criteria for assessing parallel efficiency are developed, and the results of the parallel computation are presented with a scalability analysis [29].

As a conclusion to the review of the various forms of the "problem–method–algorithm–implementation" chain, it should be noted that the least experienced developers make their first attempt to consider features of the computer architecture only during the very last stage, when analyzing the results of their first runs. However, the highest efficiency and scalability of the program can be achieved at the topmost levels, when choosing a method for solving the problem. One of the main goals for creating the AlgoWiki Open Encyclopedia of Algorithmic Features is to specifically give the user an opportunity well in advance to visualize the entire chain in full detail, from problem to implementation, and to choose the methods and algorithms that will produce the most efficient solutions.

4.5   Interconnection of Sub-chains

In fact, different sub-chains may intersect or have many common nodes and links since each method can usually be applied to the solution of various problems. Short chains can be composed of longer ones. For example, all the sub-chains considered in Subsects. 4.1–4.3 could be combined into one. Thus, the subtask of the solution of a system of linear elliptic equations (P) considered in Subsect. 4.1 can be solved by using the algorithm (A) for the solution of linear systems by the conjugate gradients method (see Subsect. 4.3), which will use as a preconditioner an incomplete triangular factorization, referred to in Subsect. 4.2. Such long chains can also arise when solving other practical problems. Another example of a complete chain description is considered in the next section.

5   An Example Description of the "Problem-Method-Algorithm-Implementation" Chain in the AlgoWiki Encyclopedia

Let us take a look at one of the full "problem-method-algorithm-implementation" chains already described in the AlgoWiki Encyclopedia—the chain that goes from the "Matrix factorization" group of problems to Householder's (reflection) method of QR factorization [25–28] of a square matrix, floating point variant (see Fig. 2).

Starting with the program (Matrix factorization), an AlgoWiki reader would see the "P" icon, indicating the problem level. In this case, however, it is not a single problem but a group of problems: Matrix factorization as a problem requires decomposing a matrix into a series of special matrices (unitary, triangular, etc.), depending on what is used in a higher-level problem (solving a system of linear equations, eigenvalue problems, etc.). Next, the reader sees that the detailed description of the problem raises another level with the same "P" sign, and again, these are not individual problems but groups of problems:

- "Triangular factorization": different versions of Gaussian decomposition, with or without transpositions, using the original matrix structures, up to compact schemes for tridiagonal matrices; a place is also prepared for pre-computed factorization of known matrices.

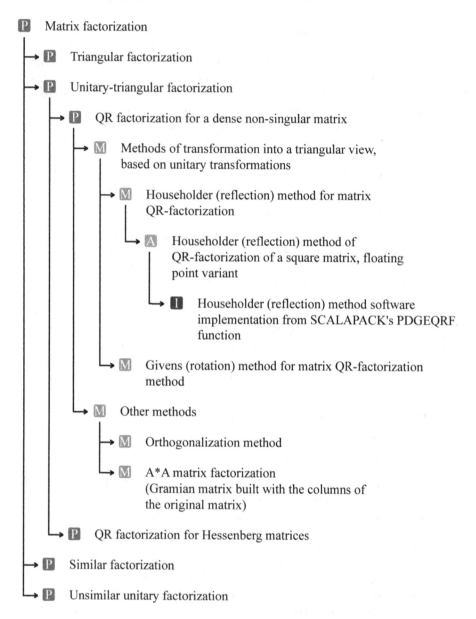

Fig. 2. Example of the full "problem–method–algorithm–implementation" chain

- "Unitary-triangular factorization": the intermediary point we seek in our chain (we will look at it in more detail later).
- "Similar factorization": reduction of the matrix by means of a two-sided similarity transform to a Hessenberg form or tridiagonal symmetric matrix form, and spectral factorization.

– "Unsimilar unitary factorization": reduction of the matrix by means of an unbounded similarity transform to a two-diagonal form, and singular factorization.

Before proceeding with the chosen factorization group, note that the pages describing variations at the lower "algorithmic" level (e.g., Gaussian method with the selection of a leading element) are quite brief, but we can say in advance that no detailed description will be available for at least two of them since they are not used in practice and are only mentioned as a tribute to the history of the algorithm development.

In the chosen group of problems, "Unitary-triangular factorization," after moving a level lower, the reader finally gets to the individual problem level. There are two of them in the classification so far: QR factorization for a dense non-singular matrix and for Hessenberg matrices. The latter only has a short description at the method level, as the actual algorithm is not used in practice (factorization is not explicitly used in the QR algorithm but as iterations with implicit shifts). The LQ decomposition is also mentioned in the description at this level as a variation of QR factorization for a matrix transposed from the original matrix.

A more detailed description of the problem "QR factorization for dense non-singular matrix" is the last step in our chain, with the "P" icon. This is why, following the rules for describing the single-problem level, a brief list of solution methods is given along with a brief comparison of the above methods.

Digressing from the chosen chain towards the Householder method, one can note that different chains do not always end at the algorithm level. For example, the last method (using Gramian matrix factorization) does not have such descriptions. The reason is that, owing to a narrow application area (in addition to requiring a non-singular matrix, the $A * A$ matrix conditioning is the square of the original matrix conditioning, so the error margin for this method is much greater), this algorithm has not been implemented by anyone.

Choosing the Householder method next, the user arrives at the method level ("M" icon on screen), specifically at the "Householder (reflection) method for matrix QR-factorization". Naturally, at this level the user is presented with just a basic mathematical concept for the method in question, as the actual algorithm is described in more detail at the algorithm-level page ("Householder (reflection) method of QR-factorization of a square matrix, floating point variant"). Other versions of the algorithm have not been described yet, but they are mentioned in the method description: "In addition to the classical point version, the method has many other implementations such as the block version."

Following the link to the algorithm, the user sees the "A" icon and a detailed description of all features for the chosen algorithm. Among other things, the general and mathematical descriptions of the algorithm contain not just general words about the Householder transformation, but also specific formulas for each step in the computation and for reducing the transformations to a sequence of scalar multiplications and weighted vector sums. These basic operations are parts of the computational core of the algorithm and are described in the respective section. A description of the algorithm macro structure shows why the main parts of scalar multiplications in the same step can be performed independently.

Looking at the implementation chart of the serial algorithm, the reader will see a description of the mathematical essence of the stages within one step. Serial complexity is presented as a formula for the number of floating point multiplications and additions/subtractions.

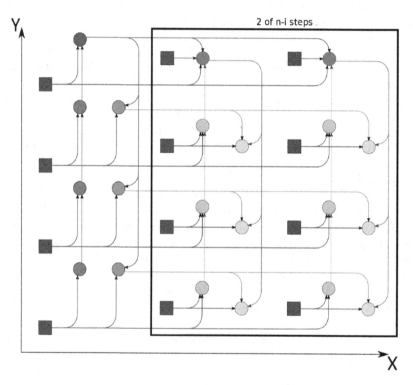

Fig. 3. Algorithm step graph (zeroing column i). Squares represent input data for this step (taken from the input data or from previous step), circles show operations. The outlined group of operations is repeated independently $n - i$ times

The information graph (see Fig. 3) is given for a single step as no parallelism is observed between individual steps. The parallelism resource is calculated assuming serial execution of scalar multiplication, but there are also assessments for other methods, namely the serial-parallel and pairing methods. Input and output data for the algorithm and its other properties are described further.

As one can clearly see, the correlation between serial and parallel complexities is *linear*, which provides a good incentive for paralleling the algorithm execution. However, the fastest level-parallel form of the graph has a *square-law* width, which indicates an imbalance between device loads in an attempt to actually program it. That is why it is more reasonable to keep the number of devices (e.g., cluster nodes) linear to the matrix size, even in the presence of a good (fast) communication network, which doubles the critical path for the level parallel form.

In this case, the algorithm's computational efficiency, expressed as the ratio of the number of operations to the total amount of input and output data, is *linear*.

The algorithm is fully determined within the selected version.

The computational error in the Householder (reflection) method grows in a *linear* manner, as well as in the Givens (rotations) method.

The next item after the description of the actual algorithm is the description of its software implementation from SCALAPACK's PDGEQRF function. The "Implementation peculiarities of the serial algorithm" section presents a fragment of the program in Fortran implementing the given algorithm.

The next session presents an analysis of the algorithm locality and its qualitative assessment.

Next, the user can see how the algorithm's dynamic characteristics change with different computation performance parameters for the same algorithm implementation.

Algorithm implementation can have relatively good performance, owing to the use of the SCALAPACK implementation of BLAS libraries, etc. Finally, at the end of the chain, the user can see some recommendations, particularly the advice to avoid using the classical version of the method and to use the block versions instead, for which numerous research works are available.

6 Conclusions

The paper considers the hierarchical approach used in the AlgoWiki Open Encyclopedia of Algorithmic Features to represent the structure of the subject area. A description is offered in the form of a "problem–method–algorithm–implementation" chain, which corresponds to numeric descriptions of the problems used in computational mathematics. In the paper, we show the interconnection between the basic concepts and analyze the descriptions of such chains obtained while working on additional content for the AlgoWiki Encyclopedia.

The proposed approach to the description of "problem–method–algorithm–implementation" chains is now being implemented in descriptions within the AlgoWiki Open Encyclopedia of Algorithmic Features. This makes it possible to describe subject areas in which computational algorithms are used more clearly and according to a single scheme.

References

1. Patwary, M.M.A., et al.: Parallel efficient sparse matrix-matrix multiplication on multicore platforms. In: Kunkel, J.M., Ludwig, T. (eds.) ISC High Performance 2015. LNCS, vol. 9137, pp. 48–57. Springer, Cham (2015). https://doi.org/10.1007/978-3-319-20119-1_4

2. Solc, R., Kozhevnikov, A., Haidar, A., Tomov, S., Dongarra, J., Schulthess, T.C.: Efficient implementation of quantum materials simulations on distributed CPU-GPU systems. In: Proceedings of the International Conference for High Performance Computing, Networking, Storage and Analysis (SC 2015), pp. 10:1–10:12. ACM, New York (2015). https://doi.org/10.1145/2807591.2807654

3. Alam, M., Khan, M., Vullikanti, A., Marathe, M.: An efficient and scalable algorithmic method for generating large-scale random graphs. In: Proceedings of the International Conference for High Performance Computing, Networking, Storage and Analysis, pp. 32:1–32:12. IEEE Press, Piscataway (2016). https://doi.org/10.1109/SC.2016.31

4. Dongarra, J., et al.: The international exascale software project roadmap. Int. J. High Perform. Comput. Appl. **25**(1), 3–60 (2011). https://doi.org/10.1177/1094342010391989

5. Arm HPC tools for SVE. https://developer.arm.com/products/software-development-tools/hpc/sve

6. Post-K computer. http://www.aics.riken.jp/en/postk/project

7. Intel Eases Use of FPGA Acceleration: Combines Platforms, Software Stack and Ecosystem Solutions to Maximize Performance and Lower Data Center Costs. https://newsroom.intel.com/news/intel-eases-use-fpga-acceleration-combines-platforms-software-stack-//ecosystem-solutions/

8. Intel Enables 5G, NFV and Data Centers with High-Performance, High-Density ARM-based Intel Stratix 10 FPGA. https://newsroom.intel.com/news/intel-enables-5g-nfv-data-centers-high-performance-high-density-arm-//based-intel-stratix-10-fpga/

9. Fu, H., et al.: The Sunway TaihuLight supercomputer: system and applications. Sci. China Inf. Sci. **59**(7) (2016). https://doi.org/10.1007/s11432-016-5588-7

10. Zhang, J., et al.: Extreme-Scale phase field simulations of coarsening dynamics on the Sunway TaihuLight supercomputer. In: Proceedings of the International Conference for High Performance Computing, Networking, Storage and Analysis (SC 2016), pp. 4:1–4:12. IEEE Press, Piscataway (2016). https://doi.org/10.1109/SC.2016.3

11. Open Encyclopedia of Parallel Algorithmic Features. https://algowiki-project.org/en

12. Voevodin, Vl., Antonov, A., Dongarra, J.: AlgoWiki: an open encyclopedia of parallel algorithmic features. Supercomput. Frontiers Innov. **2**(1), 4–18 (2015). https://doi.org/10.14529/jsfi150101

13. Antonov, A., Voevodin, V., Voevodin, Vl., Teplov, A.: A study of the dynamic characteristics of software implementation as an essential part for a universal description of algorithm properties. In: 24th Euromicro International Conference on Parallel, Distributed, and Network-Based Processing Proceedings, 17th–19th February 2016, pp. 359–363 (2016). https://doi.org/10.1109/PDP.2016.24

14. Antonov, A., et al.: Parallel processing model for Cholesky decomposition algorithm in Algowiki project. Supercomput. Frontiers Innov. **3**(3), 61–70 (2016). https://doi.org/10.14529/jsfi160307

15. Voevodin, Vl., Antonov, A., Dongarra, J.: Why is it hard to describe properties of algorithms? Procedia Comput. Sci. **101**, 4–7 (2016). https://doi.org/10.1016/j.procs.2016.11.002

16. Boyer, C.B.: The Arabic Hegemony. A History of Mathematics, Second edn. Wiley, Hoboken (1991)

17. Knuth, D.: The Art of Computer Programming. Fundamental Algorithms, vol. 1, 3rd edn. Addison-Wesley, Reading (1997)

18. Gloukhov, V.: Parallel implementation of the INM atmospheric general circulation model on distributed memory multiprocessors. In: Sloot, P.M.A., Hoekstra, A.G., Tan, C.J.K., Dongarra, J.J. (eds.) ICCS 2002. LNCS, vol. 2329, pp. 753–762. Springer, Heidelberg (2002). https://doi.org/10.1007/3-540-46043-8_76

19. Hess, R., Joppich, W.: A comparison of parallel multigrid and a fast Fourier transform algorithm for the solution of the Helmholtz equation in numerical weather prediction. Parallel Comput. **22**, 1503–1512 (1997)

20. Saad, Y.: Iterative Methods for Sparse Linear Systems, Second edn. SIAM, Philadelphia (2003). https://doi.org/10.1137/1.9780898718003

21. FFTW Home Page. http://www.fftw.org

22. Computing Cluster FFT. https://software.intel.com/node/521992

23. AccFFT. A New Parallel FFT Library. http://accfft.org

24. Davis, T.A.: Direct methods for sparse linear systems. SIAM (2006). https://doi.org/10.1137/1.9780898718881

25. Golub, G., Van Loan, C.F.: Matrix Computations, Third edn. Johns Hopkins University Press, Baltimore (1996)

26. Chu, E., George, A.: QR factorization of a dense matrix on a hypercube multiprocessor. SIAM J. Sci. Stat. Comput. **11**, 990–1028 (1990). https://doi.org/10.1137/0911057

27. Paige, C.: Some aspects of generalized QR factorization. In: Cox, M., Hammarling, S. (eds.) Reliable Numerical Computations. Clarendon Press, Oxford (1990)

28. Dongarra, J.J., D'Azevedo, E.F.: The design and implementation of the parallel out-of-core ScaLAPACK LU, QR, and Cholesky factorization routines. Department of Computer Science Technical report CS-97-347, University of Tennessee, Knoxville, TN (1997)

29. Antonov, A., Teplov, A.: Generalized approach to scalability analysis of parallel applications. In: Carretero, J., et al. (eds.) ICA3PP 2016. LNCS, vol. 10049, pp. 291–304. Springer, Cham (2016). https://doi.org/10.1007/978-3-319-49956-7_23

A Toolkit for the Development
of Data-Driven Functional Parallel
Programmes

Alexander I. Legalov$^{(\boxtimes)}$, Vladimir S. Vasilyev, Ivan V. Matkovskii,
and Mariya S. Ushakova

Siberian Federal University, Krasnoyarsk, Russia
legalov@mail.ru, rrrFer@mail.ru, alpha900i@mail.ru, ksv@akadem.ru

Abstract. In the article a technology is considered which aims at creating architecture-independent parallel programmes based on the data-driven functional paradigm. A proposed toolkit provides the translation, execution, debugging, optimisation and verification of programmes. A programme in a data-driven functional parallel language is translated into the data-flow graph (which describes the data dependencies of an implemented algorithm) of the programme. On the basis of this representation, the control-flow graph (which defines the organisation of computations) is generated. Both graphs allow to carry out various optimising transformations. The resulting data-flow graph is also used for the formal verification of the programme. A computation process is considered as a cooperation of the control-flow graph and the data-flow graph. The execution of data-driven functional parallel programmes is carried out by a special interpreter (event machine), which consist of a number of event processors controlled by a special manager.

Keywords: Data-driven functional parallel programming
Software development toolkit · Parallel-programmes translation
Parallel-programmes optimisation · Parallel-programmes verification

1 Introduction

Parallel computing have outgrown the application in high-performance computing long ago. It is widely used for solving problems in different areas. Nowadays, the main feature of parallel programming is the source-code dependence on the architecture of the target computation system. So, to port a programme to another architecture, it should be completely rewritten or appreciably modified. The reason is the intention to increase the efficiency of parallel programmes, which results in software being heavily tied to particular hardware characteristics.

Computational resources and their communications are the main characteristics of a computation system which should be considered during the development in order to increase the programmes efficiency. So, in addition to solving

The research is supported by the RFBR (research project No. 17-07-00288).

L. Sokolinsky and M. Zymbler (Eds.): PCT 2018, CCIS 910, pp. 16–30, 2018.
https://doi.org/10.1007/978-3-319-99673-8_2

an applied problem, we need to explicitly manage computations and resolve resource conflicts among parallel processes. That is why parallel programming is hard [1] and requires non-trivial analysis of programme correctness taking different approaches, for instance, model checking [2] for formal verification.

It should be pointed out that a dependency on a particular parallel hardware precludes writing truly parallel algorithms at the initial stage of the development. This leads to the reduction of problem parallelism according to hardware resources, which prevents from applying more effective solutions when a modification of the programme is needed. At the same time, the development of hardware-dependent programmes is the mainstream of parallel programming. The existing approaches have very different ideology of parallelisation. The most widespread approaches are: parallelisation with message passing [3], multithread and multi-core programming for systems with shared memory [4], graphical processing unit programming [5], and also the mixture of these three approaches in different combinations for systems with heterogeneous and distributed architecture [6–9].

Though the concept of unlimited parallelism is not widespread in parallel programmes development nowadays [10], it has some prospects as a basis of programming system that provide for subsequent transformations of programmes into resource-limited and architecture-dependent parallel programmes. So it is topical to develop a language and a toolkit to provide for creation of parallel programmes which are initially independent of peculiarities of a specific parallel computer system. Porting a programme to a particular system can be done after the processes of verification, testing and debugging.

The proposed approach is based on the concept of architecture-independent parallel programming. Its key ideas are exclusion of resource conflicts and implicit control over computations from within the programme being developed. It is supposed that a virtual machine which executes the programme has unlimited resources and a programming language allows to define solely data dependencies between the executed functions. An interaction between functions takes place on data readiness. This allows to create programmes with maximal achievable parallelism, which is compressed to special computing resources at the stage of the intermediate representation after verification and debugging of the source code. This allows to increase the efficiency of parallel-programme development process. For example, it is possible to create a generic library of functions adaptable to different existing and prospective architectures. The subsequent transformations of such programmes can be done with formal methods by changing the control-flow graph to fit the target architecture, preserving the correspondence with the data-flow graph (DFG).

The goal of our research is the development of architecture-independent parallel-programming technology based on the data-driven functional parallel paradigm [11]. To achieve the goal we solve the following problems:

- the development and further improvement of the data-driven functional parallel (DDFP) computing model, on whose basis a programming language is defined; it allows the creation of architecture-independent parallel programmes;

- the development of a toolkit to provide translation, testing, debugging and execution of data-driven functional parallel programmes;
- the development of methods for programme verification and optimisation at the level of the programme DFG;
- the development of control-flow graph transformation methods that allow to change the programme parallelism and, in the future, to transform programmes for particular parallel architectures.

2   Problems of Imperative Paradigm Employment in Parallel Programming

Wide application of the imperative programming paradigm introduces certain difficulties in the development of parallel programmes. A programmer has to explicitly or implicitly form relations between programme objects [12]. The possible relations are:

- data relations which specify the DFG of the programme; this graph defines dependencies between operations and operands being processed;
- control relations which set the order of execution of operations; these relations are associated with the DFG of the programme in order to ensure the right logic of transformations of operands;
- relations between computing resources (memory, processor units) that are used during the operation execution.

In most cases, a programme developer has to explicitly take into account the dependencies between these relations in an attempt to avoid any logical contradictions leading to an erroneous execution. In the ubiquitously used imperative programming, the relations between the data and the control are kept in the programmer's mind but are not explicitly expressed in the programme. For instance, let us consider the factorial function over the range 1 to n.

```
int fact1n(int n) {
    int r = 1;    int i = 1;
loop:
    if(i <= n) {
        r *= i;
        i++;
        goto loop;
    }
    return r;
}
```

It is evident that the only explicit relation is the relation of programme objects order in the source code or (after translation and loading for execution) in the system memory. But it does not specify the exact order of computing. A graphical representation of the given function explicitly represents all kinds of relations and

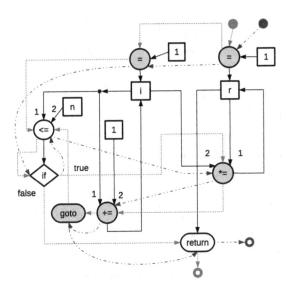

Fig. 1. A graphical representation of the relations in the factorial function (the solid, dashed and dash-dotted lines show data relations, control relations and order relations, respectively)

shows different trajectories of its execution which the programmer has to keep in mind to form an overall understanding of the programme (Fig. 1). In some cases, the control and order relations coincide (operations marked with light grey), which facilitates the programme understanding and allows to employ programme counters instead of straight-forwardly transferring the flow of execution to the address. In most cases, however, these relations are connected by implication rather than by the order of the operations in the programme.

Frequently, the relation of order can be ignored by employing a graphical representation. Particularly, flowcharts, activity diagrams, automaton graphs speed up the development of programme algorithm and allow to represent the logic of the operational behaviour clearly. This is done by explicitly defining the control and data relations on the basis of the developer's intuitive algorithm understanding.

The situation becomes more complicated if we turn to parallel programmes development. In this case, additional control operations for splitting and synchronisation appear in the control-flow graph. What is more, all available resources are to be distributed for the simultaneous execution of parallel source-code fragments. This results in a new relation between the programme and the resources. This relation can be explicitly represented by the resource graph (graph of system resources). The probability of conflicts arises, which could lead to incorrect computations even if the programme worked correctly in the sequential case.

Various parallel systems employ different computing control methods (strategies) [12]. A programme can be represented by a data-control-resource graph (we

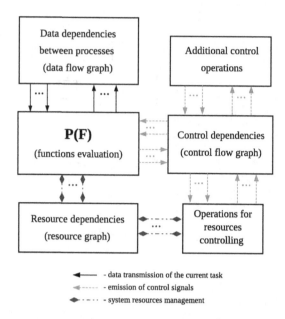

Fig. 2. A DCR-net describing the process execution within the computational resources

call it a DCR-net) in which processes P execute the operations F defined by a programmer. The execution of these operations are initiated by control signals that are emitted under certain conditions in the control-flow graph. These conditions emerge from the data dependencies of the programme, peculiarities of computational resources and some additional factors (Fig. 2). The correctness of the computing process depends on certain prerequisites for each operation. In the general case, an operation execution within the resources of the computing system is possible only if the following conditions are satisfied before the execution:

1. The condition of data readiness (Data, D-condition). Before the process execution start, all the required data have to be at the process input. The process execution in the absence of any required data leads to a wrong result.
2. The condition of resources allocation (Resource, R-condition). The process requires certain resources to be executed within them, and these resources should be allocated and provided before the process execution.
3. The condition of acknowledgement (Acknowledge, A-condition). Resources utilised by a process can be freed or reused only after the acknowledgement that the output results of computations have been received by all the processes that take them as input.

The control of readiness conditions can be performed by different means. On the one hand, a programmer can control processes directly. On the other hand, a computing system undertake many control functions. Let us distinguish the following control modes:

1. Explicit (human) control. A programmer sets the logic of generating and checking the readiness conditions in the source code.
2. Implicit control. In this case, it is assumed that the processes are executed correctly without any control. This assumption may follow, for example, from special organisation of resources in the computing system, automatically maintaining the data readiness condition (automatic control). Another possibility is when the readiness conditions are always true due to the system peculiarities, and hence no control is needed (empty control).

If a programmer uses explicit control, then he should code the readiness conditions checks. It increases the software development costs.

3 Features of the Computing Model and the Language of Data-Driven Functional Parallel Programming

The basic approach to architecture-independent parallel programming is the development of a language and a toolkit to provide the implicit control at the level of computing model. We propose a model of data-driven functional parallel computing [11], in which every function is represented as a DFG whose nodes are operators and whose arcs are data connections between operators. Any connection is marked with a value which is both the output of the operator in the beginning of the arc and the input of the second operator. There are several types of operators in the graph: the operator of interpretation and data-grouping operators.

The **operator of interpretation** is the only operator that applies a function to the function arguments. This operator has two inputs: the first one takes a function (functional input), while the second one takes an argument for the function (data input) (Fig. 3). The output of the interpretation operator is the result of the function application to the argument. Functions are either elementary predefined operations or programmer-defined. The interpretation operator semantics is defined by the axioms of the computing model and its transformation algebra [11].

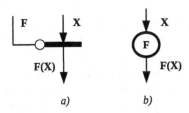

Fig. 3. Graphical symbols of the interpretation operator (*a*—the general case, *b*—the case when the function on the first input is predefined)

Data-grouping operators provide various ways of grouping operands in different structures (lists). The idea of data-grouping operators goes back to the functional forms introduced in [13]. In our case, however, it is the variety of such structures that is the principal way to increase the flexibility in writing parallel programmes and implement non-conventional ideas of parallel-algorithms development. Extending the set of such operators is one of the main approaches to the further development of the computing model and the language of data-driven functional parallel programming. This language is used to try out various approaches targeting the efficiency of expressing different types of parallelism. For instance, the usage of asynchronous lists [14] allows to develop algorithms with dynamically modifiable parallelism according to the rates of data incoming and processing. The core set of data-grouping operators is shown in Fig. 4.

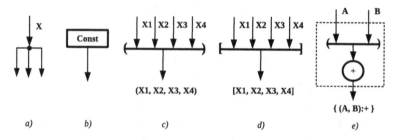

Fig. 4. Graphical symbols of data-grouping operators (*a*—data copying, *b*—constant assignment, *c*—grouping in a data list, *d*—grouping in a parallel list, *e*—delay-list creation)

The **copy operator** (Fig. 4*a*) carries out data replication. In our language, replication is done by assigning a name to a connection (marked later with a value during the execution), and then this name is used in other positions of the programme to refer to this connection (and the corresponding value). We use the prefix and postfix notation for assigning a name to the connection:

$$\text{value} >> \text{name, or name} << \text{value}.$$

The **constant operator** has no inputs (Fig. 4*b*). It has only one output that is always marked with the predefined value. In our language, the constant operator is defined by the value of a certain type.

The **data-list grouping operator** (Fig. 4*c*) has several inputs and one output. It performs structuring and ordering of the values that are transmitted through arcs from different sources. Each input has its number from 1 to N. The position of data in the resulting list equals the number of the input it has come from. In the source code, the list elements are put in parentheses "(" and ")". For example:

$$(\text{x1,x2,x3,x4}).$$

The **parallel-list grouping operator** (Fig. 4*d*) groups elements in a similar way as in a data list. However, its output is a multiple connection whose multiplicity equals the number of operator inputs. If an operator of interpretation is executed having a parallel list on its data input, then a function is applied to each individual element of the parallel list independently and in parallel. In the source code, the elements are put in square brackets "["and "]". For example:

$$[x,y,z]:sin \equiv [x:sin,y:sin,z:sin].$$

Similarly, in cases when a parallel list of functions comes to the functional input of the interpretation operator, each function is applied to the argument in parallel:

$$(x,0):[<,=,>] \equiv [(x,0):<, \ (x,0):=, \ (x,0):>].$$

The transformation algebra of the language describes all the cases of equivalent transformations of parallel lists.

The **delay-list grouping operator** (Fig. 4*e*) delays the execution of operators corresponding to some subgraph. This subgraph is considered as a single node of the DFG until the delay list is released. This node has several inputs and one output. The connections coming from outside the subgraph are the inputs of the operator, and the result produced in the subgraph comes to the operator output. The specific feature of this operator is that the delayed operators are not executed even on data readiness until the delay is not released. The release from the delay takes place if a delay list becomes an input of the interpretation operator. Delay lists allows to construct the conditional branches of the programme. In the graphical representation, a dashed line surrounding the delayed operations is used to denote the delay list. In our language, the list of delayed computations is defined by putting operators in braces "{" and "}".

On the basis of the described model, we develop the Pifagor language for data-driven functional parallel programming. The source code of the above-mentioned factorial function in the Pifagor language is the following:

```
fact1n << funcdef n {
    n1<< (n,1);
    [(n1:[<=,>]):?]^ (
        1,
        {(n, n1:-:fact1n):*}
    ):. >>return
}
```

The function is free from explicit computations control. Only the data dependencies between operators are defined. The DFG of this function is shown in Fig. 5.

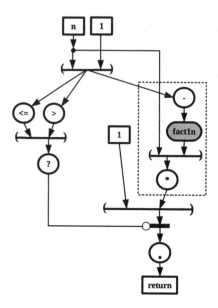

Fig. 5. Data-flow graph of the factorial function

4   A Toolkit for Architecture Independent Parallel Programming

We develop a toolkit to support the data-driven functional parallel programming paradigm in order to try out the proposed ideas and their further development on the basis of experiments. The general scheme of the toolkit is shown in Fig. 6. It includes the following subsystems:

- a translator from the language of data-driven functional parallel programming to the intermediate representation, called the reverse data-flow graph (RDFG);
- a generator of the control-flow graph (CFG), which constructs the graph for controlling computations;
- an event machine, which supports execution of data-driven functional parallel programmes by utilising RDFG and CFG;
- RDFG optimisation tools;
- CFG optimisation tools;
- tools for the DDFP programmes formal verification.

4.1   Translation of Data-Driven Functional Parallel Programmes

The translator accepts source code files in Pifagor language, each containing one or more functions. It also provides separate compilation of functions stored in a special repository. The translator generates a RDFG for each function. These

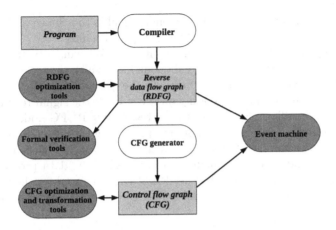

Fig. 6. The toolkit for architecture-independent parallel programming

RDFGs are saved in the repository in text format. The choice of the text format is due to the fact that an internal representation in the computer memory can be easily constructed by means of simple translators. Moreover, the developer can easily read and analyse the translated functions, considering the text form of a graph to be an analogue of the assembly language. The translator also generates auxiliary files with debug information binding the nodes of the RDFG to the function source-code lines.

A reverse data-flow graph generated by the translator allows to generate a control-flow graph that controls the function execution. Each node of the CFG is associated with the corresponding node of the RDFG and controls the moment when the operation starts executing. Each node of the CFG is a finite automaton whose states are controlled by the input signals. These signals notify the automaton of the event of the data having been prepared for the associated RDFG node. The computations on the RDFG node are initiated on certain state switches in the automaton. As the RDFG node execution completes, the readiness signal is transmitted through the output arc of the CFG to the next automaton. Before the execution, the CFG arcs are marked with initial signals. As the execution begins, the signals are transmitted along the arcs and change the states of the receiving nodes. A special utility programme forms the CFG. It is saved in the repository in text form.

4.2 Parallel Event Machine

At the current development stage, the execution of data-driven functional parallel programmes is done by a special interpreter (event machine), which consists of a set of event processors (EP) and a special event-machine scheduler controlling the EPs. Each EP handles only one function, which is run in a separate thread. Currently, the execution of operators inside the function is performed sequentially. At the present moment, our main goal is to achieve a stable functioning of the event machine rather than high performance.

Functioning of an EP (Fig. 7) is carried out in the following way. Initial signals of the CFG are added to the EP's signal queue from which they are transmitted to the handler of control signals according to the queue discipline. The handler analyses an incoming event. Depending on the state of the signal recipient node of the CFG, the handler might query the corresponding RDFG node (associated with the CFG node) if the operation of data processing is to be executed. In case it is, the handler of the RDFG nodes is called. It performs all needed functional transformations and saves intermediate results. After the data processing, the control node changes its state and, if needed, it emits a signal for the next node. The latter signal is again added to the queue of control signals, and so on.

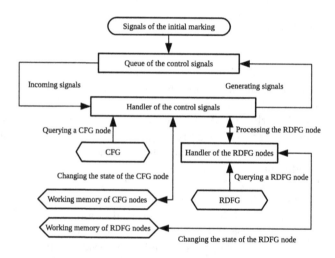

Fig. 7. Event-processor general structure

Before the event machine launch, a linker assembles separate functions from the repository into a programme. The linker checks the presence of all components that are listed in the section of the external links of the RDFG. If any required component is absent, the interpretation is reported impossible. Each required function is also linked. The event machine scheduler stores a table with RDFGs and CFGs loaded by the linker.

The process of interpretation starts with the creation of the first (initial) EP. It receives the RDFG and CFG of the function which is the first to execute. The EP saves the data in the working memory of the RDFG nodes, while automaton states are stored in the working memory of CFG nodes.

The states of the CFG nodes automata connected with the RDFG constant operators are initially set to new-signal generation. These signals are transmitted through the CFG connections and activate receiving automata. The process of signal transmission through the connections lasts until the "return" node of the corresponding RDFG is processed (in this case, the function is considered completed), or until the event queue is empty. In the latter case, the EP switches to sleep mode, sending a signal about this to the event-machine scheduler. This situation occurs when all inner signals are processed and there is no incoming control signals notifying of returned results from the called functions.

4.3   Optimisation of Data-Driven Functional Parallel Programmes

Within the system of data-driven functional parallel programming, we have developed a number of optimising transformations that take advantage of the peculiarities of our computing model.

1. Dead-code elimination (removal of code that does not affect the programme results). The optimiser starts at the "return" node, traverses the DFG and marks all the reachable nodes. The rest of the nodes are removed.
2. Optimisation within iterative calculations. Traditionally, compilers carry out this kind of transformations for loops. In our case, similar transformations are applied to recursive functions and parallel lists defined in the language model. In particular, calculations inside a recursive function that remain constant during the recursive calls are moved to a new auxiliary function, whose result is passed to the recursive function as an additional parameter; in functions applied to parallel lists, all computations that are independent from the function parameters are moved to the calling function.
3. Inline substitution of simple functions. If a function is sufficiently small (the number of nodes is below a predefined limit), then the function-call overhead is substantial compared to the overall cost of the function. As a result, the function body is better to be inserted at the place of the function call.
4. Duplicate-code elimination. If the DFG subgraphs perform the same operations on the same arguments and also are in one and the same delay list or in hierarchically nested delay lists, then they can be merged, thereby eliminating redundant computations.
5. Optimisation based on equivalent transformations. The RDFG is searched for certain subgraphs that can be transformed to a more computationally simple but equivalent form.
   In particular, the model admits the following equivalent transformations: simplification of single-element parallel list; unwrapping of directly nested parallel lists into a single parallel list; preliminary simplification of parallel lists whose size is known at compilation time.
6. Redundant control-dependencies removal. An RDFG describes data dependencies, and the CFG is created on its basis according to the data-readiness control strategy. However, in several cases, some control relations are redundant, and their removal would not affect the order of programme-operator execution.
7. Generation of a CFG that defines a sequential traversal of the RDFG nodes. This removes the overhead of data-readiness analysis.

4.4 Formal Verification of Data-Driven Functional Parallel Programmes

The proposed paradigm eases formal verification of programmes owing to the absence of resource limitations and to the fact that a programme defines only data dependencies. The main problems in this area are to study the application of formal correctness-proof methods to the proposed language and to develop a toolkit to assist formal verification.

For the correctness proof, we employ the axiomatic approach based on Hoare Logic [15]. The specification of the programme is expressed in a special formal language (specification language). A Hoare triple is represented by a data-flow graph of the programme whose input and output arcs are marked with formulas in the specification language (called a precondition and a postcondition, respectively). The process of proving the programme correctness consists in marking the graph arcs with formulas, graph modifications and folding. As a result, we obtain a number of RDFGs with all arcs marked with formulas. Each of these RDFGs can be transformed into a logic formula. If all these formulas are identically true, then the programme is correct [16].

The process of proving is quite complicated since it requires taking into account a great number of graphs and their transformations. That is why we have developed basic concepts of a toolkit for supporting formal verification of DDFP programmes [17]. The system takes a DFG and programme pre- and postcondition as its input. It searches for unmarked arcs of the graph and assists in selecting appropriate axioms and theorems for marking the arcs. The proof process of a programme correctness is considered as a tree in which each node is a partially marked DFG of the programme. The construction of the proof tree finishes when all its leaves are totally marked DFGs. Thereafter, a logic formula is generated for each DFG in the leaves. The programme correctness is proved if we manage to prove that all these formulas are identically true.

5 Overview of Related Works

In the area of languages and support tools for parallel programming, the current focus is on the creation and development of architecture-dependent systems. The difference between these and our approaches has been discussed above. There exist unconventional methods and tools for parallel programming, but usually, they are being developed by small groups of developers. The development often finishes at the stage of an experimental solution, which does not make it more popular. An exception is special-purpose systems, which target specific object domains and have a considerable optimisation potential for existing architectures owing to the limited number of tasks to solve. For instance, the non-procedural language NORMA [18] targets problems of mathematical physics and translates into parallel programmes for different architectures.

Dataflow programming is implemented in the LabVIEW system [19]. The graphical programming language named "G" is designed to target the problems of the automation of scientific researches and production processes. The language is oriented towards large-blocks programming and, in fact, describes an interaction of different resources. On the contrary, our approach targets unlimited resources and parallelism at the level of elementary operations.

Sisal is one of the universal functional languages of parallel programming that has been developed for a long time. The first version of this language was released in 1985. In recent times, the Institute of Informatics Systems of the Siberian Branch of the Russian Academy of Sciences has been developing this language, and its latest release is Sisal 3.2 [20]. It should be pointed out that the main goal of the project is to provide application programmers with convenient environment for functional programme development, with the subsequent execution of the programmes on a parallel computing system available via telecommunication networks. This is the main difference from our goals: we seek, investigate and implement operators that allow for efficient expression of unlimited parallelism in architecture-independent parallel programmes. In our view, this allows to rethink the process of development, analysis and transformation of parallel programmes. In particular, it is demonstrated in [21] how to deduce known methods of sorting by imposing different constraints on an algorithm with initially unlimited parallelism.

6   Conclusions

The toolkit being developed allows to create architecture-independent parallel programmes whose execution may be controlled using different strategies without changing the programme logic. Nothing prevents us from performing preliminary optimisation, testing and verification of the DFG in the architecture-independent manner. Further transformations of intermediate programme representations to programmes for real computing systems can be carried out on the already debugged source code by means of formal methods, which would increase programme reliability. Also, it is possible to perform additional optimisations, for instance, to increase the efficiency of memory usage.

It should be pointed out that all transformations are done only after a correctly functioning programme code is written. In the meantime, we have solved only the first part of the problem—programme execution on the emulator of the event machine. The next stage of our development is programme transformations for existing computing systems. Besides making the developed tools more convenient to use, we intend to create an integrated development environment (IDE) that additionally supports function repository, translating, verification and execution of programmes.

References

1. McKenney, P.E.: Is Parallel Programming Hard, And, If So, What Can You Do About It? www.kernel.org/pub/linux/kernel/people/paulmck/perfbook/perfbook.html

2. Karpov, Y.G.: Model Checking. Verification of Parallel and Distributed Program Systems. BHV-Petersburg, Saint Petersburg (2010). (in Russian)
3. Korneev, V.D.: Parallel programming in MPI. Institute of Computational Mathematics and Mathematical Geophysics, Siberian Branch of the Russian Academy of Sciences, Novosibirsk (2002). (in Russian)
4. Akhter, S., Roberts, J.: Multi-core Programming Increasing Performance through Software Multithreading. Intel Press, Santa Clara (2006)
5. Cheng, J., Grossman, M., McKercher Ty.: Professional CUDA Programming. Wiley, Indianapolis (2014)
6. Tay, R.: OpenCL Parallel Programming Development Cookbook. Packt Publishing Ltd., Birmingham (2013)
7. Lastovetsky, A.L.: Parallel Computing on Heterogeneous Networks. Willey, Hoboken (2003). https://doi.org/10.1002/0471654167
8. Maad, S. (ed.): Grid Computing – Technology and Applications, Widespread Coverage and New Horizons. InTech, Rijeka (2012). https://doi.org/10.5772/2290
9. Gaster, B.R., Howes, L., Kaeli, D.R., Mistry, P., Schaa, D.: Heterogeneous Computing with OpenCL. Advanced Micro Devices, Inc., Elsevier Inc., Santa Clara (2013)
10. Voevodin, V.V., Voevodin, Vl.V.: Parallel Computations. BHV-Petersburg, Saint Petersburg (2002). (in Russian)
11. Legalov, A.I.: The functional programming language for creating architecture-independent parallel program. Comput. Technol. **10**(1), 71–89 (2005). (in Russian)
12. Legalov, A.I.: Managing computation in parallel systems and programming languages. Sci. Bull. NSTU **3**(18), 63–72 (2004). (in Russian)
13. Backus, J.: Can programming be liberated from von Neuman style? A functional stile and its algebra of programs. CACM **21**(8), 613–641 (1978). https://doi.org/10.1145/359576.359579
14. Legalov, A.I., Redkin, A.V., Matkovskii, I.V.: Data driven functional parallel programming with data coming asynchronously. In: PACT 2009, pp. 573–578. South Ural State University, Chelyabinsk (2009). (in Russian)
15. Hoare, C.A.R.: An axiomatic basis for computer programming. CACM **12**(10), 576–585 (1969). https://doi.org/10.1145/363235.363259
16. Kropacheva, M., Legalov, A.: Formal verification of programs in the pifagor language. In: Malyshkin, V. (ed.) PaCT 2013. LNCS, vol. 7979, pp. 80–89. Springer, Heidelberg (2013). https://doi.org/10.1007/978-3-642-39958-9_7
17. Ushakova, M.S., Legalov, A.I.: Automation of formal verification of program in the Pifagor language. Model. Anal. Inf. Syst. **22**(4), 578–589 (2015). https://doi.org/10.18255/1818-1015-2015-4-578-589
18. Andrianov, A.N., Baranova, T.P., Bugerya, A.B., Efimkin, K.N.: Nonprocedural NORMA Language and Its Translation Methods for Parallel Architectures. University News. North-Caucasian region, Technical Sciences Series, vol. 3, no. 195, pp. 5–12 (2017). https://doi.org/10.17213/0321-2653-2017-3-5-12
19. Yang, Y.: LabVIEW Graphical Programming Cookbook. Packt Publishing, Birmingham (2014)
20. Kasyanov, V.: Sisal 3.2: functional language for scientific parallel programming. Enterp. Inf. Syst. **7**(2), 227–236 (2013). https://doi.org/10.1080/17517575.2012.744854
21. Legalov, A.I.: Parallel algorithms development. Open Syst. **9**(101), 64–68 (2004). (in Russian)

Machine Learning Techniques for Detecting Supercomputer Applications with Abnormal Behavior

Alexander Bezrukov[1] , Mikhail Kokarev[2] , Denis Shaykhislamov[2] ,
Vadim Voevodin[2]([⊠]) , and Sergey Zhumatiy[2]

[1] Plekhanov Russian University of Economics, Moscow, Russia
Bezrukov.AV@rea.ru
[2] Research Computing Center, Lomonosov Moscow State University, Moscow, Russia
mikhail.kokareff@gmail.com, sdenis1995@gmail.com,
{vadim,serg}@parallel.ru

Abstract. There are different approaches that help to solve the issue of low efficiency of modern supercomputer usage. One of them is based on constant monitoring of a supercomputer job flow in order to promptly detect inefficient programs. The execution dynamics of such programs usually differs from the "normal" behavior of common programs; however, it is very difficult to establish exact criteria for determining abnormal behavior. Machine learning methods are therefore used in this study for detecting abnormal jobs. This paper deals with an important aspect of working with machine learning methods, namely data preparation. The solution proposed herein was evaluated on the Lomonosov-2 supercomputer.

The issue of optimal input data selection is one of the key steps for transferring the methods suggested in the paper to other supercomputers. The analysis described in the article has served as a starting point for developing a methodology for applying overall solutions to other supercomputers, which is also described in this paper.

Keywords: Supercomputer · High-performance computing
Task flow · Anomaly detection · Program efficiency · Machine learning

1 Introduction

High-performance computing is becoming more and more large-scale: the number of scientists from different research areas that use supercomputer technologies for solving scientific problems is constantly growing. This is definitely a positive trend which hopefully will continue in the future. But this has an unobvious

This work was partially funded by the Russian Foundation for Basic Research (grants 16-07-00972 and 17-07-00719) and a study grant from the Russian Federation President's Fund (SP 1981.2016.5).

© Springer Nature Switzerland AG 2018
L. Sokolinsky and M. Zymbler (Eds.): PCT 2018, CCIS 910, pp. 31–46, 2018.
https://doi.org/10.1007/978-3-319-99673-8_3

drawback. A lot of new scientists entering this area are usually skillful specialists in their research areas, such as computational physics, molecular dynamics, weather forecasting, drug design, etc., but they are by no means experienced enough in parallel computing. Taking also into consideration that the architecture of modern supercomputers is highly complex, it becomes really difficult to develop efficient parallel applications that consider all the peculiarities of underlying hardware. And this results in a substantial amount of parallel programs with really low execution efficiency [1].

One can say that more people involved in parallel computing means efficient off-the-shelf application packages being developed, and this is partially true. New ready-to-use packages appear as well as existing packages enhance their functionality, although unfortunately they are often neither very scalable nor efficient in practice. But such packages play significant role in forming the overall efficiency of using supercomputer centers, so their behavior should be monitored and analyzed, which is the goal of another research being conducted at the Research Computing Center of the Lomonosov Moscow State University (RCC MSU) [2].

There are many different approaches as to how the efficiency of a particular parallel program can be analyzed and optimized. Various profilers, trace analyzers, and debuggers have been developed and successfully used to address this task. But before an application can be studied thoroughly, we would have to become aware that this application has low execution efficiency and that it needs to be analyzed. And it turns out that in many cases not only users but also system administrators do not know that an application has some performance issues. This means that a constant monitoring of all programs running on a supercomputer is needed in order to find inefficient applications with possible performance issues.

This work is aimed at solving this particular task. The main goal is to detect abnormal applications, i.e. applications with abnormal behavior which significantly differentiates from the standard behavior of the tasks in a supercomputer job flow. The behavior of applications is described using system monitoring data. Owing to the fact that it is currently almost impossible to precisely establish criteria for abnormal behavior, machine learning (ML) methods are used for that purpose. But tuning machine learning techniques to maximize its performance can be quite tricky. One of the main difficulties that are encountered on this path is to correctly select and prepare needed input data, which can drastically influence the overall accuracy of machine learning methods.

The main contribution of this paper is a description of methods for determining a suitable input data set which can lead to improvements in classification accuracy. These methods were implemented and evaluated using an anomaly detection method developed previously. Furthermore, the conducted data preparation analysis served as an entry point for developing a methodology for applying overall anomaly detection approaches to other supercomputer centers. This newly developed methodology is also presented in this paper.

The paper is organized as follows. Section 2 briefly describes the work that was previously done within this research, as well as related studies. Section 3 is devoted to the problem of data preparation for the machine learning method used for anomaly detection. A methodology for applying the developed solution for anomaly detection to other supercomputers is described in detail in Sect. 4. Section 5 contains the conclusions made as well as plans for future research.

2   Background and Related Work

The main goal of this work is to find abnormally inefficient applications in a supercomputer job flow using system monitoring data. There are several related works with similar goals that could be mentioned. Many of them are based on just static thresholds that help to determine abnormal behavior: this is how system monitoring tools like Nagios or Zabbix do. But this works well just for simple cases and it is required to accurately adjust these thresholds.

For more complex cases, machine learning methods are used. For example, in [3], ML techniques are used for program classification as well. But the authors of that paper use supervised methods for identifying specific applications (e.g., software packages like GROMACS) based on performance data. For detecting inefficient behavior, simple static thresholds were used.

Another example, which is the most related one to our research, is the paper [4]. The authors present a method for detecting performance anomalies in HPC systems. A system monitoring performance data is also used in this case for anomaly detection, although they are interested in detecting performance variations caused by resource contention or hardware/software problems on a node. A number of node-level anomalies like "orphan processes" and "hidden hardware problems" are specified, and are then detected using machine learning methods. So the main goal of this work is quite different, even though the approach used is very similar. It is interesting that Random Forest algorithm showed the best classification results, as in our study. The high performance achieved in paper [4] showed us that the methods we were planning to use in our case should lead to suitable results.

There are other works where machine learning techniques for performance analysis in the HPC area are used (for example, [5,6]), but none of them aims at solving our task. Nevertheless, it should be mentioned that studying these works helped us to determine the methods suitable in our case.

As mentioned earlier, the anomaly detection method proposed is based on analyzing data collected with monitoring systems. At the MSU Supercomputing Center, a set of proprietary tools is currently being used, but it is planned to switch in the near future to the DiMMon monitoring system [7], which is being developed at the RCC MSU for use on exascale-level supercomputers. Data from processor counters (e.g., CPU user load), memory and communication network intensity (e.g., number of L1 cache misses per second, amount of bytes sent per second): all this information is collected for each job running on the supercomputer, forming the basis needed for the performance analysis of job efficiency.

Using these dynamic characteristics for performance description, each job can be classified as normal, suspicious or abnormal. In general, a job is classified as normal if no performance issues are found. A job is called abnormal if it is definitely working incorrectly, wasting computing resources; this could happen if a program stalls or a software/hardware error has been encountered. A job is classified as suspicious if we can detect some performance issues in its behavior but we cannot be sure that this behavior is definitely incorrect (abnormal), so a more detailed analysis is needed.

The overall job classification process is organized as follows (a detailed description can be found in [8]).

Each job is represented by the values of dynamic characteristics changing during the program runtime, that is, by a number of timelines, one timeline per characteristic. For each job, these timelines are divided into time intervals. An intellectual method is used for this purpose that tries to identify substantial changes in the behavior of the program, separating different logical stages of the program execution. In this case, the behavior of each interval is quite simple and can be therefore represented accurately using integral values (e.g., max, min, median, oscillation rate). After timelines are divided into time intervals, each interval is individually classified as abnormal, suspicious or normal using integral values for the chosen dynamic characteristics.

For interval classification purposes, we use a method based on the Random Forest algorithm (Scikit-learn [9] implementation). This is a supervised method that was trained on a set of 520 manually classified intervals (270 normal, 70 abnormal and 180 suspicious intervals), leading to a classifier accuracy of ~0.93 on the Lomonosov-2 supercomputer. The accuracy is calculated as the ratio of correctly classified intervals to the number of all intervals in the set.

When each interval of a job runtime is classified, it is needed to assign a class to the job in general. It is done using a set of criteria that attempt to determine whether a substantial amount of processor time was consumed by intervals with abnormal/suspicious behavior. The results were validated on a test set of 110 applications (32 abnormal, 48 suspicious, 33 normal). The overall job classification accuracy achieved was ~0.92.

The resulting performance is quite high. However, the following should be noted. One of the most important points that influence the performance of machine-learning-based classification is data preparation. And the original selection of the feature set was based only on our initial sense of what, in our opinion, should be most useful for the classification. It was thus decided to study how much accuracy can be increased with a more intelligent approach to the choice of the input data, based on a rich existing analytical experience in this area. Within this paper, we describe a number of different methods we have tried for choosing the most suitable feature set, aimed at increasing the accuracy of the classifier we have developed.

Moreover, the approach for data preparation described in this paper makes this overall classification process much more portable, since choosing the correct input data is one of the most challenging tasks for performing an accurate classification. Following the methodology described in Sect. 4, one can try to implement the described classifier on a different supercomputer.

3 Data Preparation for the Anomaly Detection Approach

In a previous work, we selected and fixed an input feature set for the classifier according to our initial view of which data is the most important in our opinion (here and below, we will refer to this set as the "basic feature set", and, accordingly, to the classifier based on it as the "basic classifier"). However, during the working process, we figured out that this can potentially be improved with useful information about job dynamic behavior that was not used at that time. So it was decided to rethink our data preparation process.

The data preparation stage for machine-learning-based algorithms consists of three steps: selection, preprocessing and transformation [10]. Usually a lot of different types of input data are available that can be used for classification purposes, but using all data not always results in the best performance and, also, it can be very computationally complex. So the **data selection** step is aimed at choosing the right subset of data types that will lead to the best accuracy and/or classification speed.

In our case, the machine learning algorithm uses system monitoring data as input, so we can potentially use all the information that can be collected from the system counters describing the utilization of CPUs, memory subsystem, communication network, etc. It should be noted that there are always hardware restrictions in a processor (which are not the same for different processor families), which allows us to collect only a small amount of processor counters simultaneously. For each supercomputer, we heuristically chose the most suitable data. For example, the set of data being collected on our Lomonosov-2 supercomputer is the following:

- *CPU utilization:* CPU user load, other types of CPU load (system, iowait), loadavg.
- *Memory usage intensity:* number of L1/L2/L3 cache misses per second; number of load/store operations per second.
- *Communication network usage intensity:* number of bytes/packets sent/ received per second, separately for MPI and file system networks.
- *GPU utilization:* GPU user load, GPU memory load, GPU memory utilization.

Using all this data is not the best option, so we need to choose a suitable subset. This task turned out to be the most challenging in the data preparation stage; a description of methods used is further provided.

The second step in the data preparation stage is **data preprocessing**. This step includes such processes as data cleaning and formatting, which in our case is almost not needed at all: the monitoring system provides us with all the needed data in a suitable format.

The third step is **data transformation**. During this step, data scaling or decomposition as well as aggregation can be performed. In this work, a machine learning algorithm is used for time interval classification, and this is done with the Random Forest algorithm. It does not require data normalization, so no need for any data scaling in our case.

Data aggregation is done twice before passing the input feature set to the classifier. At first, 2-min approximation is used: raw data collected by the monitoring system is replaced every two minutes with integral values (max, min, average, etc.). This is necessary to reduce the amount of data that needs to be stored, and it was decided not to change this step in this work.

The next aggregation is done on our side, at the interval level. As mentioned earlier, we form time intervals in such a way that they show a simple behavior which can be accurately described using integral values. This means that each interval and each dynamic characteristic (like CPU user load) is described with only maximum, median, etc. values instead of using a time series of raw numbers (usually, there are 30 to 100 time points representing each interval). Furthermore, most jobs studied in our work are parallel, which means that we also need to aggregate the data across different processor cores in a node, as well as between nodes. So we use triple aggregation in this case: first by space (between cores in a node), then again by space (between nodes) and then by time (between time points).

The question is, what integral values should be used for interval description? We have made a list of possible variants that can be useful in practice, according to our experience. For each variant, three aggregation methods are provided (within a node/between nodes/time):

- *Minimum (within a node)/minimum (between nodes)/minimum (between time points).* Maximum is not so interesting since it is often equal to the peak.
- *Average/average/median.* We have selected median for time aggregation due to its better resistance to outliers, which leads to a more accurate description in many cases.
- *Maximum/average/average.* This value helps to detect imbalance between nodes.
- *Minimum/(average-minimum)/average.* We have selected (average-minimum) for aggregation between nodes instead of just average, since using just average leads to values very close to the minimum, which was already described earlier.
- *Average/average/oscillation_rate.* The oscillation rate is calculated as maximum-minimum range divided by the overall average. This number reflects the relative fluctuation of the max and min values of the characteristic around the average.
- *Maximum/maximum/oscillation_rate.* This helps us to describe fluctuation of characteristics as well. We do not use minimum/minimum/oscillation_rate because it is usually equal to zero.
- *(Maximum/maximum/maximum)/(average/average/average).* Another way of describing fluctuation.

We also thought about adding more integral values, such as skewness or kurtosis [11], but it seems like it would not add any new information to the list of considered values, so it was decided not to expand the list, which was already quite big.

Taking into account the aforesaid, there are two options that we can adjust in order to try to improve classification accuracy, namely what data types should be selected and what integral values should be used. But we have 20 different data types, with 7 integral values to represent each of them, which makes 2^{140} different possible feature subsets. This means that we need some intellectual method to choose one close to the optimal for our purpose.

But before starting to choose the appropriate set of features, it is necessary to understand whether it is worth doing in principle. We need to evaluate whether the possible accuracy gain is statistically significant. If no, there is no sense in performing any analysis at all; the current feature set would then be quite suitable.

For checking the statistical significance of the model accuracy ratio increase, we use Student's t-criterion (1):

$$T = max + 3 * (st\_error), \tag{1}$$

where max is the maximum accuracy obtained using the basic feature set, and st_error is the standard error of the mean for the accuracy with the basic classifier. We use the t-test as it is commonly applied to check the significance of the difference between two values, in our case, the model accuracy values. If the accuracy obtained with new feature sets is higher than this value, we can say that we found a statistically significant better result.

We ran the basic classifier 1000 times and analyzed the cross-validation accuracy in the intervals of real-life applications from the Lomonosov-2 supercomputer. The max was 0.9398, while the average was 0.9324 (most of the accuracy variation is related to the random nature of the classifier due to the use of the Random Forest algorithm). The t-criterion in this experiment is then equal to 0.94. The analysis showed that this value can be exceeded using new feature sets. For example, the average accuracy for the most complete feature set (all 140 features considered) is 0.9434, which is higher than the t-criterion. It should be noted that this set does not suit us owing to the following reasons: (1) the result obtained is possibly not the highest one; (2) using so many characteristics is likely to lead to an overfitted model (and also to issues with classification speed), so this amount should be reduced.

This analysis proved that it is worthwhile to search for more optimal feature sets for our classifier. The standard approach for solving this issue is to use discriminant function analysis. The next section describes in detail how it was used in our research.

3.1 Discriminant Function Analysis

There are three possible approaches for choosing an appropriate feature set:

1. Use all possible features and manually select the most important ones (standard method).
2. Backward stepwise.
3. Forward stepwise.

All these approaches are usually quite similar in terms of the accuracy that can be achieved as a result. Both backward and forward stepwise methods were attempted in order to quantitatively evaluate the significance of the features and obtain the preliminary list of features that remain in the model after the completion of both algorithms, as well as to gain an a priori insight into the features' influences.

Backward Stepwise Method. The main idea of this algorithm is quite simple: we perform a number of steps, and at each step, the feature leading to the least accuracy loss is removed. Scikit-learn provides its own implementation of the backward stepwise algorithm [12] but it calculates the accuracy using only one cross-validation at a time, which in our case leads to rather unstable results. So we decided to implement five cross-validations and take the average result. The overall backward stepwise algorithm looks as follows:

- While the break condition is not satisfied:
  - Temporarily remove one feature from the set A. Calculate the classification accuracy using the resulting feature set A_1.
  - Repeat the previous step for each feature in the set A. As a result, we have N accuracies for all possible sets A_n (where N is the size of the feature set A) without one feature.
  - Find the set A_i that shows the highest accuracy.
  - Use this set A_i and go to the next iteration.

There are two points that need to be clarified. The first one is how the accuracy is calculated. We use cross-validation: the training data is split into five equal parts; one part serves as the test set, the other four are chosen as the training set. There are five possible ways to do this, and the cross-validation accuracy is calculated based on this 5-fold splitting. Then we repeat cross-validation five times and take the average accuracy value. The second point to clarify is the break criterion. There are several standard ways to do this: do not stop until a particular number of features is reached or do not stop until the accuracy loss is less (or the overall accuracy is higher) than a specified threshold.

Figure 1 shows average accuracy results for the backward stepwise method. Along the X axis, we have the numbers of features left in the feature set; the Y axis corresponds to the accuracy value for the chosen feature set at each iteration of the backward stepwise algorithm.

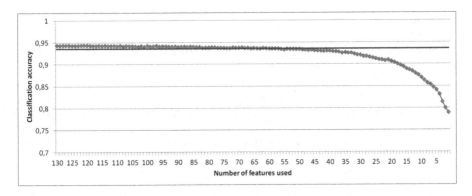

Fig. 1. Interval classification accuracy variation during the backward stepwise method. The horizontal line corresponds to the basic classifier accuracy

It can be seen that the same accuracy as for the basic classification (marked with a red horizontal line) is achieved with 50 to 70 features, whereas the basic feature set consists of 33 features. Using 70+ features is not a suitable option for us because it does not provide a substantial accuracy gain, while decreasing the classification speed and increasing the probability of the overfitting problem. This means that the use of the backward stepwise method leads to worse results than those of the basic classification.

So it was decided to switch to the second method: the forward stepwise.

Forward Stepwise Method. The forward stepwise algorithm works the other way round compared with backward stepwise: we start with a small feature set and add one feature that maximizes the accuracy at each iteration. But there are also several points that should be determined in this case: what features to start with and when to stop.

One of the most important steps in the forward stepwise algorithm is the choice of a suitable starting feature set. Usually it includes 5 to 10 features. But they can be selected in a different way, for example, randomly or based on expert knowledge. At first, we tried to rely on our experience-based assumption that the median and the oscillation rate are the most important integral values that are useful for the classification process. So we tried to use only them as the starting feature set. The break criterion was the maximum size of the feature set: not more than 35 to 40 features. However, this led to poor accuracy results, meaning that this assumption is not good enough. Using randomly chosen starting sets is not a good option either, since we know that some combinations of features must be included, otherwise important behavior peculiarities could be omitted by the classifier.

As a result of trying different variants, the following method for choosing a starting feature set and finishing the forward stepwise process was formed. We combined data types into the following groups (several groups contain only one data type):

- CPU load;
- load average;
- memory reference intensity (number of load/store operations per second);
- number of L1 cache misses per second;
- number of L2 cache misses per second;
- number of L3 cache misses per second;
- MPI network usage intensity (number of bytes/packets sent/received per second);
- file system network usage intensity (number of bytes/packets sent/received per second);
- GPU utilization (GPU user load, GPU memory utilization).

Each group represents a part of information that must be included in the resulting feature set one way or another. Each group can be represented using any integral value specified earlier; there are no restrictions on that. It should be noted that three data types, namely CPU system load, CPU iowait load and GPU memory load, were considered not important enough, so they are not included in any group and may not therefore be included in the resulting feature set.

Initially, five random features were selected from the list above. We chose the following break criteria: (1) at least one feature from each group must be included into the feature set; (2) the size of the feature set must be not less than 30. The first criterion is needed to be sure that the resulting feature set includes all the information we think is necessary. The second one guarantees that the feature set will not be too small, which, in most cases, leads to poor accuracy results.

Owing to the random nature of the classification algorithm used, the feature set obtained can be quite different each time we run the forward stepwise method. So, in order to choose the most appropriate one, we need to collect enough statistics.

We ran the forward stepwise method 50 times and obtained 50 different feature sets. Next, we needed to determine which feature set shows the best performance results. So we performed cross-validation 2000 times for each feature set and calculated the average accuracy. After that, we took the top five feature sets having highest accuracy (their results were very similar) and then selected the best one among them, using our own knowledge on what features are more important. The chosen set turned out to be also the most uniform: the number of features in each group was almost equal.

Final Results. After we determine the final feature set, it is necessary to evaluate the results achieved using the final feature set (referred further as the "final classifier") compared to the performance of the basic classifier.

The overall cross-validation accuracy of the interval classification for the Lomonosov-2 real-life applications improved from ~ 0.93 with the basic classifier to 0.95 with the final classifier. This means that the final accuracy is above the t-criterion value, which is 0.94 (calculated in the beginning of Sect. 3), so that the achieved accuracy improvement is statistically significant.

The confusion matrix (Table 1) summarizes the classification accuracy for particular classes (average values for 1000 cross-validation runs). It can be seen that our final classifier works best with the normal class, but the results for other classes are also quite high.

Table 1. Confusion matrix for interval classification

		Predicted class		
		Normal	Abnormal	Suspicious
Actual class	Normal	268	0	6
	Abnormal	2	62	5
	Suspicious	10	1	168

False-negative error is also an important measurement in our case. It is not a big issue to misclassify a few normal jobs as suspicious or abnormal since we want just to notify users about anomalies found. But it is not acceptable to miss abnormal or suspicious behavior since it leads to loss of computing resources. The results show that this error is very small for the final set: 0.027 in average.

We have also compared other classification performance metrics, such as the F-score and the false-negative error. F-score results for both the basic and the final versions are summarized in Table 2. The F-score helps to evaluate the classification results from another point of view. It is calculated using the following formula (2):

$$F = 2 * (precision * recall)/(precision + recall). \tag{2}$$

According to Table 2, the F-score also improved for each class. This is especially true for suspicious jobs, where it increased from 0.892 to 0.935.

Table 2. Comparison of F-score values for the basic and the final classifiers. F-score calculated independently for each class

	F-score (normal)	F-score (abnormal)	F-score (suspicious)
Basic classifier	0.944	0.918	0.892
Final classifier	0.966	0.93	0.935

All the described results for interval classification show that the accuracy improved compared to the basic version. But our final goal is to detect jobs with abnormal behavior, so we need to evaluate job classification results as well. The performance for the job classification was tested on previously unclassified real-life jobs from the Lomonosov-2 supercomputer. We detected 190 suspicious and 64 abnormal jobs with our final classifier based on the analysis of ~10 days of the Lomonosov-2 functioning. These results were manually validated, leading to accuracies of 0.98 and 0.95 for abnormal and suspicious jobs, respectively.

Thus, it can be seen that the final feature set obtained using discriminant analysis enabled us to improve the overall classification accuracy. Moreover, the data preparation process described in this section made the overall classification process much more portable. The next section is devoted to this topic.

4   Methodology for Applying Anomaly Detection Method to Other Supercomputers

The methodology in this section describes in detail the process of applying the solution proposed for anomaly detection to other supercomputing systems. At the top level, this methodology is quite universal and, in fact, is suitable for most machine learning based classification methods; however, more specific details of this process relate to this method in particular. A description of the sequential steps of this methodology is provided below.

1. Prepare core software tools based on the proposed methods. This can be implemented manually using the description given in this paper, or the source code that we plan to upload to GitHub in the near future. This core software should include methods that are independent of the supercomputer it is used on: method for partitioning the job timeline into intervals, interval classifier, and job classifier based on the interval classification results.

2. Determine the range of possible input data. In this step, all data that can potentially be useful for the interval classification process should be selected. As in our case, which was described in Sect. 3, this includes selection of data types and integral values used for aggregation. It is worth recalling that integral values are used for triple aggregation: by time and twice by space (between cores in a node and between nodes). We believe that both data types and integral values chosen in our case can serve as a good starting point by default, but a user may make his own changes in these lists if needed, according to his knowledge of the usual behavior of jobs on the target supercomputer.

3. Implement data collection using a monitoring system. For detection of anomalies, we need the input data that were described in the previous step. This is provided by a monitoring system, so one should be installed and configured at the target supercomputer. This step is done outside of our anomaly detection process, so we do not specify how this step should be done.

Nevertheless, several remarks should be made. Usually, it is impossible to store all the data that a monitoring system can provide for the whole supercomputer, so data aggregation is used (as it was done in our case, see Sect. 3). The frequency of data aggregation can influence the classification performance, and one should keep that in mind: if classification results are too low, a possible reason could be that data are too frequently aggregated.

The second remark is that this step can reduce the range of possible data from step 2. This is due to the limitations of the hardware being monitored.

For example, modern processors allow to simultaneously collect data from just a few hardware counters, so we need to select the most needed ones. Moreover, different processor families provide different sets of counters. So it is likely that the range of data specified in step 2 will have to be adjusted after the installation of a monitoring system.

4. Choose the initial feature set. At this point we have fixed the data that can possibly be used for classification. So now the initial feature set that will be used in the basic (reference) version of the classifier can be determined. This set is chosen on the basis of our understanding of a feature importance since, at this point, we have no analytical insights on which features are more important for classification.

Creating this basic version is done for several reasons. Based on the results of the reference version, we can decide whether the proposed approach is applicable in general in this case. Working with the basic version, we get first assumptions on the preferable size of the feature set, the time needed for classification, etc. Next, we can use reference classification results for comparison with all other versions that are going to be created in future. And also, this allows us to verify whether this implementation works correctly on this supercomputer.

5. Create the training set. This is both one of the most challenging and one of the least automated steps. We need to scan through real data (real-life supercomputer job flow) and pick out intervals we want to include in the training set. Each interval should be classified as normal, suspicious or abnormal, based on the chosen monitoring data set. Also, since our final goal is to classify applications, we need to create a second training set of classified jobs, so we must assign a class to each job according to its classified intervals. It is not necessary to use every interval from a job in the training set; only the most useful ones may be included. But if not all intervals in a job are classified, this job normally should not be included in the second training set for the job classifier (the one based on the simple criteria, not on ML techniques) since the result in this case can be incorrect.

In our experience, the training sets do not need to be very big: 500+ intervals from 100+ jobs were enough for our classifier.

There are several general rules that should be followed during this process:

- The manual interval classification for the training set should be based on exactly the same data that will be used in the main classification process.
- The training set should include as many different types of dynamic behavior as one wants the classifier to identify. If one type of behavior is not present in the training set, then the classifier is likely to misclassify it.
- The numbers of intervals in the classes of the training set should not differ significantly. Otherwise, the classification results can be incorrectly biased to the more popular class.
- A suspicious class can be divided into subclasses if desired. This can make classification results more informative but only if the division into subclasses in the training set was made accurately enough. There is usually no point

in detecting subclasses in a normal class (since there are no performance issues in such jobs) as well as in an abnormal class (since such jobs behavior normally is not so diverse). At the same time, there are many possible types of dynamic behavior for suspicious jobs. For example, on the Lomonosov-2 supercomputer, we have manually found such types of specific behavior as "1 active process on each node", "1 active process on all nodes", "stalled because of Lustre issues", etc. It should be noted that the classifier developed has not been tested in practice with subclass division, but this should work out of the box.

6. Configure proposed core software. The methods developed in the core software have a number of input parameters that can be configured. All input parameters can be used with default values but some of them depend on the target supercomputer peculiarities, so it is recommended to analyze if more suitable values can be used.

The method for partitioning the job timeline into intervals has only one parameter: the minimal number of time points in an interval. In our case, it was decided that each interval should be not less than 30 min, otherwise the number of intervals in the job could be too large. Since 2-min aggregation is used on the Lomonosov-2 supercomputer, it results in a minimum of 15 time points per interval.

The main interval classifier has two internal parameters that are used for tuning the Random Forest algorithm: (1) the number of trees in the ensemble, and (2) a measure for choosing the optimal database split (impurity). On the Lomonosov-2 supercomputer, 256 decision trees are used in production mode and 32 in test mode (using less trees speeds up the classification process significantly and leads to only a slight decrease in accuracy); increasing this value does not lead to any significant changes in classification accuracy but causes a decrease in speed. Also, Gini impurity measure is used since it tends to provide more accurate results. We believe that it is not necessary to change these parameters in most cases, but they may be adjusted if needed.

The last method developed is the job classifier based on interval classification results. It uses a set of criteria (see [8]) based on constant thresholds that generally determine how much CPU hours (and what part of the overall job) are consumed by abnormal/suspicious intervals. It is recommended to configure these parameters based on the job flow structure of the target supercomputer, since the default thresholds were specifically adjusted for the Lomonosov-2 supercomputer.

7. Run the classification using the initial feature set. All the preliminary steps are done, so now it is time to run the classification for the first time, using data from the initial feature set. If the accuracy results are unsatisfying, it may be necessary to rethink steps 2 (in the part related to the aggregation implementation), 5 or 6.

8. Search for a suitable feature set using the forward stepwise method. This step is devoted to the feature selection, which was described in detail in

Sect. 3. This is another one of the most challenging steps, along with step 5, since there is a lot of possible ways to implement it. According to the results from Sect. 3, the following steps should be carried out:

1. *Choose forward stepwise parameters.* This includes the initial set of features to start from, as well as the break criterion (at what point the forward step-wise algorithm must stop). By default, values specified in this paper may be used but these can be altered if needed. This step also includes changing the forward stepwise method to other possible methods, but we hope this will be necessary on rare occasions.
2. *Run the forward stepwise method.* As a result, a new feature set will be defined.
3. *Tune the feature set.* It is always useful to add more semantic knowledge to the classification process. If some features are known to be meaningful for classification, it is likely that they should be added to the resulting feature set. But this should be done with caution, since it is very hard, in our experience, to understand all the dependencies in the performance data collected for supercomputer jobs.
3. *Check the accuracy.* After the feature set is formed, it is necessary to check the classification accuracy. The result can be now compared to the reference results achieved with the initial feature set.
4. *Repeat if necessary.* If the classifier works poorly, return to the feature set tuning in this step. If this does not help, return to step 5 or 6.

9. Verify the results. At this step, we have developed a working classifier that shows, hopefully, high-performance results. However, this was achieved on training data, and the classification accuracy on real-life data can be slightly different owing to possible shortcomings of the training set (see step 5). So we need to verify the accuracy of the resulting classifier on unclassified real-life data. This can be done in a similar way as shown in Subsect. 3.1.

5   Conclusions

This paper describes the data preparation process for a machine learning method used for anomaly detection in a supercomputer job flow. The main goal is to determine what data types should be included in the feature set and how this data should be aggregated. Discriminant analysis methods were used for this purpose. The best results were obtained with the forward stepwise method, resulting in a new feature set that helped to increase the accuracy from 0.93 to 0.95.

The research conducted in this paper have shown that the basic classifier with the initial feature set shows a very good accuracy which can only be slightly improved. But the developed method for choosing an appropriate feature set has allowed us to make the overall anomaly detection solution much more portable. Thus, a methodology for applying this solution to other supercomputer systems has been proposed, which is also described within this paper.

In the future, we plan to further apply the proposed anomaly detection solution in practice. It is planned to implement an online job classification which would allow us to promptly notify supercomputer users about their running jobs that exhibit a suspicious behavior. Also, we are looking forward to trying our solution on other supercomputers, so we could analyze its performance and portability and make further improvements if required.

References

1. Voevodin, V., Voevodin, V.: Efficiency of exascale supercomputer centers and supercomputing education. In: Gitler, I., Klapp, J. (eds.) ISUM 2015. CCIS, vol. 595, pp. 14–23. Springer, Cham (2016). https://doi.org/10.1007/978-3-319-32243-8_2

2. Shvets, P., Voevodin, V., Zhumatiy, S.: Statistics of software package usage in supercomputer complexes. In: Proceedings of the 3rd Ural Workshop on Parallel, Distributed, and Cloud Computing for Young Scientists. CEUR Workshop Proceedings, vol. 1990, pp. 20–29 (2017)

3. Gallo, S.M., et al.: Analysis of XDMoD/SUPReMM data using machine learning techniques. In: 2015 IEEE International Conference on Cluster Computing, pp. 642–649. IEEE, September 2015. https://doi.org/10.1109/CLUSTER.2015.114

4. Tuncer, O., et al.: Diagnosing performance variations in HPC applications using machine learning. In: Kunkel, J.M., Yokota, R., Balaji, P., Keyes, D. (eds.) ISC 2017. LNCS, vol. 10266, pp. 355–373. Springer, Cham (2017). https://doi.org/10.1007/978-3-319-58667-0_19

5. Zhang, H., You, H., Hadri, B., Fahey, M.: HPC usage behavior analysis and performance estimation with machine learning techniques. In: Proceedings of the International Conference on Parallel and Distributed Processing Techniques and Applications (PDPTA), p. 1. The Steering Committee of The World Congress in Computer Science, Computer Engineering and Applied Computing (WorldComp) (2012)

6. Sidnev, A., Gergel, V.: Automatic selection of the fastest algorithm implementations. Numer. Methods Program.: Adv. Comput. **15**(4), 579–592 (2014). (in Russian)

7. Stefanov, K., Voevodin, V., Zhumatiy, S., Voevodin, V.: Dynamically reconfigurable distributed modular monitoring system for supercomputers (DiMMon). Procedia Comput. Sci. **66**, 625–634 (2015). https://doi.org/10.1016/j.procs.2015.11.071

8. Shaykhislamov, D., Voevodin, V.: An approach for detecting abnormal parallel applications based on time series analysis methods. In: Wyrzykowski, R., Dongarra, J., Deelman, E., Karczewski, K. (eds.) PPAM 2017. LNCS, vol. 10777, pp. 359–369. Springer, Cham (2018). https://doi.org/10.1007/978-3-319-78024-5_32

9. Pedregosa, F.: Scikit-learn: machine learning in Python. J. Mach. Learn. Res. **12**(Oct), 2825–2830 (2011)

10. How to Prepare Data for Machine Learning. https://machinelearningmastery.com/how-to-prepare-data-for-machine-learning/

11. Measures of Skewness and Kurtosis. http://www.itl.nist.gov/div898/handbook/eda/section3/eda35b.htm

12. Feature Elimination Implementation in Scikit-Learn. http://scikit-learn.org/stable/modules/generated/sklearn.feature_selection.RFECV.html

Role-Dependent Resource Utilization Analysis for Large HPC Centers

Dmitry Nikitenko(✉) ⓘ, Pavel Shvets ⓘ, Vadim Voevodin ⓘ,
and Sergey Zhumatiy ⓘ

Research Computing Center, Lomonosov Moscow State University, Moscow, Russia
{dan,shpavel,vadim,voevodin}@parallel.ru

Abstract. The resource utilization analysis of HPC systems can be performed in different ways. The method of analysis is selected depending primarily on the original focus of research. It can be a particular application and/or a series of application run analyses, a selected partition or a whole supercomputer system utilization study, a research on peculiarities of workgroup collaboration, and so on. The larger an HPC center is, the more diverse are the scenarios and user roles that arise. In this paper, we share the results of our research on possible roles and scenarios, as well as typical methods of resource utilization analysis for each role and scenario. The results obtained in this research have served as the basis for the development of appropriate modules in the Octoshell management system, which is used by all users of the largest HPC center in Russia, at Lomonosov Moscow State University.

Keywords: HPC center management
Application efficiency analysis · User roles · Analysis scenarios
Supercomputer

1 Introduction

1.1 The Variety of Resource Utilization Analysis Levels

Nowadays, the issue of computing resource utilization efficiency is a very hot topic. There are many points of view and research subjects that fit into this issue, such as workload efficiency, power consumption, and others. The most interesting thing is that all these aspects effect efficiency, and one has to take into consideration many of them at a time to gain a realistic overall picture. Moreover, there are many different levels of efficiency analysis, especially when

The results were obtained at the Research Computing Center of Lomonosov Moscow State University. The work was partially funded by the Russian Foundation for Basic Research (grant № 17-07-00719), and with financial support from the Russian Science Foundation (grant № 17-71-20114) in the part of the program implementation described in Sect. 4. The research was carried out on equipment of the shared research facilities of HPC resources at Lomonosov Moscow State University.

© Springer Nature Switzerland AG 2018
L. Sokolinsky and M. Zymbler (Eds.): PCT 2018, CCIS 910, pp. 47–61, 2018.
https://doi.org/10.1007/978-3-319-99673-8_4

one considers HPC and distributed computing. For instance, we can define four levels of computing resources.

First, at the top of the pyramid, a supercomputer center administration is interested in global figures and resource utilization rates with almost no need to go through all the messy details of thousands of applications. At the same time, it is natural at this level of observation to have an interest in comparing workload with resource utilization rates of supercomputers available at an HPC center.

Second, a close level is one at which one studies resource utilization and workload for a specific system. No system holder is interested in wasting costly resources for the whole system and for each system partition.

Third, at this point levels stop being mapped to any specific part of the HPC system. This is the level of research projects. Every research project can have a number of participants that run jobs on some or all HPC center computers. Of course, both system holder and project member are interested in details of the project resource utilization, at least to fit into the granted amount of resources for the project.

The last but not the least is the level of application run. Every job is interesting because it has en effect both on the whole HPC center resource utilization profile and on its own job efficiency, as it can significantly bring closer or delay the obtention of the result.

The set of these levels or layers can be extended to a more complicated hierarchy, but even at this point we can see obviously different scopes of interest with personalized accents on some specific system utilization parameters. Moreover, every level requires its own access permissions, so we see different roles of users at each level of abstraction.

As soon as we speak about resource utilization, one of the most common techniques of getting all required information is system monitoring. There is a diversity of various monitoring systems that are focused on specific targets. In this paper, we keep to the tools that have been developed or adopted and widely used in our practice at Lomonosov Moscow State University HPC center.

1.2   The Paper Structure

The "Background" section describes the current state at MSU regarding the paper topic, which served as the basis for the research. The next section, "The Proposed Approach Principles", provides a description of selected key roles and scenarios of resource utilization study. The "Implementation" section provides technical solution details. The "Evaluation" section describes our experience in using the methods developed during our research. The "Conclusions" section lists further steps of research and development. The "References" section ends the paper.

2   Background

Understanding resource utilization profiles regarding both machines and research projects has always been an element of primary importance at every HPC center [1]. The larger the HPC center is, the more important that element is. This has always been a hot topic for the Lomonosov State University HPC center as the largest of this kind in Russia [2].

There is an impressive number of various approaches to performance and efficiency analysis for HPC applications [3–7]. Nonetheless, the peculiarities of running a large academic supercomputing center drove MSU to develop a set of mutually reinforcing and complimentary tools and methodologies. Every part of this toolkit has originally been developed as an open-source tool. Figure 1 gives a short overview of the tools hierarchy.

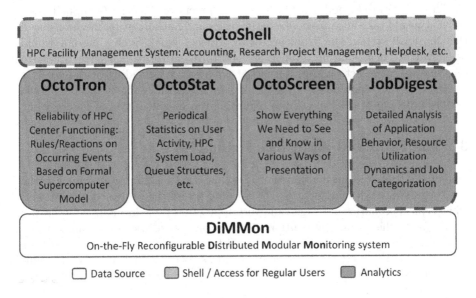

Fig. 1. JobDigest and OctoShell system as a part of the MSU HPC center toolkit (Color figure online)

The work described in this paper extends the interaction of JobDigest [8,9], a detailed application analysis tool, with Octoshell [10,11], a general management system. These two blocks are shown enclosed in dashed red boxes in Fig. 1.

The JobDigest[1] approach provides details on resource utilization for every application. This can be done in various ways [12,13], but generally the JobDigest reports can also be superfluous and some extra lightweight forms, the mini

---

[1] JobDigest® is a Russian registered trademark. An application for the creation of the JobDigest approach was filed and the corresponding patent was granted.

JobDigest reports, may be required. JobDigest was originally developed as a precise tool that can be used both by experienced users and by beginners, by users and by administrators. Nevertheless, the important issue of private and business data isolation from third party or unauthorized users has not been thoroughly studied yet.

As a result, this analysis tool, though perfect at its main objective, really needs to implement access privilege techniques for sharing collected data for jobs, systems and components between authorized users only, and the development of mini JobDigest is required for quick job reviews.

At this point, the authors are quite happy with having Octoshell, a modular management system, at their disposal. This system was originally developed to serve as a connecting link between managed objects of totally different kinds: accounts, users, projects, quotas, and so on. Notably, a set of user roles are present in the system basics.

These facts are a good basis for the development of a special OctoShell module that would take advantage of existing roles and authorization mechanisms of the OctoShell system and grant access to projects logics. Such a module would be aimed at encapsulating JobDigest reports in all forms, securing access to sensitive user, project or application data, as well as providing a user-friendly interface to the available resource utilization submodules for every user of the center according to the access level regarding the specified level of observation.

3 The Proposed Approach Principles

One of the first, and most important questions that should be answered in the very beginning of development is who is going to use the proposed services and what typical scenarios of usage can every type of user go through.

Actually, the main contribution of the paper is the way we combine these two things. From one side, there is a set of typical user role definitions. On the opposite side, there are typical scenarios and usage cases that are often encountered in resource utilization studies based on system monitoring and resource management data.

3.1 Levels of Analysis

As it has already been mentioned in the introduction, we can go down from the level of overall system observation to the level of detailed job analysis.

Here we emphasize the following levels of abstraction and observation:

1. Overall job run states.
2. Integral job characteristics.
3. Detailed job information.
4. Heuristics and ML-based reports.
5. HW/SW failure influence.
6. Other custom levels.

Overall Job Run States. Overall job run statistics is the top level of abstraction, which represents the actual resource utilization by the whole system or its part for a specified period of time. In our opinion, it is reasonable to limit the bottom of this level to the system partition level. The log files of most resource managers allow for grouping finished and running jobs into categories by job state and, what is more, for summarizing utilized CPU or core hours. System monitoring integration allows calculating sums for any other resource amounts utilized and/or granted. So, in a very similar way, one can observe the distribution of jobs according to their states and utilized resources for every partition, system or the whole HPC center.

It is quite useful to have an option to quickly jump to more detailed information for specified partition jobs, or even for jobs with a certain state, say, "TIMEOUT". In other words, an option to go deeper into details, going down to the next level of observation.

Integral Job Characteristics. At the integral job characteristics level, one starts seeing the details of jobs. At his step, job details are presented as basic information from the resource manager, supplied with average rates of dynamic job characteristics, such as CPU_user, load average, etc., and tags for every job. Integral job characteristics can be highlighted according to some rules or thresholds. At this step, the user can see all the available jobs with easy-to-understand general job information: was it resourceful in terms of memory, CPU or GPU usage, did it finish normally, and so on.

This step is obviously expected to have a possibility to proceed to the details of a chosen job. The detailed job information is the next to the bottom level of abstraction.

Detailed Job Information. Detailed job information can be provided in various ways. The most natural for us is the JobDigest approach. The JobDigest reports provide basic job information received from the resource manager and dynamical job characteristics from the monitoring system, in the form of heat maps, diagrams or raw data for export and further analysis. It also provides tags for every job, i.e. some automatically (and/or manually) assigned categories based on thresholds or more complicated rules.

The problem is that such a report sometimes is redundant, that is why we have introduced the lightweight version containing no diagrams but showing all important information: the so-called mini Job Digest. It is specially designed to be provided to every user of the HPC center, and for every job. If required, it can also contain unique links to the full JobDigest report version.

Heuristics and ML-Based Information. Of course, thresholds are still of a significant value to identify many categories of jobs, but recently a number of methodologies have evolved that provide efficient methods for class revealing, similarity study, and so on [14, 15]. There is an interesting direction of research at the MSU HPC center devoted to such techniques. It allows revealing anomalies both in job profiles and in job queues [16].

Even though it does not have a user interface yet, the results obtained are already quite promising and we expect a special module to appear and become available soon.

HW/SW Failure Influence. It is quite natural that one of the root causes of drop-downs in the efficiency of an HPC system and its applications are failures of system hardware, such as interconnect interface, or software, such as problems with schedulers. Sometimes, such problems are found and fixed almost immediately. Nevertheless, the influence of such factors can be on occasions critical for the result or accuracy.

That is why we keep in mind a tool that would allow matching jobs with known problems all over the system, based on resilience system logs. In our case, the OctoTron system [17].

Other Custom Levels. As we realistically look at the problem, we understand that we should support extending this set of levels with new ones as soon as it is wanted and developed.

3.2 User Roles

Regular User. Actually, regular user is the most important role, just because all these systems are originally designed to perform actions that allow achieving real-life research goals by a scientist or an engineer. And that explains the scope of interest of most users. Some regular users do not care much about efficiency of applications, but if a user has some limitations like disk quota or limited CPU time, that becomes critical as it can prevent from obtaining the results in time or at all. Users who run their codes or packages regularly usually feel more interested in the efficiency and execution time of the routines. Moreover, most users still understand that efficient resource utilization is beneficial for everybody: both for application owners and for system holders.

Anyway, the variety of users determines the scope covered by the analysis.

The important thing is to provide means to collaborate in job efficiency and study overall stats regarding workgroup activity.

Another important task is to secure job-related data from being accessed by any other regular user outside the workgroup. One can configure the system in a way to limit job-detail access rights either to the set of jobs run by the owner or to the set of jobs run by the workgroup which the user is a member of.

Project Manager. Project manager is almost a regular user with one key difference: responsibility for the workgroup actions, being the official representative of the workgroup. So, in any case, it is quite natural for such role to have access to all personal jobs and to job stats of the workgroup members. The main difference from the user is a more concentrated focus on overall statistics, as the project manager is more concerned about keeping the project to the granted amount of resources.

Administrator. Going to the other side, system administrators are originally targeted at running HPC systems and helping users to overcome difficulties while using these machines. That implies covering all possible levels of observation in all possible combinations.

System holders or HPC center managers have almost the same rights that administrators have, but like project managers are more focused on overall stats on system usage, and certain workgroup or account activity.

Expert. There can be supervisors with some reduced scopes of analysis. For example, a role that can be used for real-time open demonstrations of what is going in the center right now, but only for some special events regarding a selected workgroup or partition.

As for the MSU HPC center, we actively use the Expert role for annual project expertise. This allows experts to see the job history and details only of those sets of accounts that belong to the project that has been assigned to the expert for review.

3.3   Jumps Between Levels of Analysis

The described levels of analysis, as noted, are interrelated. In order to develop a more convenient and effective tool, we consider the following requirements for quick links between levels.

- Jump from overall job run states to a list of jobs with more detailed, integral characteristics for a selected set of logins (i.e. projects), for a certain state, for a certain queue, for a certain system, or for combinations thereof.
- Jump from the list of jobs to a sublist of jobs (specification of the list of jobs by tags, dates, etc.).
- Jump from the job list to detailed job info, mini JobDigest for a selected job by its ID.
- Jump from a mini JobDigest to a full-format external JobDigest for a selected job by its ID.

3.4   Functional Description of the Interface

We consider the following basic functional features for each one of the proposed levels of the prototype.

Overall Job Run States. Purpose: granted resource-utilization rate assessment by user applications and an estimation of conformity of resources utilization to allocated limits.

Content: average and total amounts of CPUh, GPUh, disk usage for multiple logins grouped by whole systems, partitions, job states.

Filtration: by system, by partition, by job states, by time interval, by project, by login.

Features: job data access segregation: user (own logins), project manager (own logins and managed projects' logins), expert (logins of additional projects assigned to an expert), administrator (all logins, all projects).

Additional features: comparison with allocated quotas for a project; comparison with the same period preceding a displayed interval, quick jump to a job list corresponding to the selected group (for example, all completed jobs or all successfully completed jobs in a specified section).

Integral Job Characteristics (Job List). Purpose: qualitative assessment of resource utilization by jobs, search for abnormal launches, comparison of application runs.

Content: a list of jobs with characteristics and color markup.

Filtration: by system, by partition, by job states, by time interval, by project, by login, by values range for each characteristic, by tags.

Features: job data access segregation: user (own logins), project manager (own logins and managed projects' logins), administrator (all logins, all projects).

Detailed Job Information. Purpose: qualitative assessment of resource utilization by a job.

Content: reduced version of JobDigest: integrated characteristics and data from the resource manager.

Features: job data access segregation: user (own logins), project manager (own logins and managed projects' logins), administrator (all logins, all projects).

Additional functions: unique link to the full JobDigest report.

4   Implementation

Let us now describe the technical implementation. All the tools are implemented as a module of the OctoShell system, which allows using the built-in roles separation mechanism, while users get access to a generally familiar interface, so as to expect a more successful and frequent use of the development by the users.

All necessary data is stored locally in the system and is obtained from a third-party tool operating in 24/7 mode, which builds a full-format JobDigest, allocates categories, and so on. The Octoshell job service retrieves all data from an external supercomputer job data storage and processing service.

Data access is performed using ORM technique, and Ruby on Rails web app development framework is used. All general data are stored in a database table with a structure as shown in Table 1. Fields `id`, `login`, `start_time`, `end_time` are used for indexing. It allows to speed up the most common requests for user job querying in a selected time interval.

Integral job characteristics are stored using three tables in the database.

The first table contains three fields: `id`, `name`, `type`. The `name` field holds the name of the characteristic, and the `type` field its type (numeric or text). This table is used to identify what kind of characteristics are available and what kind of data they present.

Table 1. The structure of the general job information storage table

Attribute	Description
id	Entry ID
job_id	Job ID
login	System user name
partition	Supercomputer partition
account	Accounting user name
submit_time	Submit time of the job
start_time	Start time of the job
end_time	End time of the job
timelimit	Time limit of the job
job_name	Name of the job
state	State of the job
priority	Priority of the job
req_cpus	Number of requested cores
alloc_cpus	Number of allocated cores
nodelist	List of allocated nodes

The other two tables have the same structure, as shown in Table 2.

Those two tables are used for storing actual integral characteristics data. The only difference between them is the type of their value field.

The `id` and `task_id` fields are used for indexing. To obtain the integral characteristics for a task, one should query the characteristics metainfo from the first table and query actual data from the corresponding characteristic table.

The service allows displaying the short version of the JobDigest with optional access to the full JobDigest as an external service.

The structure of the short JobDigest is stored using a table similar to the one described previously (`id`, `name`, `type`). That table stores the description of the monitoring sensors used in JobDigest. Values are stored in a table with a structure as show in Table 3.

Table 2. The structure of the job integral characteristics storage table

Attribute	Description
id	Entry ID
name	Name of the characteristic
task_id	Job entry ID (see Table 1)
value	Value

Table 3. The structure of the job dynamic characteristics storage table

Attribute	Description
name	Name of the characteristic
task_id	Job entry ID (see Table 1)
time	Time
value	Value

The id and task_id fields are used for indexing. The access to the full Job-Digest is granted with a unique URL which is stored as usual job characteristics of text type.

The service allows using tags assigned to a job. Tags are stored in a table with a structure as show in Table 4.

Table 4. The structure of the job tags storage table

Attribute	Description
id	Entry ID
name	Tag name
task_id	Job entry ID (see Table 1)

All fields are used for indexing.

Updates are performed using external POST requests.

The first request is used to update general JSON information. If a job is not present, then a new entry is added.

The second type of request inserts data about the integral characteristics into the database, and the data is transmitted in the body of a POST request in JSON format.

The third type of request inserts the tag data into the database, and the data is sent in the body of a POST request in JSON format.

The fourth type of request adds to the database data about a series of changes in the value of the sensor during job operation. The data is sent in the request body in CSV format.

The overall system workflow is shown in Fig. 2.

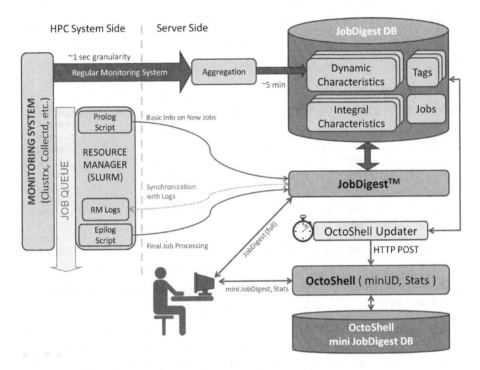

Fig. 2. General OctoShell mini JobDigest DB workflow

5   Evaluation

The implemented prototype is available for users of the MSU HPC center. At present, we are collecting feedback that should aid us in further approach elaborations. We hereby thank one of the workgroups at the MSU HPC center for depicting the interface. All presented data correspond to real research [18,19].

Figure 3 illustrates the interface for the level of overall job run states. We can see that the "Lomonosv-2" system was used by the project only during the 2017Q2. It is quite nice to see that users did really use the test partition for testing. The majority of resources have been spent for successfully completed jobs in the compute partition during the period.

| Start date | 2017-06-01 |
| End date | 2017-09-01 |

Show

Total

System	Total tasks	Total CPU*Hours
Lomonosov-1	0	0.00
Lomonosov-2	3150	1283027.17
TOTAL	3150	1283027.17

Lomonosov-1

Partition	State	Count	Cores*Hours

Lomonosov-2

Partition	State	Count	Cores*Hours
compute	TOTAL	2717	1278331.02
	COMPLETED	1672	1004594.98
	CANCELLED	774	51094.17
	FAILED	210	41484.33
	TIMEOUT	44	70845.87
	NODE_FAIL	17	110311.68
test	TOTAL	433	4696.15
	COMPLETED	205	815.45
	FAILED	88	264.77
	CANCELLED	81	648.31
	TIMEOUT	59	2967.62

Fig. 3. States of selected project jobs for a certain period of time with total resources utilization example

Figure 4 provides the details for the compute partition run jobs. One can see job IDs, allocated amount of resources, and actually spent CPU time. This list can be easily enriched with general integral job characteristics, such as average CPU_user, Load Average, network usage, etc.

Figure 5 shows a prototype of the mini JobDigest tool. We can see all general data on the job, including command line and node list. Note that we can also see the average resource utilization rates highlighted with colors based on thresholds. The job tag corresponding to the job category of jobs with poor cache data stats has been imported also from the full size JobDigest report.

This type of short but informative report seems to be sufficient for most regular users for an initial job analysis. Nevertheless, the set of characteristics in such a brief report is subject to investigation and will be updated based on users' feedback.

Job table query interface

From	To	Cluster
20.10.2017	27.10.2017	Lomonosov-1 ▼

Owned logins	Involved logins
vurdizm	emel_251996

States	Partitions
All	All
Completed	compute
Failed	gpu
Cancelled	regular4

Show

Results (limited to 100)

cluster	job_id	login	partition	start_time	end_time	state	num_cores	duration(h)	cores*hours	cpu load	gpu load	loadavg	ib receive	ib sent
lomonosov-2	460043	vurdizm	compute	10/21/17 03:06:13	10/21/17 06:38:50	COMPLETED	14	3.5	49.6	50.5	0.0	15.0	117.2	117.2
lomonosov-2	460042	vurdizm	compute	10/21/17 02:58:09	10/21/17 06:31:08	COMPLETED	14	3.5	49.7	49.6	0.0	15.0	0.0	0.0
lomonosov-2	459417	vurdizm	compute	10/20/17 05:31:41	10/20/17 23:25:03	CANCELLED	140	17.9	2504.5	47.0	0.0	15.0	0.0	0.0
lomonosov-2	459415	vurdizm	compute	10/20/17 04:00:09	10/20/17 23:24:54	COMPLETED	56	19.4	1087.1	79.4	0.0	26.1	3.8	0.0

Fig. 4. List of selected project jobs for a certain period of time in the specified section with example of details

Job info for459415

/mnt/scratch/users/vurdizm/students/shupanov/science/cub/X=0.0/System 0/without obr/t=0.001/polymer.sh

General info

cluster	lomonosov-2
Job id	459415
Login	vurdizm
State	COMPLETED
Partition	compute
Num cores	56
Submit time	10/20/17 00:42:26
Start time	10/20/17 04:00:09
End time	10/20/17 23:24:54

Job performance

Metric	Value	Ranking
Average CPU load (%)	79.35	average
Average Loadavg	26.14	good
Average GPU load (%)	0.00	low
Average IB receive data (MB/s, total MPI + FS)	3.85	low
Average IB send data (MB/s)	0.04	low

Fig. 5. Example of mini JobDigest report (Color figure online)

6   Conclusions

In the near future, our plans have a strong focus on usability for regular users. At the same time, there are at least two levels of analysis to be added to the prototype. The first is a machine-learning-based module for anomaly detection, and the second is a role-sensitive situational screen based on earlier research, known as OctoScreen or TentaView [20].

We would also like to encourage all interested HPC users to contact the authors if additional implementation and functional details are required.

References

1. Voevodin, V., Voevodin, V.: Efficiency of exascale supercomputer centers and supercomputing education. In: Gitler, I., Klapp, J. (eds.) ISUM 2015. CCIS, vol. 595, pp. 14–23. Springer, Cham (2016). https://doi.org/10.1007/978-3-319-32243-8_2
2. Voevodin, V., et al.: Practice of "Lomonosov" supercomputer. Open Syst. J. **7**, 36–39 (2012)
3. Gunter, D., Tierney, B., Jackson, K., Lee, J., Stoufer, M.: Dynamic monitoring of high-performance distributed applications. In: Proceedings of the 11th IEEE International Symposium on High Performance Distributed Computing, pp. 163–170 (2002). https://doi.org/10.1109/hpdc.2002.1029915
4. Mellor-Crummey, J., Fowler, R.J., Marin, G., Tallent, N.: HPCVIEW: a tool for top-down analysis of node performance. J. Supercomput. **23**(1), 81–104 (2002). https://doi.org/10.1023/A:1015789220266
5. Jagode, H., Dongarra, J., Alam, S., Vetter, J., Spear, W., Malony, A.D.: A holistic approach for performance measurement and analysis for petascale applications. In: Allen, G., Nabrzyski, J., Seidel, E., van Albada, G.D., Dongarra, J., Sloot, P.M.A. (eds.) ICCS 2009. LNCS, vol. 5545, pp. 686–695. Springer, Heidelberg (2009). https://doi.org/10.1007/978-3-642-01973-9_77
6. Adhianto, L., et al.: HPCTOOLKIT: tools for performance analysis of optimized parallel programs. Concurr. Comput.: Pract. Exper. J. **22**(6), 685–701 (2009). https://doi.org/10.1002/cpe.1553
7. Kluge, M., Hackenberg, D., Nagel, W.E.: Collecting distributed performance data with dataheap: generating and exploiting a holistic system view. Procedia Comput. Sci. J. **9**, 1969–1978 (2012). https://doi.org/10.1016/j.procs.2012.04.215
8. Nikitenko, D., et al.: JobDigest - detailed system monitoring-based supercomputer application behavior analysis. In: Voevodin, V., Sobolev, S. (eds.) RuSCDays 2017. CCIS, vol. 793, pp. 516–529. Springer, Cham (2017). https://doi.org/10.1007/978-3-319-71255-0_42
9. JobDigest components. https://github.com/srcc-msu/job_statistics
10. Nikitenko, D., Voevodin, V., Zhumatiy, S.: Resolving frontier problems of mastering large-scale supercomputer complexes. In: ACM International Conference on Computing Frontiers (CF 2016), pp. 349–352. ACM, New York (2016). https://doi.org/10.1145/2903150.2903481
11. Nikitenko, D., Voevodin, V., Zhumatiy, S.: Octoshell: large supercomputer complex administration system. In: Russian Supercomputing Days International Conference, Moscow, Russia, CEUR Workshop Proceedings, vol. 1482, pp. 69–83 (2015)

12. Nikitenko, D., Stefanov, K., Zhumatiy, S., Voevodin, V., Teplov, A., Shvets, P.: System monitoring-based holistic resource utilization analysis for every user of a large HPC center. In: Carretero, J., et al. (eds.) ICA3PP 2016. LNCS, vol. 10049, pp. 305–318. Springer, Cham (2016). https://doi.org/10.1007/978-3-319-49956-7_24

13. Nikitenko, D.A., et al.: Supercomputer application integral characteristics analysis for the whole queued job collection of large-scale HPC systems. In: 10th Annual International Scientific Conference on Parallel Computing Technologies, PCT 2016, Arkhangelsk, Russian Federation, CEUR Workshop Proceedings, vol. 1576, pp. 20–30 (2016)

14. Movchan, A., Zymbler, M.: Time series subsequence similarity search under dynamic time warping distance on the Intel many-core accelerators. In: Amato, G., Connor, R., Falchi, F., Gennaro, C. (eds.) SISAP 2015. LNCS, vol. 9371, pp. 295–306. Springer, Cham (2015). https://doi.org/10.1007/978-3-319-25087-8_28

15. Rechkalov, T., Zymbler, M.: Accelerating medoids-based clustering with the Intel many integrated core architecture. In: Proceedings of the 9th International Conference on Application of Information and Communication Technologies (AICT 2015), 14–16 October 2015, Rostov-on-Don, Russia, pp. 413–417. IEEE (2015). https://doi.org/10.1109/ICAICT.2015.7338591

16. Voevodin, V., Voevodin, V., Shaikhislamov, D., Nikitenko, D.: Data mining method for anomaly detection in the supercomputer task flow. In: Numerical Computations: Theory and Algorithms, The 2nd International Conference and Summer School, Pizzo calabro, Italy, 20–24 June 2016, AIP Conference Proceedings, vol. 1776, pp. 090015-1–090015-4 (2016). https://doi.org/10.1063/1.4965379

17. Antonov, A., et al.: An approach for ensuring reliable functioning of a supercomputer based on a formal model. In: Wyrzykowski, R., Deelman, E., Dongarra, J., Karczewski, K., Kitowski, J., Wiatr, K. (eds.) PPAM 2015, Part I. LNCS, vol. 9573, pp. 12–22. Springer, Cham (2016). https://doi.org/10.1007/978-3-319-32149-3_2

18. Rudyak, V., Krakhalev, M., Sutormin, V.: Electrically induced structure transition in nematic liquid crystal droplets with conical boundary conditions. Phys. Rev. E. **96**, 052701-1–052701-5 (2017). https://doi.org/10.1103/PhysRevE.96.052701

19. Guseva, D., Rudyak, V., Komarov, P., et al.: Crosslinking mechanisms, structure and glass transition in phthalonitrile resins: insight from computer multiscale simulations and experiments. J. Polym. Sci. Part B: Polym. Phys. (2017). https://doi.org/10.1002/polb.24548

20. Nikitenko, D., Zhumatiy, S., Shvets, P.: Making large-scale systems observable – another inescapable step towards exascale. Supercomput. Front. Innov. J. **3**(2), 72–79 (2016). https://doi.org/10.14529/jsfi160205

High-Performance Reconfigurable Computer Systems with Immersion Cooling

Ilya Levin, Alexey Dordopulo$^{(\boxtimes)}$, Alexander Fedorov, and Yuriy Doronchenko

Scientific Research Center of Supercomputers and Neurocomputers (LLC),
106, Italyanskiy alley, Taganrog, Russia 347900
{levin,doronchenko}@superevm.ru, scorpio@mvs.tsure.ru, ss24@mail.ru

Abstract. In the paper, we review the design principles and architecture of reconfigurable computer systems with immersion cooling. We prove that systems with immersion cooling are the most promising for the design of high-performance computer complexes. We give selection criteria and design results for the principal components of the immersion cooling system. We demonstrate the design of our computational module prototype, based on advanced Xilinx UltraScale FPGAs and give testing results for the principal technical solutions. We prove that the designed immersion cooling system has a high power efficiency and power reserve for designing advanced reconfigurable computer systems on the basis of new UltraScale+ FPGAs and other next-generation FPGAs. We suggest new design solutions for the case of our computational module, as well as for the layout of the main computational board and other components of the computational module for use of Xilinx UltraScale+ FPGAs.

Keywords: Immersion cooling system · Liquid cooling
Reconfigurable computer systems · FPGAs
High-performance computer systems · Energy efficiency

1 Introduction

Having considerable advantages in real performance and energetic efficiency in comparison with cluster-like multiprocessor computer systems, reconfigurable computer systems (RCS) containing an FPGA computational field of large logic capacity are used for the implementation of computationally laborious tasks from various domains of science and technique. An RCS provides adaptation of its architecture to the structure of any task. In this case, a special-purpose computer device is created. It hardwarily implements all the computational operations of the information graph of the task with the minimum delays. Here we have a contradiction between the implementation of the special-purpose device and its general-purpose use for solving tasks from various problem areas. It is possible to solve this contradiction by combining the creation of a special-purpose computer

© Springer Nature Switzerland AG 2018
L. Sokolinsky and M. Zymbler (Eds.): PCT 2018, CCIS 910, pp. 62–76, 2018.
https://doi.org/10.1007/978-3-319-99673-8_5

device with a wide range of solvable tasks within a concept of reconfigurable computer systems based on FPGAs which are used as principal computational resource [1].

A practical experience of maintenance of large RCS-based computer complexes proves that air cooling systems have reached their heat limit. The continuous increase of both the circuit complexity and the clock rate of each new FPGA family leads to a considerable growth of the power consumption and maximal operating temperature of the chip. So, for the XC6VLX240T-1FFG1759C FPGAs of a computational module (CM) Rigel-2, the maximum overheat of the FPGAs relative to an environment temperature of 25 °C in operating mode, and with a power of 1255 W consumed by the CM, is 33.1 °C, i.e. the maximum temperature of the FPGA chip in the CM Rigel-2 is 58.1 °C. For the XC7VX485T-1FFG1761C FPGAs of the CM Taygeta, the maximum overheat of the FPGAs relative to an environment temperature of 25 °C in operating mode, and with a power of 1661 W consumed by the CM, is 47.9 °C, i.e. the maximum temperature of the FPGA of the CM Taygeta is 72.9 °C. If we take into account that the permissible temperature of an FPGA functioning, providing high reliability of the equipment during a long operation period, is 65...70 °C, then it is evident that the CM Taygeta maintenance requires a decrease in environment temperature.

According to the obtained experimental data, the conversion from the FPGA family Virtex-6 to the next family, Virtex-7, leads to an increase of the FPGA maximum temperature by 11...15 °C. Thus, further development of FPGA production technologies and conversion to the next FPGA family, Virtex UltraScale (with a power consumption of up to 100 W for each chip), will lead to an additional increase in FPGA overheat by 10...15 °C. This will shift the range of their operating temperature limit (80...85 °C), which has a negative influence on their reliability when the workload on the chips reaches up to 85–95% of the available hardware resource. This circumstance requires a quite different cooling method which provides for keeping the performance growth rates of advanced RCS.

2   Liquid Cooling Systems for Reconfigurable Computer Systems

The development of computer technologies leads to the design of computer technique providing higher performance and, hence, more heat. Dissipation of released heat is provided by a system of electronic element cooling which transfers heat from the more heated object (the cooled object) to the less heated one (the cooling system). If the cooled object is constantly heated, then the temperature of the cooling system grows and, for some period of time, will be equal to the temperature of the cooled object. So heat transfer stops and the cooled object will get overheated. The cooling system is protected from overheat with the help of a cooling medium (a heat-transfer agent). Cooling efficiency of the heat-transfer agent is characterized by the heat capacity and heat dissipation. As a rule, heat transfer is based either on the principles of heat conduction, which requires a physical contact of the heat-transfer agent with the cooled object, or on the principles of convective heat exchange with the heat-transfer agent, which consists in the physical transfer of the freely circulating heat-transfer agent. To organize heat transfer to the heat-transfer agent, it is necessary to provide heat contact between the cooling system and the heat-transfer agent. Various *radiators* – facilities for heat dissipation in the heat-transfer agent are used for this purpose. Radiators are set on the most heated components of computer systems. To increase efficiency of heat transfer from an electronic component to a radiator, a *heat interface* is set between them. The heat interface is a layer of heat-conducing medium (usually multicomponent) between the cooled surface and the heat dissipating facility, used for reduction of heat resistance between two contacting surfaces. Modern processors and FPGAs need cooling facilities with as low as possible heat resistance, because at present even the most advanced radiators and heat interfaces cannot provide necessary cooling if an air cooling system is used.

To organize heat transfer to the heat-transfer agent, it is necessary to provide heat contact between the cooling system and the heat-transfer agent. Various *heat-sinks*-facilities for heat dissipation in the heat-transfer agent – are used for this purpose. Heat-sinks are set on the most heated components of computer systems. To increase the efficiency of heat transfer from an electronic component to a heat-sink, a *heat interface* is set between them. The heat interface is a layer of heat-conducing medium (usually multicomponent) between the cooled surface and the heat dissipating facility, used for reduction of heat resistance between two contacting surfaces. Modern processors and FPGAs need cooling facilities with as low as possible heat resistance, because at present even the most advanced heat-sinks and heat interfaces cannot provide the required cooling if an air cooling system is used.

Before 2013, air cooling systems were used quite successfully for cooling supercomputers. But due to a growth of performance and circuit complexity of microprocessors and FGAs, used as components in supercomputer systems, air cooling systems have practically reached their limits for advanced supercomputers of that time, including hybrid computer systems. The majority of vendors of computer technique therefore consider liquid cooling systems as an alternative solution to the cooling problem. Today, liquid cooling systems constitute the most promising design area for cooling modern intensively operating electronic components in computer systems.

A considerable advantage of all liquid cooling systems is the heat capacity of liquids, which is better than that of air (from 1500 to 4000 times), and a higher heat-transfer coefficient (up to 100 times higher). To cool one modern FPGA chip, 1 m^3 of air or 0.00025 m^3 (250 ml) of water per minute is required. Much less electric energy is required to transfer 250 ml of water than to transfer 1 m^3 of air. Heat flow, transferred by similar surfaces at the conventional velocity of the heat-transfer agent, is 70 times more intensive in the case of liquid cooling than in the case of air cooling. An additional advantage is the use of traditional, rather reliable and cheap, components such as pumps, heat exchangers, valves, control devices, etc. For corporations and companies dealing with equipment with high packing density of components operating at high temperatures, liquid cooling is in fact the only possible solution to the problem of cooling modern computer systems. One more option to increase liquid cooling efficiency consists in improving the initial parameters of the heat-transfer agent: increasing velocity, decreasing temperature, creating turbulent flow, increasing heat capacity, reducing viscosity.

The heat-transfer agent in liquid cooling systems used in computer technique is a liquid such as water, or any dielectric liquid. Heated electronic components transfer heat to the permanently circulating heat-transfer agent - a liquid which, after being cooled in the external heat exchanger, is used again for cooling heated electronic components. There are several types of liquid cooling systems. In closed-loop liquid cooling systems, there is no direct contact between liquid and electronic components of the printed circuit boards [6,7]. In open-loop cooling systems (liquid immersion cooling systems), electronic components are immersed directly into the cooling liquid [8,9]. Each type of liquid cooling systems has its own advantages and disadvantages.

In closed-loop liquid cooling systems all heat-generating elements of the printed circuit board are enclosed by one or several flat plates with a channel for liquid pumping [10,11]. So, for example, the cooling system of the SKIF-Avrora supercomputer [12] is based on the principle "one cooling plate, one printed circuit board". The plate, of course, has a complex surface relief to provide tight heat contact with each chip. In the IBM Aquasar supercomputer, cooling is based on the principle "one cooling plate, one (heated) chip". In each case, the channels of the plates are joined by manifolds into a single loop connected to a common heat-sink (or another heat exchanger), usually placed outside the computer case and/or rack, or even the computer room. With the help of the pump, the heat-transfer agent is pumped through the plates and dissipates, by means of the heat exchanger, the heat generated by the computational elements. In such systems, it is necessary to provide access of the heat-transfer agent to each heat-generating element of the calculator, which means a rather complex "piping system" and a large number of pressure-tight connections. Besides, if it is necessary to provide maintenance of the printed circuit boards without any significant demounting, then the cooling system must be equipped with special liquid connectors providing pressure-tight connections and simple mounting/demounting of the system.

In closed loop liquid cooling systems all heat-generating elements of the printed circuit board are covered by one or several flat plates with a channel for liquid pumping. So, for example, cooling of a supercomputer SKIF-Avrora is based on a principle "one cooling plate for one printed circuit board". The plate, of course, had a complex surface relief to provide tight heat contact with each chip. Cooling of a supercomputer IBM Aquasar is based on a principle "one cooling plate for one (heated) chip". In each case the channels of the plates are united by collectors into a single loop connected to a common radiator (or another heat exchanger), usually placed outside the computer case and/or rack or even the computer room. With the help of the pump the heat transfer agent is pumped through the plates and dissipates heat, generated by the computational elements, by means of the heat exchanger. In such system it is necessary to provide access of the heat transfer agent to each heat-generating element of the calculator, what means a rather complex "piping system" and a large number of pressure-tight connections. Besides, if it is necessary to provide maintenance of the printed circuit boards without any serious demounting, then the cooling system must be equipped with special liquid connectors which provide pressure-tight connections and simple mounting/demounting of the system.

In closed-loop liquid cooling systems, it is possible to use water or glycol solutions as the heat-transfer agent. However, leak of the heat-transfer agent can lead to possible ingress of electrically conducting liquid to unprotected contacts of the printed circuit boards of the cooled computer, and this, in its turn, can be fatal for both separate electronic components and the whole computer system. To eliminate failures, the whole complex must be stopped, and the power supply system must be tested and dried up. The control and monitoring systems of such computers always contain many internal humidity and leak sensors. Cooling systems with liquid at negative pressure are frequently used to solve the leak problem. In these systems, water is not pumped in under pressure but instead is pumped out, thus practically excluding leaks of liquid. If the air-tightness of the cooling systems is damaged, then air enters the system but no leak of liquid happens. Special sensors are used for detection of leaks, while modular design allows maintenance without stoppages of the whole system. However, all these capabilities considerably complicate the design of the hydraulic system.

Another issue affecting closed-loop liquid cooling systems is the dew point problem. In the section of data processing, the air is in contact with the cooling plates. It means that if some parts of these plates are too cold and the air in the section of data processing is warmer and not very dry, then moisture can condense out of the air on the plates. The consequences of this process are similar to leaks. This problem can be solved either by hot-water cooling, which is not effective, or by controlling and keeping at the required level the temperature and humidity parameters of the air in the section of data processing, which is complicated and expensive.

The design becomes even more complicated when it is necessary to cool several components with a water flow that should be proportional to their heat generation. In addition to branched pipes, it is necessary to use complex control devices (simple T-branches and four-ways are not enough). An alternative approach is to use an industrial device with flow control, but in this case, the user cannot considerably change the configuration of the cooled computational modules.

In open-loop liquid cooling systems, the principal component is the heat-transfer agent, which is a dielectric liquid based, as a rule, on a white mineral oil that provides a much higher heat storage capacity than the air does in the same volume. According to their design, these systems contain printed circuit boards, servers of computational equipment, and a bath that is filled with heat-transfer liquid and placed into a computer rack. The heat generated by the electronic components is dissipated by the heat-transfer agent, which flows within the whole bath. We can mention here some advantages of immersion liquid cooling systems: simple design and capability of adaptation to the changing geometry of printed circuit boards, simplicity of mani-folds and liquid connectors, no problems with control of liquid flows, no dew point problem, high reliability and low cost of the product.

The main problem of open-loop liquid cooling systems is the chemical composition of the used heat-transfer liquid which must fulfil strict requirements of heat transfer capacity, electrical conduction, viscosity, toxicity, fire safety, stability of the main parameters and reasonable cost.

Considering the given advantages and disadvantages of the two types of liquid cooling systems, we can affirm that open-loop cooling systems for electronic components of computer systems have more weighty advantages. In this connection, when dealing with advanced RCS, it is reasonable to use direct immersion of heat-generating system components into a mineral oil-based liquid heat-transfer agent.

At present, the technology for liquid cooling of servers and separate computational modules is being developed by many vendors, and some of them have achieved success in this direction [9–11]. These technologies, however, are intended for cooling computational modules containing only one or two microprocessors. All attempts to adapt this technology for cooling computational modules, which contain a large number of heat-generating components (an FPGA field of eight chips), failed due to a number of shortcomings.

The main disadvantages of existing technologies of immersion liquid cooling [10–14] for computational modules containing FPGA computational fields are:
- poor adaptation of the cooling system for placement into standard computer racks;
- inefficiency of cooling of electronic component chips with considerable (over 50 W) heat generation;
- the thermal paste between FPGA chips and heat-sinks is washed out during long-term maintenance;
- the system of cooling-liquid circulation inside the module is designed for one or two chips but not for an FPGA field, and this fact leads to considerable thermal gradients.

In the systems based on the IMMERS technology [9], all cooling liquid is circulating within a closed loop through the chiller, and this fact leads to some problems:
- complex maintenance stoppages are necessary to remove separate components and devices;
- it is necessary to use a power specialized pump and hydraulic equipment adapted to the cooling liquid;
- a complex system for the control of cooling-liquid circulation, which causes periodic failures;
- high cost of the cooling liquid, produced by only one manufacturer.

These disadvantages can be considered as an inseparable part of other existing open-loop liquid cooling systems since the cooling of RCS computational modules containing not less than eight FPGA chips has some specific features compared with the cooling of a single microprocessor.

The special feature of the RCS produced at the Scientific Research Center of Supercomputers and Neurocomputers is the number of FPGAs, which is not less than six to eight chips on each printed circuit board, and high packing density. This considerably increases the number of heat-generating components compared with microprocessor modules, making more complicated the application of the IMMERS direct liquid cooling technology along with other end solutions of immersion systems, and requires additional technical and design solutions for an effective cooling of RCS computational modules.

The use of open liquid cooling systems is efficient owing to the heat-transfer agent characteristics and the design and specification of the used FPGA heat-sinks, pump equipment, and heat-exchangers.

The heat-transfer agent must have the best possible dielectric strength, high heat transfer capacity, the maximum possible heat capacity, and low viscosity.

The heat-sink must have the maximum possible surface of heat dissipation, must allow the circulation of the heat-transfer agent turbulent flow through itself, and manufacturability. Specialists at SRC SC & NC have performed heat engineering research and suggested a fundamentally new design of a heat-sink with original solder pins which create a local turbulent flow of the heat-transfer agent. The used thermal interface cannot be deteriorated or washed out by the heat-transfer agent. Its coefficient of heat conductivity can remain permanently high. SRC SC & NC specialists have created an effective thermal interface that fulfills all specified requirements, and additionally, its coating and removal technology has also been improved.

The pumping equipment, that is to say, is not the least of the components of a CM cooling system. The principal criteria that must be met are the following:
- performance parameters;
- overall dimensions and coordinated placement of the input and the output fittings;
- the pump must be suitable for interaction with oil products with a specified viscosity and chemical composition;
- continuous maintenance mode;
- minimal vibrations;
- the pump must have the minimal permissible positive suction head;
- the protection class of the pump electric motor must be not less than IP-55.

The heat exchanger is also an important component of the cooling system. Its design must be compact and must provide an efficient heat exchange. Research performed by the SRC SC & NC scientific team has proved that the most suitable design of the heat exchanger is a plate-type one designed for cooling mineral oil in hydraulic systems of industrial equipment.

The liquid cooling system must have a control subsystem containing sensors of level, flow, and temperature of the heat-transfer agent, and a temperature sensor for cooling components.

3 "SKAT" Reconfigurable Computer System Based on Xilinx UltraScale FPGAs

The SRC SC & NC scientific team has actively developed since 2013 the creation of next-generation RCS on the basis of their original liquid cooling system for computational circuit boards with high packing density and large number of heat-generating electronic components. The design criteria of computational modules (CM) of next-generation RCS with an open-loop liquid cooling system are based on the following principles:
- the RCS configuration is based on a computational module with a 3U height and 19 width, and self-contained circulation of the cooling liquid;
- one computational module can contain 12 to 16 computational circuit boards (CCB) with FPGA chips;
- each CCB must contain up to eight FPGAs, with a dissipating heat flow of about 100 W from each FPGA;
- a standard water cooling system based on industrial chillers must be used for cooling the liquid.

The principal element in the modular implementation of an open-loop immersion liquid cooling system for electronic components of computer systems is a new generation reconfigurable computational module (see design in Fig. 1-a). The new-generation CM casing consists of a computational section and a heat-exchange section. The casing, which is the base of the computational section, contains a hermetic container with dielectric cooling liquid, and electronic components with elements that generate heat during operation. The electronic components can be computational modules (not less than 12 to 16), control boards, RAM, power supply blocks, storage devices, daughter boards, etc. The computational section is closed with a cover.

The computational section adjoins the heat exchange section which contains a pump and a heat exchanger. The pump moves the heat-transfer agent in the CM through a closed loop: from the computational module, the heated heat-transfer agent passes into the heat exchanger and is cooled there. From the heat exchanger, the cooled heat-transfer agent again passes into the computational module and cools the heated electronic components there. As a result of heat dissipation, the agent becomes heated and again passes into the heat exchanger, and so on. The heat exchanger is connected to the external heat-exchange loop via fittings and is intended for cooling the heat-transfer agent with the help of a secondary cooling liquid. A plate heat exchanger in which the first and the second loops are separated can be used as a heat exchanger. So, as the secondary cooling liquid, it is possible to use water cooled by an industrial chiller. The chiller can be placed outside the server room and can be connected to the reconfigurable

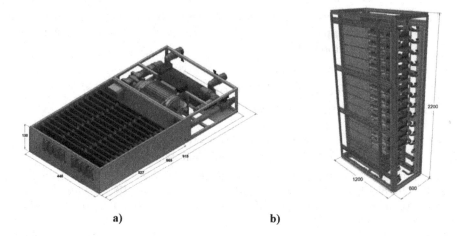

Fig. 1. The design of a computer system based on liquid cooling (a - design of a new generation CM, b - design of the computer rack)

computational modules by means of a stationary system of engineering services. The design of the computer rack with mounted CMs is shown in Fig. 1-b.

The computational and the heat exchange sections are mechanically inter-connected into a single reconfigurable computational module. Maintenance of the reconfigurable computational module requires its connection to the source of the secondary cooling liquid (by means of valves), to a power supply block or to a hub (by means of electrical connectors).

In the casing of the computer rack, the CMs are placed one over another. Their number is limited by the dimensions of the rack, by technical capabilities of the computer room, and by the engineering services.

Each CM of the computer rack is connected to the source of secondary cooling liquid with the help of supply and return manifolds through fittings (or balanced valves) and flexible pipes; the connection to both the power supply and the hub is performed via electric connectors.

The supply of cold secondary cooling liquid and the extraction of the heated one into the stationary system of engineering services connected to the rack are made via fittings (or balanced valves).

For the purpose of testing technical and technological solutions, and deter-mining the expected technical and economical characteristics and service perfor-mance of the designed high-performance reconfigurable computer system with liquid cooling, we designed a number of models, experimental and technological prototypes. Figure 2 shows the prototype of a new-generation "SKAT" CM. A new-design CCB with high packing density was created for this CM.

The CCB of the advanced computational module contains eight Kintex Ultra-Scale XCKU095T FPGAs; each FPGA has a specially designed thermal interface and a low-height heatsink for heat dissipation.

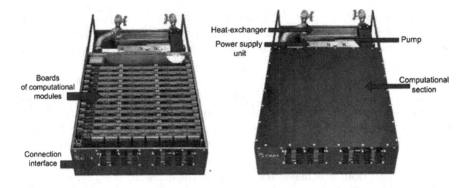

Fig. 2. The prototype of the new-generation CM

We have designed an immersion power supply unit providing DC/DC 380/12 V transducing with the power up to 4 kW for four CCBs.

The computational section of the "SKAT" CM contains 12 CCBs with a power of up to 800 W each, and three power supply units. In addition, all boards are completely immersed into an electrically neutral liquid heat-transfer agent.

To achieve an effective immersion cooling system, we developed a dielectric heat-transfer agent with the best possible dielectric strength, high heat transfer capacity, the maximum possible heat capacity and low viscosity.

The heat exchange section contains pump components and the heat exchanger, both providing the effective flow and cooling of the heat-transfer agent. The design height of the CM is 3U.

The performance of a next-generation "SKAT" CM is increased in 8.7 times in comparison with the "Taygeta" CM. Original design solutions provide more than triple increasing of the system packing density. Clock frequency and logic capacity of the FPGAs are also increased. As a result, all this provides such qualitative increasing of the system specific performance.

Experimental tests of the developed solutions of the immersion liquid cooling system proved that the temperature of the heat-transfer agent does not exceed 30 °C, and the power consumed by each FPGA in operating mode equals 91 W (8736 W for the whole CM). In addition, the maximum FPGA temperature during heat experiments did not exceed 55 °C. All this proves that the designed immersion liquid cooling system has a reserve and can provide effective cooling for the designed RCS based on the advanced Xilinx UltraScale+ FPGA family.

4 "SKAT+" Advanced Reconfigurable Computer System Based on Xilinx UltraScale+ FPGAs

The use of UltraScale+ FPGAs based on the 16FinFET Plus 16-nm technology and produced by Xilinx since 2017 will provide a three time increase in computational performance due to an increase in clock frequency and FPGA circuit complexity, whereas the size of the computer system will still remain unchanged. However, despite the reduction of relative energetic consumption due to new technological standards of FPGA manufacturing and also to a certain power reserve of the designed liquid cooling system, it is expected that FPGA operating temperatures will approach again their critical values.

In addition, the new FPGAs of the UltraScale+ family have larger geometric sizes. The size of the FPGAs in the "SKAT" RCS is 42.5 × 42.5 mm. The size of the FPGAs that will be placed into the "SKAT+" RCS amounts to 45 × 45 mm. Owing to this circumstance, it is impossible to use the existing CCB design since the width of the printed circuit board will become larger and will not fit in a standard 19 rack.

In this connection, it is necessary to modify the designed open liquid cooling system and the CCB design, which will lead to a modification of the whole CM.

At present, the SRC of SC & NC scientific team is working on the design of an advanced RCS based on Xilinx UltraScale+ FPGAs. Due to these works related to the modification of the cooling system, we are going to solve the following problems:

1. Increase the effective surface of heat-exchange between FPGAs and the heat-transfer agent.
2. Increase the performance of the heat-transfer agent supply pump.
3. Increase the reliability of the liquid cooling system by means of immersed pumps.
4. Experimentally improve the heat-sink optimal design.
5. Experimentally improve the technology of thermal interface coating.

We have designed a prototype of an advanced computational module with a modified immersed cooling system (Fig. 3). Some distinctive features of the new design are immersed pumps and a considerable reliability increase of the CM due to a reduction of the number of components and simplification of the cooling system. According to our plans, the heat-exchange section will house only the heat exchanger. We are working on experimental pump equipments that can operate in the heat-exchange agent. During modification of the CCB design,

Fig. 3. A prototype of a computational module with a modified immersed cooling system

we have created a prototype of an advanced board, shown in Fig. 4. The CCB contains eight UltraScale+ FPGAs of high circuit complexity. To provide room for the new CCB into a 19 rack, it is necessary to exclude its CCB controller from its structure. The CCB controller was always implemented as a separate FPGA and used to provide access to the FPGA computational resources of the CCB, FPGA programming, and monitoring of the CCB resources.

FPGAs are rather small, but their resource constantly grows with each new family. At the same time, the variety of functions of the CCB controller grows only slightly. As a result, the resources required at present for the implementation of all the CCB controller functions amount to only some percent of the logic capacity of the FPGAs currently used. In this connection, further implementation of the CCB controller as a separate FPGA is considered unnecessary. A single FPGA in the computation field will be able to perform all the functions of the controller. We need a system of hydraulic balancing for the heat-transfer agent flow within each hydraulic loop. Such a system will provide an equal flow of the heat-exchange agent in each computational module inside the computer rack during servicing of one or several computational modules. The additional subsystem will considerably complicate the cooling system. To simplify the hydraulic balancing of the heat-transfer system, the SRC SC & NC scientific team (LLC, Russia) has suggested an engineering solution for balancing the heat-transfer agent flow in the heat-exchange system (see Fig. 5).

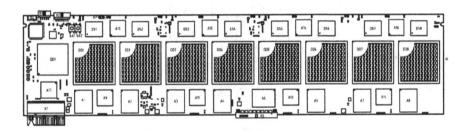

Fig. 4. The prototype of the CCB modified packing

The designed heat-exchange system includes a pump 1, a chiller 2 (that is, a cooling machine), several circulation loops 3, with heat-exchangers 15, intended for the heat exchange process between the primary heat-transfer agent (water) and the secondary one (oil MD-4.5), which is circulating in the computational modules 4. The circulation of the secondary heat-transfer agent (oil MD-4.5) in computational modules 4 and heat-exchangers 15 of each circulation loop is ensured by an additional pump (not shown in Fig. 1) connected to each circulation loop. Heat-exchangers 15 are connected by parallel tubes to supply manifold 6 and return manifold 5. The inlets and outlets of the first (No. 1), second (No. 2), third (No. 3), etc., circulation loops are arranged along the heat-transfer agent flow 7 and near the inlets of supply manifold 8 and return manifold 9. The inlets

and outlets of the last circulation loop No. 6 in Fig. 1) are situated near the outlets of supply manifold 10 and return manifold 11. The return pipe 12 connects
the outlet of return manifold 11 with the chiller 2, the pump 1, and the inlet
of supply manifold 8. Besides, each circulation loop may be complemented with
a balancing valve for finer balance-tuning. The heat-exchange system is filled
with the primary heat-transfer agent (water, antifreeze, etc.), then decreased,
and the pump 1 is switched on. The primary heat-transfer agent is supplied to
the inlet 8 of the supply manifold 6 and then through the circulation loops 3
into the heat-exchangers 15, where heat is transferred from the primary heat-
transfer agent to the secondary one, which is circulating in the computational
modules 4, where the secondary heat-transfer agent (oil MD-4.5) dissipates the
heat from heating electronic components. The primary heat-transfer agent gets
warm and enters the return manifold. There is a return pipe 12 at outlet 11
of the return manifold. Through return pipe 12 and the chiller 2, the primary
heat-transfer agent is again transferred to pipe 1, and then to inlet 8 of supply
manifold 6, and then flows along the closed loop. In the chiller 2, the heated
primary heat-transfer agent is chilled.

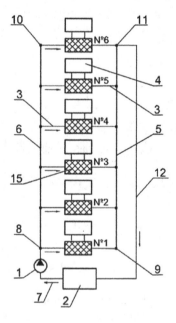

Fig. 5. The layout of hydraulic balancing of the heat-exchange system for computational modules in a computer rack

If a circulation loop in any computational module fails, then the heat-transfer
agent flow is evenly changed in the rest of modules, since the closed trajectory of
the heat-transfer agent flow is similar for all loops, and the distance between each
loop and the pump is the same: pump – inlet of the supply manifold – supply

manifold – circulation loop – return manifold – outlet of the return manifold – return pipe – chiller – pump. The described engineering solution makes it possible to balance the hydraulic resistance in all the circulation loops when the heat-transfer agent flow is pumped through them. No additional hydraulic balancing system is needed here.

Thanks to breakthrough technical solutions that we have found while designing the "SKAT" RCS with an immersed liquid cooling system, we are now able to develop this direction of high-performance RCS design, and after some design improvements, we will be able to create a computer system providing a new level of computational performance.

5   Conclusions

The use of air cooling systems in the design of supercomputers has practically reached its limit, since cooling effectiveness decreases as the rate of consumed and dissipated power grows at the same time as circuits in microprocessors and other chips become more and more complex. This explains why the use of liquid cooling in modern computer systems is considered as a priority direction for the improvement of cooling systems and has excellent prospects of further development. Liquid cooling of RCS computational modules containing not less than eight FPGAs of high circuit complexity has some specific features compared with the cooling of microprocessors and requires the development of a specialized immersion cooling system. The original liquid cooling system that has been designed for a new-generation RCS computational module provides high maintenance characteristics, such as a maximum FPGA temperature not exceeding 55 °C, while keeping the heat-transfer agent temperature below 30 °C in operating mode. Thanks to breakthrough solutions found for the immersion liquid cooling system, it is now possible to mount not less than 12 new-generation CMs, with a total performance above 1 PFlops, in a single 47U computer rack. The power reserve of the liquid cooling system for the new-generation CMs ensures an effective cooling not only for the existing but also for future FPGA families (Xilinx UltraScale+ and UltraScale 2).

FPGAs, as principal components of reconfigurable supercomputers, provide a stable, practically linear growth of the RCS performance. This makes the performance of an RCS based on Xilinx Virtex UltraScale FPGAs similar to that of the world best cluster supercomputers and opens new possibilities for the design of super-high performance supercomputers.

References

1. Perkowski, M.: FPGA computer architectures. Northcon/93. In: Conference Record, 12–14 October 1993. https://doi.org/10.1109/northc.1993.505038. ISBN: 0-7803-9972-2
2. Tripiccione, R.: Reconfigurable computing for statistical physics. The weird case of JANUS. In: IEEE 23rd International Conference on Application-Specific Systems, Architectures and Processors (ASAP) (2012). https://doi.org/10.1109/asap.2012.38
3. Baity-jesi, M., et al.: The Janus project: boosting spin-glass simulations using FPGAs. In: IFAC Proceedings Volumes, Programmable Devices and Embedded Systems, vol. 12, no. 1 (2013). https://doi.org/10.3182/20130925-3-cz-3023.00039
4. Shaw, D.E., et al.: Anton, a special-purpose machine for molecular dynamics simulation. Commun. ACM 51(7), 91–97. https://doi.org/10.1145/1364782.1364802
5. Kalyaev, I.A., Levin, I.I., Dordopulo, A.I., Slasten, L.M.: Reconfigurable computer systems based on Virtex-6 and Virtex-7 FPGAs. IFAC Proc. Volumes Programmable Devices Embed. Syst. 12(1), 210214 (2013). https://doi.org/10.3182/20130925-3-cz-3023.00009
6. Coolitsystems. http://www.coolitsystems.com/index.php/data-center/liquid-cooling-options.html
7. Asetek Data Center Liquid Cooling. http://www.asetek.com/data-center/oem-data-center-coolers
8. Data Center Cooling: Liquid Immersion - Green Revolution Cooling. http://www.grcooling.com/carnotjet
9. Immers.ru. http://www.immers.ru/sys/immers660
10. Eurotech HPC. http://www.eurotech.com/aurora
11. RSC Technology. http://www.rscgroup.ru
12. T-platforms. http://www.t-platforms.ru/products/hpc/a-class/cooling.html
13. Iceotope. http://www.iceotope.com/product.php
14. LiquidCool Solutions - Liquid cooled servers for a CRAC - free future. http://www.liquidcoolsolutions.com

Hybrid Supercomputer Desmos with Torus Angara Interconnect: Efficiency Analysis and Optimization

Nikolay Kondratyuk[1,2], Grigory Smirnov[1], Ekaterina Dlinnova[3], Sergey Biryukov[4], and Vladimir Stegailov[1(✉)]

[1] Joint Institute for High Temperatures of the RAS, Moscow, Russia
stegailov.vv@mipt.ru
[2] Moscow Institute of Physics and Technology, Dolgoprudny, Russia
[3] National Research University Higher School of Economics, Moscow, Russia
[4] JSC NICEVT, Moscow, Russia

Abstract. The paper describes the first experience of practical deployment of the hybrid supercomputer Desmos at the Joint Institute for High Temperatures of the Russian Academy of Sciences (JIHT RAS). We consider job scheduling statistics, energy efficiency, case studies of GPU acceleration efficiency and benchmarks of the distributed storage with a parallel file system.

Keywords: Job accounting and statistics · Energy efficiency
GPU acceleration · Parallel I/O

1 Introduction

Desmos is a supercomputer targeted to molecular dynamics (MD) calculations that was installed in the JIHT RAS in December 2016. Desmos is the first application of the Angara interconnect for a GPU-based MPP system [1,2].

Modern MPP systems can combine up to 10^5 nodes for solving one computational problem. For this purpose, MPI is the most widely used programming model. The architecture of the individual nodes can differ significantly and is usually selected (co-designed) for the main type of MPP system deployment. The most important component of MPP systems is the interconnect. The interconnect properties have a major influence on the scalability of any MPI-based parallel algorithm. In this work, we describe the Desmos supercomputer, which is based on cheap 1CPU+1GPU nodes connected by the original Angara interconnect.

The JIHT team was supported by the Russian Science Foundation (grant No. 14-50-00124). The Desmos supercomputer is a part of the Supercomputer Centre of JIHT RAS. The authors acknowledge the Shared Resource Center "Far Eastern Computing Resource" IACP FEB RAS (http://cc.dvo.ru) for granting access to the IRUS17 supercomputer.

© Springer Nature Switzerland AG 2018
L. Sokolinsky and M. Zymbler (Eds.): PCT 2018, CCIS 910, pp. 77–91, 2018.
https://doi.org/10.1007/978-3-319-99673-8_6

The Angara interconnect is a Russian-designed communication network with a torus topology. The interconnect ASIC was developed by JSC NICEVT and manufactured by TSMC using the 65 nm process. The Angara architecture uses some principles of both the IBM Blue Gene L/P and the Cray Seastar2/Seastar2+ torus interconnects. The torus interconnect developed by EXTOLL is a similar project [3]. The Angara chip supports deadlock-free adaptive routing based on bubble flow control [4], direction ordered routing [5,6] and initial and final hops for fault tolerance [5].

The results of the benchmarks confirmed the high efficiency of commodity GPU hardware for MD simulations [2]. The scaling tests for electronic structure calculations also showed the high efficiency of MPI-exchanges over the Angara network.

In this paper, we combine the results of the Desmos supercomputer performance analysis. These results pave the way to optimizations of the supercomputer efficiency and could be relevant for other HPC systems.

2   Related Work

Job scheduling determines the efficiency of a supercomputer practical deployment and is a very important topic in parallel systems (see, e.g., [7]). The everyday work of supercomputer centers shows a need for separation of cloud-like jobs (which do not require a high-bandwidth low-latency interconnect between nodes) from regular parallel jobs. Such a separation is a way for increasing efficiency of supercomputer deployment [8]. There have been some attempts of statistical analysis of supercomputers operation in Russian HPC centers (see, e.g., [9]).

The increase of power consumption and heat generation of computing platforms is a significant problem. Measurement and presentation of the results of performance tests of parallel computer systems become more and more often evidence-based [10], including the measurement of energy consumption, which is crucial for the development of exascale supercomputers [11].

Nowadays, partial use of single precision in MD calculations with consumer-grade GPUs cannot be regarded as a novelty. The results of such projects as Folding@Home confirmed the broad applicability of this approach. Recent developments in optimized MD algorithms include the validation of the single-precision solver (see, e.g., [12]). The authors of [13] give very instructive guidelines for achieving the best performance at minimal cost in 2015.

The success of the TeraChem package [14] illustrates the amazing perspectives of GPU usage for quantum chemistry.

The ongoing increase of data generated by HPC calculations leads to the requirement of a parallel file system for rapid I/O operations. However, benchmarking parallel file system is a complicated (and usually expensive!) task, which is why accurate results of particular case studies are quite rare (see, e.g. [15]).

3   Statistical Data of Desmos Deployment

The batch system for user jobs scheduling of Desmos is based on Slurm, which is an open-source workload manager designed for Linux clusters of any size [16]. It is used in many HPC centers worldwide (the paper [16] has been cited more than 500 times). Slurm has the following main features:

- allocates exclusive and/or non-exclusive access to resources (Compute Nodes) to users for some time so they can perform a work;
- provides a framework for starting, executing, and monitoring work (normally a parallel job) on the set of allocated nodes;
- arbitrates conflicting requests for resources by managing a queue of pending work.

The SlurmDB daemon stores data into a MySQL database. The SlurmDB daemon runs on the management node. In September 2017, the SlurmDB database was activated on Desmos, giving us the possibility of detailed analysis of supercomputer load statistics. The default Slurm tool *sreport* has quite limited functionality. That is why we use SQL-queries for accessing the SlurmDB for statistical analysis. For example, the following command retrieves and calculates the duration of allocated jobs:

```
select timestampdiff(second, from_unixtime(time_start),
from_unixtime(time_end)) as running,
timestampdiff(second, from_unixtime(time_submit),
from_unixtime(time_start)) as waiting,t.*
from desmos_job_table;
```

Figure 1 shows the distribution of jobs over the number of cores used and over running time t_R. GPU floating point performance is not taken into account when drawing the iso-levels of $R_{peak} * t_R$ constant value. This quantity corresponds to the total number of floating-point operations that CPUs deployed for the particular job are able execute theoretically during time t_R.

Parallel algorithms can be executed either slowly on a modest number of cores (nodes) or quickly if their parallel scalability justifies using a large number of processing elements efficiently. Two iso-levels separate three regions of total number floating-point operations corresponding to individual jobs: less than 10 PFlos, between 10 and 100 PFlops, and above 100 PFlops. The percent values shown in the blue boxes correspond to the share of each region in the Desmos total workload since the beginning of SlurmDB logging. We see that the major part of all the jobs executed on Desmos have been essentially supercomputer-type jobs.

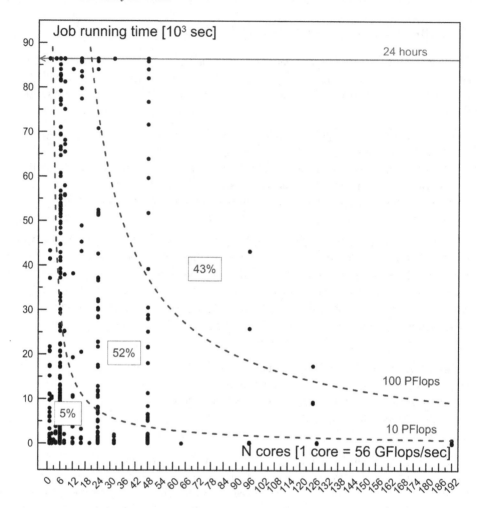

Fig. 1. Job running time vs. job size. Each point corresponds to one job (Color figure online)

At the same time, we see that there are jobs that were executed on six cores or less, i.e. on a single node. This type of jobs can be easily moved away from the supercomputer either to the cloud or to a personal workstation.

The efficiency of the supercomputer job scheduling policy can be evaluated by such type of graphs. The more points we see on the right side of the graph, the more efficient is the end-user collective deployment of the supercomputer. Users should be motivated to use scalable codes and to choose larger number of nodes for speeding calculations up. The following Slurm batch system partitions have been created on the Desmos supercomputer:

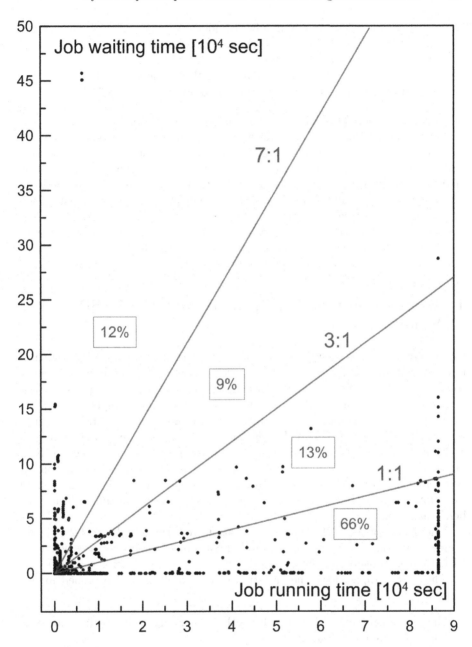

Fig. 2. Job waiting time vs. job running time. Each point corresponds to one job. The percent values shown in the red boxes correspond to the share of each region in the supercomputer total workload (Color figure online)

- test: max time = 15 min, any number of nodes;
- max1n: max time = 1440 min, min/max number of nodes = 1;
- max8n: max time = 1440 min, min/max number of nodes = 4/8;
- max16n: max time = 720 min, min/max number of nodes = 4/16;
- max32n: max time = 360 min, min/max number of nodes = 4/32.

This policy motivates users to deploy higher numbers of nodes. Also, it prevents overloading the supercomputer with small one-node or two-node jobs.

Another aspect of job scheduling is the waiting time for the job to start calculation after being submitted to the job queue. Figure 2 shows the correlation of the job running time t_R to the job waiting time t_W (i.e. the time between the moment of job submission into the batch queue and the moment of job execution). Three levels of ratios t_W/t_R are shown in Fig. 2. Fortunately, the majority of jobs (66%) fall into the category with $t_W < t_R$. Obviously, jobs with $t_W > t_R$ should be regarded as inefficient. Diminishing the number of such jobs is another criterion of supercomputer efficient usage.

4   Energy Consumption Optimization: The VASP Case Study

Our recent studies of energy consumption at frequency variation [17–19] show that a variation of CPU frequency can have a positive effect on reducing energy consumption for memory-bound algorithms. Here we extend this type of analysis from the level of a single CPU to the level of a whole supercomputer.

VASP 5.4.1 was compiled for Desmos using gfortran 4.8, Intel MPI and linked with Intel MKL for BLAS, LAPACK and FFTW calls. The model represents a GaAs crystal consisting of 80 atoms in the supercell. All 32 nodes are used for the benchmark runs. Each run corresponds to one complete iteration for electron density optimisation that consists of 35 steps. We use digital the logging capabilities of the UPS for digital sampling of the power consumed during the benchmark runs.

The results of the CPU frequency variation from the 3.5 GHz to 1.2 GHz are presented in Fig. 3. We see how the level of consumed power decreases when the CPU frequency decreases. At the same time, the time-to-solution increases.

The lower graph shows the variation of the total energy consumed in two cases:

- The real benchmark of Desmos shows that no energy saving regime can stem from CPU frequency variation.
- The hypothetical case with all fans in the chassis switched off shows a shallow minimum of total energy consumed. The total power consumption of the chassis fans working at full speed is about 4 kW. If we subtract the fan-determined power draw from the total power level (e.g., it could be the case if liquid immersion cooling would be used). This minimum corresponds to saving about 3.4% of energy at the cost of about 3.8% longer calculations.

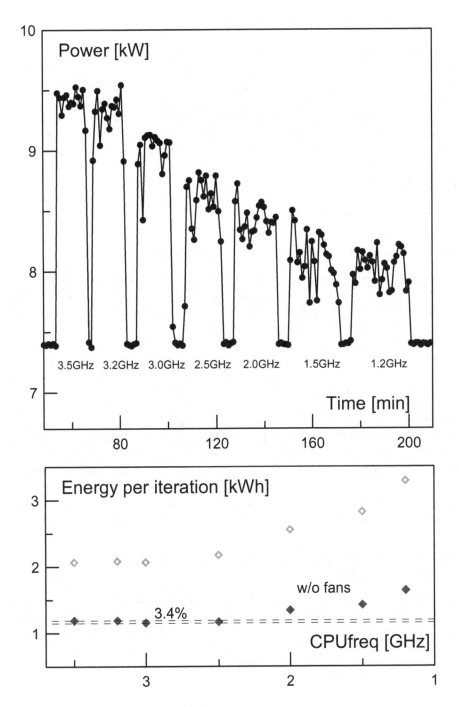

Fig. 3. Power variation and consumed energy variation on the same VASP test benchmark at different CPU frequencies

5   Case Studies of GPU Efficiency

NVIDIA CUDA technology was released in 2007. The past decade became a time of gradual adoption of this programming paradigm. Nowadays, the CUDA-enabled software ecosystem is quite mature. The GPU usage in HPC is motivated not only by energy efficiency but by cost efficiency as well. Consumer cards with teraflops performance in single precision represent an attractive option for cheap computational acceleration. The deployment of such commodity GPU accelerators in the Desmos supercomputer was a carefully planned decision [2]. However, the absence of double-precision capabilities narrows the spectrum of potential problems that can be solved using this hardware.

In this context, we present benchmarks showing the efficiency of the Desmos supercomputer for certain workloads.

5.1   Classical Molecular Dynamics with Gromacs

Classical molecular dynamics is an important modern scientific tool (see, e.g., [20–24]). In Fig. 4, we can see the results of the Intel Xeon-based super-computer IRUS17 and the Desmos supercomputer on two detailed biomolecular benchmarks [13] from the Gromacs package.

The cost of each node in Fig. 4 consists of the price of the computational resources with the corresponding infrastructure excluding the costs of intercon-nect. The prices of single nodes are estimated according to the price lists from the ThinkMate.com website at the end of November 2017.

A Desmos node without GPU costs about $ 2600, while a Desmos node with one GTX 1070 costs $ 3100. An IRUS17 node with two Intel Xeon E5-2698 v4 costs $ 11 000 (IRUS17 consists of dual-node blades in an enclosure), and an IRUS17 node with two Intel Xeon E5-2699 v4 costs $ 13 000. The labels in Fig. 4 show the amount of nodes. The cost is multiplied by the corresponding number of nodes.

We see that Desmos is ahead of IRUS17 for these benchmarks, in terms of both maximum attainable speed of calculation (ns/day) and cost-efficiency.

5.2   Quantum Molecular Dynamics with TeraChem and GAMESS-US

Quantum chemistry and electronic structure calculations are among the major consumers of HPC resources worldwide (see, e.g. [25–30]). The TeraChem pack-age is a rare example of CUDA-based software that deploys very efficiently single-precision floating point operations of NVIDIA GPU accelerators. In this work, we compare the performance of TeraChem with the well-know quantum chemistry package GAMESS-US.

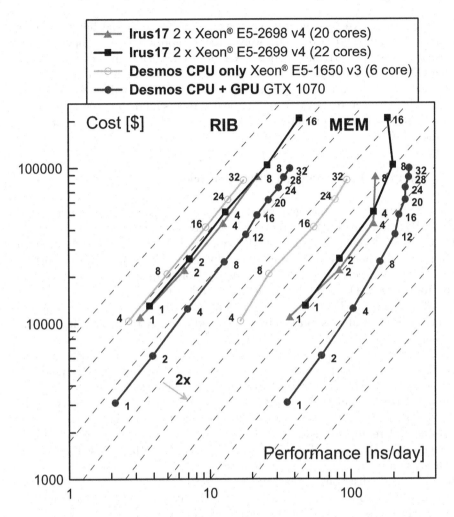

Fig. 4. Comparison of the supercomputers Desmos and IRUS17, on two biomolecular benchmarks (RIB: 2 million atoms, MEM: 82 thousand atoms; see [13])

The test model is the *ab initio* DFT molecular dynamics of the molecule of malondealdehyde $CH_2(CHO)_2$. The 6–31g basis is used together with the B3LYP exchange-correlation functional.

TeraChem is not MPI-parallelized and runs on a single node of Desmos (on a single core with a GTX 1070 accelerator). This hardware gives 0.5 s per one MD step in this test benchmark for TeraChem. The same level of performance we see in the CPU-only MPI-parallelized GAMESS-US calculation on 12 Desmos nodes (0.5 s per one MD step).

It is instructive to compare the peak performance of the hardware under consideration in these two tests. Twelve Desmos nodes have 4 TFlops of double-precision peak performance and 540 GB/s DRAM total memory bandwidth. One GTX 1070 accelerator has 6 TFlops of single-precision peak performance and 256 GB/s DRAM memory bandwidth. These numbers allow us to conclude that, with respect to GAMESS-US, the Desmos supercomputer is equivalent to a 128-TFlop supercomputer (= 32 nodes × 4 TFlops) based on Intel Xeon Broadwell CPUs.

6   Parallel File System Benchmarks

Many scientific HPC codes generate huge amounts of data. For example, in classical MD, the limits of the system size are trillions of atoms [31]. Desmos allows for GPU-accelerated modeling of MD systems with up to 100 million atoms. On-the-fly methods of data processing help considerably but cannot substitute post-processing completely. Another unavoidable requirement is the saving of control (or restart) points during or at the end of the calculation.

All 32 nodes of Desmos have been equipped with fast SSD drives, and the BeeGFS parallel file system has been installed in order to use all these disks as one distributed storage.

For comparison, we consider the Angara-K1 supercomputer, located at JSC NICEVT. This cluster is based on the Angara interconnect as well. Angara-K1 has a dedicated storage server (hardware RAID-adapter Adaptec 5405z, RAID level: 6 Reed-Solomon, HDD: 8 × 2 TB SATA2, FS: Lustre 2.10.1, FS type: ext3).

The schemes of the Desmos and Angara-K1 supercomputers and relevant parameters are given in Fig. 5.

The standard Lennard-Jones benchmark was run with the LAMMPS molecular dynamics package (the benchmark is based on the model "melt" from the LAMMPS distribution package, the model corresponds to a f.c.c. crystal of Lennard-Jones particles and has been replicated to 16 million particles).

LAMMPS has two variants for output of large amounts of data. It is possible to use either standard output methods or MPI-IO capability.

Figure 6 depicts the results of the benchmarks for different sizes of the MD model. We see that the absolute values of the calculation time are higher for Angara-K1 than for Desmos. However, the performance degradation due to storing large files is more pronounced for Desmos. The MPI-IO output gets the evident benefits of the distributed storage of Desmos.

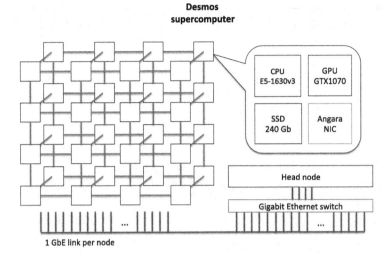

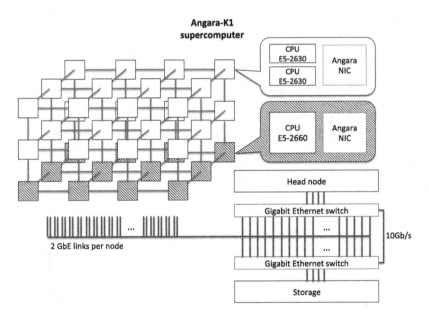

Fig. 5. The schemes of the supercomputers Desmos and Angara-K1

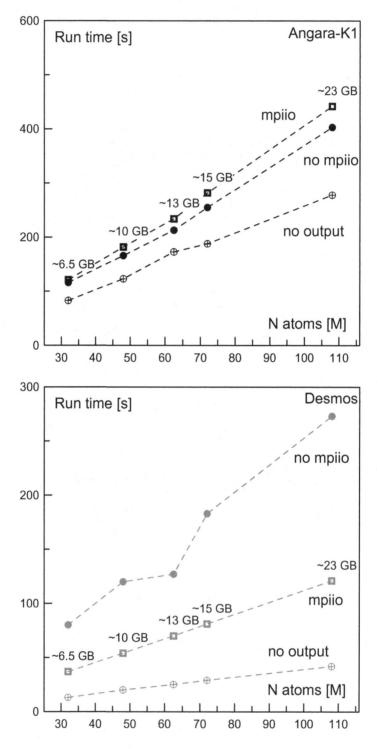

Fig. 6. Parallel output benchmarks based on the LAMMPS test model for Angara-K1 and Desmos supercomputers

7   Conclusions

The paper presents the results of efficiency and performance analyses of the Desmos supercomputer.

- The job accounting statistics of the Desmos supercomputer were reviewed. Two methods of quantitative efficiency monitoring were proposed.
- A variation of CPU frequency was attempted for energy optimization. The effect of lower energy consumption does indeed show up but the figures promise no practical benefits.
- GPU-accelerated classical MD with Gromacs runs faster and is more cost effective on supercomputers similar to Desmos than on wide-spread super-computers based on expensive Intel Xeon multi-core CPUs.
- GPU-accelerated quantum MD can be effectively computed on Desmos nodes using single precision. A comparison with the GAMESS-US package shows that TeraChem is able to efficiently substitute double-precision CPU performance with single-precision GPU performance for solving *ab initio* problems.
- It is shown that BeeGFS effectively combines the distributed storage units located on the Desmos nodes into a single drive. MPI-IO shows a very good speed in storing data from the LAMMPS MD calculation on the Desmos parallel file system. However, LAMMPS MPI-IO shows no benefits in the case of a conventional storage benchmarked on the Angara-K1 supercomputer.

References

1. Stegailov, V., et al.: Early performance evaluation of the hybrid cluster with torus interconnect aimed at molecular-dynamics simulations. In: Wyrzykowski, R., Dongarra, J., Deelman, E., Karczewski, K. (eds.) PPAM 2017 Part I. LNCS, vol. 10777, pp. 327–336. Springer, Cham (2018). https://doi.org/10.1007/978-3-319-78024-5_29
2. Vecher, V.S., Kondratyuk, N.D., Smirnov, G.S., Stegailov, V.V.: Angara-based hybrid supercomputer for efficient acceleration of computational materials science studies. In: Proceeding of International Conference Russian Supercomputing Days 2017, pp. 557–571 (2017)
3. Neuwirth, S., Frey, D., Nuessle, M., Bruening, U.: Scalable communication architecture for network-attached accelerators. In: 2015 IEEE 21st International Symposium on High Performance Computer Architecture (HPCA), pp. 627–638 (2015). https://doi.org/10.1109/HPCA.2015.7056068
4. Puente, V., Beivide, R., Gregorio, J.A., Prellezo, J.M., Duato, J., Izu, C.: Adaptive bubble router: a design to improve performance in torus networks. In: Proceedings of the 1999 International Conference on Parallel Processing, pp. 58–67 (1999). https://doi.org/10.1109/ICPP.1999.797388
5. Scott, S.L., Thorson, G.M.: The Cray T3E network: adaptive routing in a high performance 3D torus. In: HOT Interconnects IV. Stanford University, 15–16 August 1996 (1996)
6. Adiga, N.R., et al.: Blue Gene/L torus interconnection network. IBM J. Res. Dev. **49**(2), 265–276 (2005). https://doi.org/10.1147/rd.492.0265

7. Gómez-Martín, C., Vega-Rodríguez, M.A., González-Sánchez, J.L.: Fattened backfilling: an improved strategy for job scheduling in parallel systems. J. Parallel Distrib. Comput. **97**(Suppl. C), 69–77 (2016). https://doi.org/10.1016/j.jpdc.2016.06.013

8. Kraemer, A., Maziero, C., Richard, O., Trystram, D.: Reducing the number of response time SLO violations by a Cloud-HPC convergence scheduler. In: 2016 2nd International Conference on Cloud Computing Technologies and Applications (CloudTech), pp. 293–300 (2016). https://doi.org/10.1109/CloudTech.2016.7847712

9. Mamaeva, A.A., Voevodin, V.V.: Methods for statistical analysis of large supercomputer job flow. In: Proceeding of International Conference Russian Supercomputing Days 2017, pp. 788–799 (2017)

10. Hoefler, T., Belli, R.: Scientific benchmarking of parallel computing systems: twelve ways to tell the masses when reporting performance results. In: Proceedings of the International Conference for High Performance Computing, Networking, Storage and Analysis, SC 2015, pp. 73:1–73:12. ACM, New York (2015). https://doi.org/10.1145/2807591.2807644

11. Scogland, T., Azose, J., Rohr, D., Rivoire, S., Bates, N., Hackenberg, D.: Node variability in large-scale power measurements: perspectives from the Green500, Top500 and EEHPCWG. In: Proceedings of the International Conference for High Performance Computing, Networking, Storage and Analysis, SC 2015, pp. 74:1–74:11. ACM, New York (2015). https://doi.org/10.1145/2807591.2807653

12. Höhnerbach, M., Ismail, A.E., Bientinesi, P.: The vectorization of the Tersoff multibody potential: an exercise in performance portability. In: Proceedings of the International Conference for High Performance Computing, Networking, Storage and Analysis, SC 2016, pp. 7:1–7:13. IEEE Press, Piscataway (2016). https://doi.org/10.1109/SC.2016.6

13. Kutzner, C., Pall, S., Fechner, M., Esztermann, A., de Groot, B.L., Grubmuller, H.: Best bang for your buck: GPU nodes for gromacs biomolecular simulations. J. Comput. Chemis. **36**(26), 1990–2008 (2015). https://doi.org/10.1002/jcc.24030

14. Luehr, N., Ufimtsev, I.S., Martínez, T.J.: Dynamic precision for electron repulsion integral evaluation on graphical processing units (GPUs). J. Chem. Theory Comput. **7**(4), 949–954 (2011). https://doi.org/10.1021/ct100701w

15. Mills, N., Alex Feltus, F., Ligon III, W.B.: Maximizing the performance of scientific data transfer by optimizing the interface between parallel file systems and advanced research networks. Futur. Gener. Comput. Syst. **79**(Part 1), 190–198 (2018). https://doi.org/10.1016/j.future.2017.04.030

16. Yoo, A.B., Jette, M.A., Grondona, M.: SLURM: simple linux utility for resource management. In: Feitelson, D., Rudolph, L., Schwiegelshohn, U. (eds.) JSSPP 2003. LNCS, vol. 2862, pp. 44–60. Springer, Heidelberg (2003). https://doi.org/10.1007/10968987_3

17. Vecher, V., Nikolskii, V., Stegailov, V.: GPU-accelerated molecular dynamics: energy consumption and performance. In: Voevodin, V., Sobolev, S. (eds.) RuSCDays 2016. CCIS, vol. 687, pp. 78–90. Springer, Cham (2016). https://doi.org/10.1007/978-3-319-55669-7_7

18. Stegailov, V., Vecher, V.: Efficiency analysis of intel and AMD x86_64 architectures for Ab initio calculations: a case study of VASP. In: Voevodin, V., Sobolev, S. (eds.) RuSCDays 2017. CCIS, vol. 793, pp. 430–441. Springer, Cham (2017). https://doi.org/10.1007/978-3-319-71255-0_35

19. Stegailov, V., Vecher, V.: Efficiency analysis of Intel, AMD and Nvidia 64-Bit hardware for memory-bound problems: a case study of Ab Initio calculations with VASP. In: Wyrzykowski, R., Dongarra, J., Deelman, E., Karczewski, K. (eds.) PPAM 2017 Part II. LNCS, vol. 10778, pp. 81–90. Springer, Cham (2018). https://doi.org/10.1007/978-3-319-78054-2_8

20. Smirnov, G.S., Stegailov, V.V.: Anomalous diffusion of guest molecules in hydrogen gas hydrates. High Temp. **53**(6), 829–836 (2015). https://doi.org/10.1134/S0018151X15060188

21. Orekhov, N.D., Stegailov, V.V.: Simulation of the adhesion properties of the Polyethylene/Carbon nanotube interface. Polym. Sci. Ser. A **58**(3), 476–486 (2016). https://doi.org/10.1134/S0965545X16030135

22. Pavlov, S.V., Kislenko, S.A.: Effects of carbon surface topography on the electrode/electrolyte interface structure and relevance to li-air batteries. Phys. Chem. Chem. Phys. **18**, 30830–30836 (2016). https://doi.org/10.1039/C6CP05552D

23. Antropov, A.S., Fidanyan, K.S., Stegailov, V.V.: Phonon density of states for solid uranium: accuracy of the embedded atom model classical interatomic potential. J. Phys.: Conf. Ser. **946**(012094), 94 (2018). https://doi.org/10.1088/1742-6596/946/1/012094

24. Logunov, M.A., Orekhov, N.D.: Molecular dynamics study of cavitation in carbon nanotube reinforced polyethylene nanocomposite. J. Phys.: Conf. Ser. **946**(1), 2044 (2018). https://doi.org/10.1088/1742-6596/946/1/012044

25. Stegailov, V.V., Orekhov, N.D., Smirnov, G.S.: HPC hardware efficiency for quantum and classical molecular dynamics. In: Malyshkin, V. (ed.) PaCT 2015. LNCS, vol. 9251, pp. 469–473. Springer, Cham (2015). https://doi.org/10.1007/978-3-319-21909-7_45

26. Aristova, N.M., Belov, G.V.: Refining the thermodynamic functions of scandium triflouride SCF3 in the condensed state. Russ. J. Phys. Chemis. A **90**(3), 700–703 (2016). https://doi.org/10.1134/S0036024416030031

27. Kochikov, I.V., Kovtun, D.M., Tarasov, Y.I.: Electron diffraction analysis for the molecules with degenerate large amplitude motions: intramolecular dynamics in arsenic pentafluoride. J. Mol. Struct. **1132**, 139–148 (2017). https://doi.org/10.1016/j.molstruc.2016.09.064

28. Stegailov, V.V., Zhilyaev, P.A.: Warm dense gold: effective ionioninteraction and ionisation. Mol. Phys. **114**(3–4), 509–518 (2016). https://doi.org/10.1080/00268976.2015.1105390

29. Minakov, D.V., Levashov, P.R.: Melting curves of metals with excited electrons in the quasiharmonic approximation. Phys. Rev. B **92**, 224102 (2015). https://doi.org/10.1103/PhysRevB.92.224102

30. Minakov, D., Levashov, P.: Thermodynamic properties of LiD under compression with different pseudopotentials for lithium. Comput Mat. Sci. **114**, 128–134 (2016). https://doi.org/10.1016/j.commatsci.2015.12.008

31. Eckhardt, W., et al.: 591 TFLOPS multi-trillion particles simulation on SuperMUC. In: Kunkel, J.M., Ludwig, T., Meuer, H.W. (eds.) ISC 2013. LNCS, vol. 7905, pp. 1–12. Springer, Heidelberg (2013). https://doi.org/10.1007/978-3-642-38750-0_1

Performance of Elbrus Processors for Computational Materials Science Codes and Fast Fourier Transform

Vladimir Stegailov[1,2], Alexey Timofeev[1,2(✉)], and Denis Dergunov[1,2]

[1] Joint Institute for High Temperatures of the Russian Academy of Sciences, Izhorskaya st. 13 Bd.2, Moscow 125412, Russia
stegailov@gmail.com, timofeevalvl@gmail.com

[2] National Research University Higher School of Economics, Myasnitskaya st. 20, Moscow 101000, Russia

Abstract. Modern Elbrus-4S and Elbrus-8S processors provide a level of floating-point performance close to that of widespread x86_64 CPUs that are predominantly used in high-performance computing (HPC). The uniqueness of the software ecosystem of Elbrus processors requires special attention in the case of their deployment for execution of mainstream computational codes. In this paper, we consider the performance of one widely used code for computational materials science (VASP), as well as FFT libraries. The results for the Elbrus processors are embedded into the context of performance of modern x86_64 CPUs.

Keywords: Elbrus architecture · VASP · Fourier transform

1   Introduction

A large share of HPC resources installed during the last decade is based on Intel CPUs. However, the situation is gradually changing. In March 2017, AMD released the first processors based on the novel x86_64 architecture, called Zen. In November 2017, Cavium presented the server-grade 64-bit ThunderX2 ARMv8 CPUs, which are to be deployed in new Cray supercomputers. The Elbrus microprocessors stand among the emerging types of high-performance CPU architectures [1,2].

The diversity of CPU types significantly complicates the choice of the best variant for a particular HPC system. The main criterion is certainly the time-to-solution of a given computational task or a set of different tasks, which represents the envisaged workload of a system under development.

The work was supported by the grant No. 14-50-00124 of the Russian Science Foundation. The authors acknowledge Joint Supercomputer Centre of Russian Academy of Sciences (http://www.jscc.ru) for the access to the supercomputer MVS1P5. The authors acknowledge JSC MCST (http://www.msct.ru) for the access to the servers with Elbrus CPUs. The authors are grateful to Vyacheslav Vecher for the help with calculations based on hardware counters.

© Springer Nature Switzerland AG 2018
L. Sokolinsky and M. Zymbler (Eds.): PCT 2018, CCIS 910, pp. 92–103, 2018.
https://doi.org/10.1007/978-3-319-99673-8_7

Computational materials science provides an essential part of the deployment time of HPC resources worldwide. The VASP code [3–6] is among the most popular programs for electronic structure calculations. It makes it possible to calculate materials properties using non-empirical (so called *ab initio*) methods. *Ab initio* calculation methods based on quantum mechanics are important modern scientific tools (see, e.g., [7–11]). According to recent estimates, VASP alone consumes from 15 to 20% of the world's supercomputing power [12,13]. Such an unprecedented popularity has led to a special attention directed towards the optimization of VASP for both existing and novel computer architectures (see, e.g., [14]).

The computation of Fourier transforms accounts for a significant part of the calculation time in software packages for computational materials science. One of the most time consuming components in VASP is 3D-FFT [15]. FFT libraries were tested on the Elbrus processor in order to determine the most optimal tool for computing fast Fourier transforms. The EML library, developed by the manufacturer of the Elbrus processor, and the most popular FFTW library are under consideration.

In this work, we present an efficiency analysis of Elbrus CPUs compared with Intel Xeon Haswell CPUs, using a typical VASP workload example. Here we also give the results of the test of FFT libraries on Elbrus processors.

2 Related Work

HPC systems are notorious for operating at a small fraction of their peak performance. The deployment of multi-core and multi-socket compute nodes further complicates performance optimization. Many attempts have been made to develop a more or less universal framework for algorithm optimization that takes into account essential properties of the hardware (see, e.g., [16–18]). The recent work of Stanisic *et al.* [19] emphasizes many pitfalls encountered while trying to characterize both the network and the memory performance of modern machines.

A fast Fourier transform is used in computational modeling programs for calculations related to quantum computations, Coulomb systems, etc., and takes a significant part of the program's running time [20], especially in the case of VASP [15]. A detailed optimization of the computation of 3D-FFT in VASP to prepare the code for an efficient execution on multi- and many-core CPUs as Intel's Xeon Phi is considered in [15]. In this article, the threading performance of the widely used FFTW library (Cray LibSci) and Intel's MKL on the Cray-XC40 with Intel Haswell CPUs and the modern Cray-XC30 Xeon Phi (Knights Corner, KNC) system is evaluated. Recently, several 64-bit x86_64 and Armv8 CPUs have been compared using a VASP benchmark test with the focus on the memory bandwidth [21,22].

At the moment, Elbrus processors are ready for use [1,2], so we decided to benchmark them using one of the main HPC tools applied in materials science studies (VASP) and the library that determines the performance of this code (FFT). The architecture of the Elbrus processors [1,2] allows us to expect that, during the execution of the FFT, the butterfly computation occurs in a smaller number of cycles than it does on such CPUs as Intel's Xeon Phi.

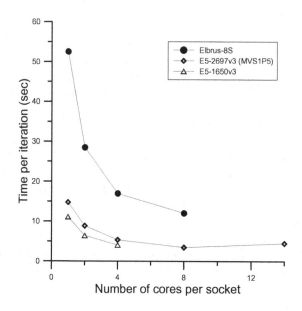

Fig. 1. Dependence between the first iteration time in the liquid-Si model test and the number of cores per socket

3   Methods and Software Implementation

3.1   Test Model in VASP

VASP 5.4.1 is compiled for Intel systems using Intel Fortran, Intel MPI and linked with Intel MKL for BLAS, LAPACK and FFT calls. For the Elbrus-8S system, lfortran compatible with gfortran ver.4.8 is used together with MPICH, EML BLAS, Netlib LAPACK and FFTW libraries.

Our test model in VASP represents a liquid-Si system consisting of 48 atoms in the supercell. The Perdew–Burke–Ernzerhof model for the xc-functional is used. The calculation protocol corresponds to molecular dynamics. We use the first iteration time of the electron density optimization τ_{iter} as the target parameter of the performance metric.

The τ_{iter} values considered in this work range from 5 to 50 s approximately and correspond to a single CPU performance. At the first glance, these times are not sufficiently long to be accelerated. However, *ab initio* molecular dynamics usually requires 10^4 to 10^5 time steps and larger system sizes. That is why decreasing τ_{iter} by several orders of magnitude is an actual problem for modern HPC systems targeted at materials science computing.

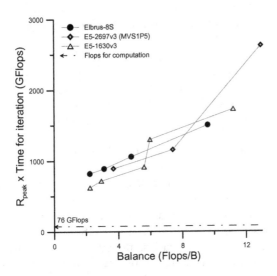

Fig. 2. Dependence between the first iteration time in the liquid-Si model test and the number of cores per socket, for reduced parameters $R_{\text{peak}}\tau_{\text{iter}}$ and balance B (R_{peak} is the total peak performance of all the cores used; the balance B corresponds to the total bandwidth for a single/dual-socket server)

The choice of a particular test model has a certain influence on the benchmarking results. However, our preliminary tests of other VASP models show that the main conclusions of this study do not depend significantly on a particular model.

3.2   Fast Fourier Transform

FFTW 3.3.6 is compiled using lcc, the analogue of gcc for Elbrus systems. As an input array for the Fourier transforms, a sinusoidal signal, white, pink and brown noise are used. In this article, we report the results for white noise.

The usual pattern when calling FFT (or MKL through its FFTW interface) is as follows:

1. Preparation stage: creates plans for FFT computations, e.g., via `fftw_plan p=fftw_plan_dft(..)` for FFTW, and via `eml_Signal_FFTInit(...)` for EML.
2. Execution stage: performs FFT computations using the plan created, e.g., via `fftw_execute_dft(p,in,out)` for FFTW, and via `eml_Signal_FFTFwd(...)` for EML.
3. Clean up.

We consider the work of the first two stages as they are the most time consuming. Preparation takes the main time when one starts the Fourier transform once for a fixed size of the input array. When the Fourier transform is repeatedly started, the running time of the program can determine the execution time of the Fourier transform itself.

So, for these two stages, we compare the FFTW and EML libraries on the processors Elbrus-4S and Elbrus 8S. For the moment, the EML library has fewer useful functions than the FFTW library. In particular, the size of the input array can only be a power of two, so the preparation stage has to be partially implemented by the user. The number of functions in the EML library is much smaller than that in the FFTW library.

Plan creation with FFTW can be done by planner schemes that differ in their costs: FFTW_ESTIMATE (cheap), FFTW_MEASURE (expensive), FFTW_PATIENT (more expensive) and FFTW_EXHAUSTIVE (most expensive). Except for FFTW_ESTIMATE , plan creation involves testing different FFT algorithms together with runtime measurements to achieve the best performance on the target platform. On servers with Elbrus-4S and Elbrus-8S processors, the authors, owing to lack of libraries, managed to compile FFTW only in FFTW_ESTIMATE mode, in which the preparation time is short and the execution time is long.

To average the operating time values and obtain the dispersion of the results, calculations were repeated 30 to 1000 times. The dispersion of the results was within 1%, and sometimes did not exceed 0.001%.

4   Results and Discussion

4.1   VASP Benchmark on Elbrus-8S and Xeon Haswell CPUs

VASP is known to be both a memory-bound and a compute-bound code [14]. Figure 1 shows the results of the liquid-Si model test runs.

Performance comparison of different CPUs usually resembles a comparison of "apples and oranges". To compare CPUs with different frequencies and different peak Flops/cycle values, it is better to use the reduced parameter $R_{peak}\tau_{iter}$ [7,23].

Another reduced parameter that characterizes the memory subsystem is the so-called balance B, which is the ratio of R_{peak} to the CPU memory bandwidth (in this work, we measure the latter quantity using the STREAM benchmark).

Figure 2 shows the same data as Fig. 1 but in reduced coordinates. This allows to eliminate the differences in floating-point performance and memory bandwidth between dissimilar CPU cores. In these reduced coordinates, the scatter of data points is much smaller, and there is an evident common trend.

The test model considered fixes the total number of arithmetic operations (Flops) required for its solution. An increase in $R_{peak}\tau_{iter}$ (that is proportional to the number of CPU cycles) leads to an increase in overhead due to the limited memory bandwidth. More CPU cycles are required for the CPU cores involved in computations to get data from DRAM.

We calculated the number of floating-point operations that corresponds to τ_{iter}. We used a system with Intel Core i7 640UM CPU. This CPU does not support AVX instructions and the performance counters work unambiguously. The resulting value of $N_{FP} = 76$ GFlops is shown in Fig. 2 as a dashed-dotted horizontal line. The ratio $R_{\mathrm{peak}}\tau_{\mathrm{iter}}/N_{FP}$ indicates the overhead of CPU cycles that are not deployed for computations because the required data from DRAM are not available. We should notice that the overall trend in Fig. 2 corresponds quite well to the limiting case $R_{\mathrm{peak}}\tau_{\mathrm{iter}} \to N_{FP}$ when $B \to 0$.

4.2   Fast Fourier Transform on Elbrus CPUs: EML vs. FFTW

We split the Fourier transformation process into two stages: the preparation of the algorithm (Figs. 3, 4, 5 and 6), and the execution of the transformation (Figs. 7, 8, 9 and 10). The preparation takes the main amount of time when one starts the Fourier transform once. The algorithm execution time can determine the total running time of the Fourier transform in situations when the Fourier transform is started many times for a fixed size of the input array.

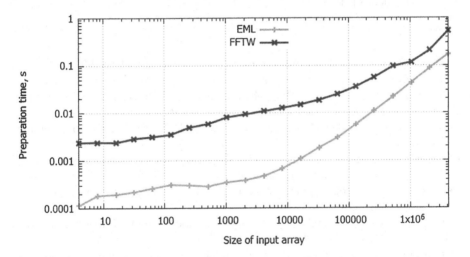

Fig. 3. Dependence between the FFT preparation time and the size of the input array, for Elbrus-4S

The preparation time of the FFT algorithm for Elbrus-4S appears to be an order of magnitude smaller when using the EML library than it is when using FFTW, for array sizes smaller than 2^{15} (Figs. 3 and 4). For larger array sizes, the preparation time is only 2 to 3 times smaller with EML than it is with FFTW. All points have an error less than 1%. As Figs. 5 and 6 show, the difference in preparation time is even greater for the Elbrus-8S. For array sizes smaller than 2^{15}, the preparation time when using EML is 10 to 20 times less than it is when using FFTW. For larger array sizes (up to 2^{17}), the preparation time when using EML is 50 to 90 times less than it is in the case of FFTW.

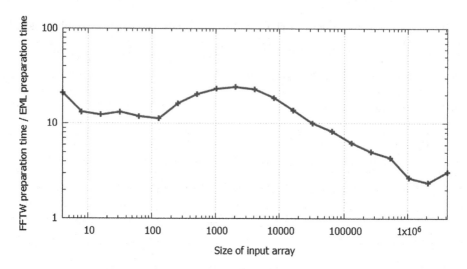

Fig. 4. Dependence between the ratio of FFT preparation time with FFTW to that with EML and the size of the input array, for Elbrus-4S

We can thus make an interim summary: single launches of the FFT on Elbrus-4S and Elbrus-8S are more efficient when using the EML library because the preparation of the FFT algorithm when using EML is faster (2 to 20 times for Elbrus-4S, and 10 to 90 times for Elbrus-8S) than it is when using FFTW.

And now we consider the second stage of the FFT implementation, namely the execution of the algorithm. The execution stage of the algorithm takes from one to several orders of magnitude less time than its preparation stage, so it has a significant effect only if the algorithm is run multiple times after a single preparation. This often happens when we need to execute an FFT on a set of arrays of the same size.

For array sizes less than 2^{11}, the execution time of the FFT algorithm using EML turns out to be from 1 to 10 times greater than it is when using FTTW (Figs. 9 and 10). For larger array sizes, the situation reverses, and the ratio of the execution time with FFTW to that with EML increases from 1 to 6 for array sizes between 2^{14} and 2^{22}. Figures 9 and 10 show that the difference in preparation time is smaller for the Elbrus-8S than for the Elbrus-4S. For arrays smaller than 2^{12}, the execution time when using EML is close to that when using FFTW. For larger arrays (up to 2^{18}), the ratio of execution time with FFTW to that with EML ranges from 1.4 to 1.9.

On Elbrus-4S, multiple starts (more than 1000) of FFT for small arrays (less than 2^{11}) are more efficient when using FFTW than they are when using EML. On Elbrus-4S, the execution time when using FFTW is 1 to 10 times faster than it is when using the EML library. On Elbrus-8S, FFT for arrays of almost all sizes is more efficient when using the EML library, but the ratio of the execution time for FFTW to that for EML is less than 2.

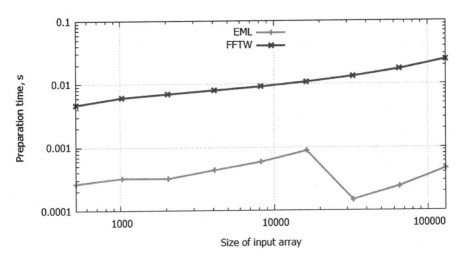

Fig. 5. Dependence between the FFT preparation time and the size of the input array, for Elbrus-8S

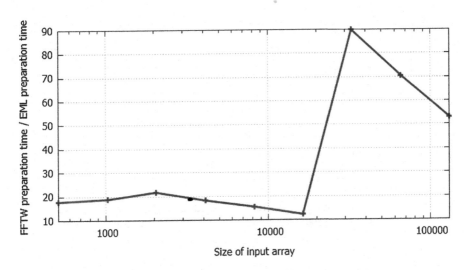

Fig. 6. Dependence between the ratio of FFT preparation time with FFTW to that with EML and the size of the input array, for Elbrus-8S

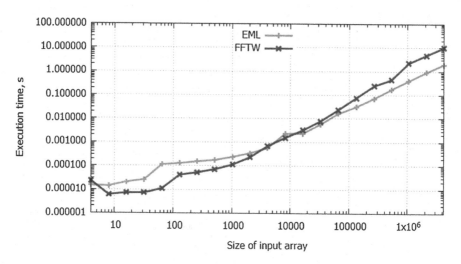

Fig. 7. Dependence between the FFT execution time and the size of the input array, for Elbrus-4S

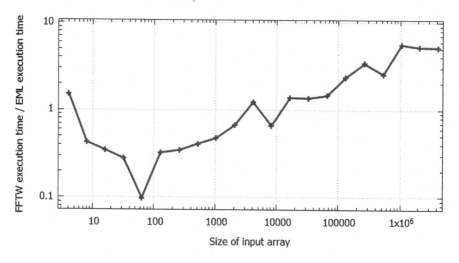

Fig. 8. Dependence between the ratio of FFT execution time with FFTW to that with EML and the size of the input array, for Elbrus-4S

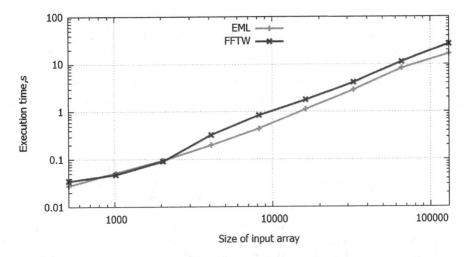

Fig. 9. Dependence between the FFT execution time and the size of the input array, for Elbrus-8S

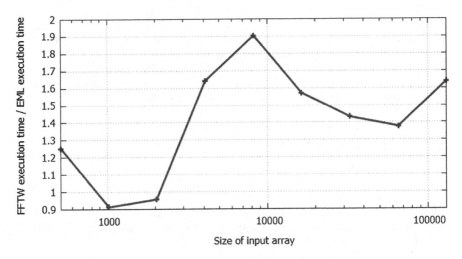

Fig. 10. Dependence between the ratio of FFT execution time with FFTW to that with EML and the size of the input array, for Elbrus-8S

5   Conclusions

We performed test calculations for the VASP model on Intel Xeon Haswell and Elbrus-8S CPUs with the best choice of mathematical libraries available. Elbrus-8S shows larger time-to-solution values, but there is not a large gap between the Elbrus-8S performance and that of Xeon Haswell CPUs. The major target for optimization, which could significantly speed up VASP on Elbrus-8S, is the FFT library.

We tested the native EML library and an unoptimized FFTW library on the Elbrus-4S and Elbrus-8S processors. Single launches of the FFT on both Elbrus-4S and Elbrus-8S are more efficient when using the EML library. Nevertheless, for small arrays (less than 4000), multiple starts (more than 1 000) of FFT are more efficient with FFTW than they are with EML. On Elbrus-8S, FFT for arrays of any sizes is more efficient when running with the EML library.

References

1. Kozhin, A.S., et al.: The 5th generation 28nm 8-core VLIW Elbrus-8C processor architecture. In: Proceedings - 2016 International Conference on Engineering and Telecommunication, EnT 2016, pp. 86–90 (2017). https://doi.org/10.1109/EnT.2016.25
2. Tyutlyaeva, E., Konyukhov, S., Odintsov, I., Moskovsky, A.: The Elbrus platform feasibility assessment for high-performance computations. In: Voevodin, V., Sobolev, S. (eds.) RuSCDays 2016. CCIS, vol. 687, pp. 333–344. Springer, Heidelberg (2016). https://doi.org/10.1007/978-3-319-55669-7_26
3. Kresse, G., Hafner, J.: Ab initio molecular dynamics for liquid metals. Phys. Rev. B **47**, 558–561 (1993). https://doi.org/10.1103/PhysRevB.47.558
4. Kresse, G., Hafner, J.: Ab initio molecular-dynamics simulation of the liquid-metal-amorphous-semiconductor transition in germanium. Phys. Rev. B **49**, 14251–14269 (1994). https://doi.org/10.1103/PhysRevB.49.14251
5. Kresse, G., Furthmuller, J.: Efficiency of ab-initio total energy calculations for metals and semiconductors using a plane-wave basis set. Comput. Mater. Sci. **6**(1), 15–50 (1996). https://doi.org/10.1016/0927-0256(96)00008-0
6. Kresse, G., Furthmüller, J.: Efficient iterative schemes for ab initio total-energy calculations using a plane-wave basis set. Phys. Rev. B **54**, 11169–11186 (1996). https://doi.org/10.1103/PhysRevB.54.11169
7. Stegailov, V.V., Orekhov, N.D., Smirnov, G.S.: HPC hardware efficiency for quantum and classical molecular dynamics. In: Malyshkin, V. (ed.) Parallel Computing Technologies. LNCS, vol. 9251, pp. 469–473. Springer, Heidelberg (2015). https://doi.org/10.1007/978-3-319-21909-7_45
8. Aristova, N.M., Belov, G.V.: Refining the thermodynamic functions of scandium triflouride SCF3 in the condensed state. Russ. J. Phys. Chem. A **90**(3), 700–703 (2016). https://doi.org/10.1134/S0036024416030031
9. Kochikov, I.V., Kovtun, D.M., Tarasov, Y.I.: Electron diffraction analysis for the molecules with degenerate large amplitude motions: intramolecular dynamics in arsenic pentafluoride. J. Mol. Struct. **1132**, 139–148 (2017). https://doi.org/10.1016/j.molstruc.2016.09.064
10. Minakov, D.V., Levashov, P.R.: Melting curves of metals with excited electrons in the quasiharmonic approximation. Phys. Rev. B **92**, 224102 (2015). https://doi.org/10.1103/PhysRevB.92.224102
11. Minakov, D., Levashov, P.: Thermodynamic properties of LiD under compression with different pseudopotentials for lithium. Comput. Mater. Sci. **114**, 128–134 (2016). https://doi.org/10.1016/j.commatsci.2015.12.008
12. Bethune, I.: Ab initio molecular dynamics. Introduction to Molecular Dynamics on ARCHER (2015)
13. Hutchinson, M.: VASP on GPUs. When and how. GPU technology theater, SC15 (2015)

14. Zhao, Z., Marsman, M.: Estimating the performance impact of the MCDRAM on KNL using dual-socket Ivy Bridge nodes on Cray XC30. In: 2016 Proceedings of the Cray User Group (2016)
15. Wende, F., Marsman, M., Steinke, T.: On enhancing 3D-FFT performance in VASP. In: CUG Proceedings, p. 9 (2016)
16. Burtscher, M., Kim, B.D., Diamond, J., McCalpin, J., Koesterke, L., Browne, J.: Perfexpert: an easy-to-use performance diagnosis tool for HPC applications. In: Proceedings of the 2010 ACM/IEEE International Conference for High Performance Computing, Networking, Storage and Analysis, SC 2010, pp. 1–11. IEEE Computer Society, Washington (2010). https://doi.org/10.1109/SC.2010.41
17. Rane, A., Browne, J.: Enhancing performance optimization of multicore/multichip nodes with data structure metrics. ACM Trans. Parallel Comput. 1(1), 3:1–3:20 (2014). https://doi.org/10.1145/2588788
18. Mantovani, F., Calore, E.: Performance and power analysis of HPC workloads on heterogeneous multi-node clusters. J. Low Power Electron. Appl. 8(2), 13 (2018)
19. Stanisic, L., Mello Schnorr, L.C., Degomme, A., Heinrich, F.C., Legrand, A., Videau, B.: Characterizing the performance of modern architectures through opaque benchmarks: pitfalls learned the hard way. In: IPDPS 2017 – 31st IEEE International Parallel and Distributed Processing Symposium (RepPar Workshop), Orlando, United States, pp. 1588–1597 (2017)
20. Baker, M.: A study of improving the parallel performance of VASP. Ph.D. thesis, East Tennessee State University (2010)
21. Stegailov, V., Vecher, V.: Efficiency analysis of Intel and AMD x86_64 architectures for ab initio calculations: a case study of VASP. In: Voevodin, V., Sobolev, S. (eds.) RuSCDays 2017. CCIS, vol. 793, pp. 430–441. Springer, Heidelberg (2017). https://doi.org/10.1007/978-3-319-71255-0_35
22. Stegailov, V., Vecher, V.: Efficiency analysis of Intel, AMD and Nvidia 64-bit hardware for memory-bound problems: a case study of ab initio calculations with VASP. In: Wyrzykowski, R., Dongarra, J., Deelman, E., Karczewski, K. (eds.) PPAM 2017. LNCS, vol. 10778, pp. 81–90. Springer, Cham (2018). https://doi.org/10.1007/978-3-319-78054-2_8
23. Nikolskiy, V.P., Stegailov, V.V., Vecher, V.S.: Efficiency of the Tegra K1 and X1 systems-on-chip for classical molecular dynamics. In: 2016 International Conference on High Performance Computing Simulation (HPCS), pp. 682–689 (2016). https://doi.org/10.1109/HPCSim.2016.7568401

Performance and Energy Analysis of Nighttime Satellite Image Archive Processing Module

Ekaterina Tyutlyaeva[1(✉)], Sergey Konyukhov[1], Igor Odintsov[1], Alexander Moskovsky[1], and Mikhail Zhizhin[2]

[1] ZAO RSC Technologies, Moscow, Russia
{xgl,s.konyuhov,igor_odintsov,moskov}@rsc-tech.ru
[2] Institute of Space Research, University of Colorado, Denver, USA
Mikhail.zhizhin@colorado.edu

Abstract. The main goal of this work is to analyze the behavior of a nighttime image processing module and find out basic estimates of required computational time and energy consumption for processing large data archives.

As part of this work, we have performed the code refactoring of the most computing-intensive module in a system for detecting fishing boat lights.

The algorithm is capable of detecting isolated bright spikes that are sharply visible on the sea surface at night. The refactored module has been optimized for effective usage of multi- and many-core Intel Xeon architectures. In the paper, we describe the algorithmic complexity for all computational stages of the module. Also, we have collected detailed statistic data for two data sets, different input parameter sets, and three test beds: Intel® Xeon® E5-2697A (codename Broadwell), Intel® Xeon® Gold 6148 (Skylake), and Intel® Xeon Phi® 7250 (KNL).

Key correlations between module behavior and energy consumption are also included in the paper. The results of the study were used for calculations of the estimate time and energy requirements for a whole year archive of day/night band (DNB) images from the Visible Infrared Imaging Radiometer Suite (VIIRS). Moreover, driving factors, including price and legacy software systems, are presented for discussion.

Keywords: Nighttime imaging processing
Energy consumption analysis · Nighttime image processing module
Archive processing analysis

1 Introduction

The module studied herein is the most computationally intensive part of an automatic system for detecting fishing boat lights in nighttime images from the VIIRS multispectral radiometer [1]. The original version was implemented using the MATLAB programming language.

© Springer Nature Switzerland AG 2018
L. Sokolinsky and M. Zymbler (Eds.): PCT 2018, CCIS 910, pp. 104–115, 2018.
https://doi.org/10.1007/978-3-319-99673-8_8

The system is able to detect isolated bright spikes that are sharply visible on the sea surface at night. In the moonlight, the interference of clouds and the glint of the moon are taken into account as well.

In our previous work [2], we studied a nighttime infrared remote sensing algorithm based on the Nelder–Mead method.

Another module based on direct Fourier transformation in a moving window has been refactored and studied in this work. Contrary to the previous algorithm, the current module is based on an archive processing approach and processes multiple images at a time.

The processing algorithms are changed and upgraded periodically, and archive data require to be re-computed to correct or define the results in compliance with the new implementations. In this regard, archive processing is regularly encountered in practice.

We have implemented and optimized the module under study using Intel® performance libraries and made an analysis of the execution time and energy consumption depending on the number of cores used and DNB images processed.

2   The Hardware

The codenames and specifications of the studied test beds are listed in Table 1.

Table 1. Test beds specifications

Codename	CPU	# Cores	Memory	GB per core
Broadwell	Intel® Xeon® E5-2697A v4	2 × 16	8x DRAM Samsung 16 GB DDR4/2133 MHz	4
Skylake	Intel® Xeon® Gold 6148	2 × 20	12 × 16 GB DRAM DDR4-2400 MHz	4.8
KNL	Intel® Xeon Phi® 7250	68	MCDRAM Intel® 16 GB + 6x DRAM Micron 32 GB DDR4/2133 MHz	2.8

3   The Algorithm and Implementation Details

In terms of the computation behavior, the algorithm can be divided broadly into four stages:

- **Input.** At this stage, we use Day/Night Band (DNB) observation data from nighttime satellite imagery collected by the Visible Infrared Imaging Radiometer Suite (VIIRS) onboard the Suomi National Polar-Orbiting Partnership (Suomi NPP). The data were locally stored in HDF5 format. The HDF5-1.8.19 Technology suite was used for data reading.
- The **Preparation** stage includes logarithmic transformation of the brightness histogram (stretch), applying the Wiener filter [3] and computing the Spike Median Index (SMI).
- The **Processing** stage is the most computationally intensive in the module. The Sharpness Index (SI) [4] is computed in a $size_{blk} \times size_{blk}$ moving window. This routine repeatedly performs direct Fourier transforms and solves overdetermined real linear systems.
- **Output.** The result is locally stored in ENVI format.

We implemented the module using C++ and MPI. We used LAPACK and FFTW optimized primitives for direct Fourier transforms and the solution of overdetermined real linear systems.

Each MPI rank processes its own images independently, so there are minimum communications between processes.

A hybrid (MPI + OpenMP) parallelization scheme was used to effectively utilize the multi-core Intel® architectures. Preliminary tests indicate that the most time/energy effective pinning pattern for this module is 1 MPI rank per physical core and $N_{Hyperthreads}$ OpenMP threads per each MPI, where $N_{Hyperthreads}$ equals two for the Broadwell test bed and four for the KNL test bed (hyperthreading support is turned on for all test beds).

Manual and compiler-supported code vectorization for Intel architectures were also applied before the analysis stage. The "vectorization" term describes the use of the Intel® SSE instruction set, which is an extension to the x86 architecture [10].

The following libraries were used:

- Intel® MKL Library (2018 Studio);
- Intel® C++ Compilers 2018;
- Intel® MPI Library Version 2018;
- HDF5-1.8.19 Technology suite.

4   Study of Performance and Energy Consumption

We have conducted a series of test runs on each of the available hardware test beds to assess the execution time. Figure 1 presents the median results of the test.

We vary the values of the input variable $size_{blk}$ and the number of processor cores (N_{cores}) used on the instrumented runs, where

- N_{cores} is the number of physical cores used. The number of MPI ranks used is equal to the number of physical cores used[1]. Explicit pinning for MPI processes was used to define a set of processor cores on which the program is allowed to run.
- $size_{blk}$ is the moving window size for the Spectral and Spatial Sharpness Measure algorithm.

It is important to outline that the number of cores used is equal to the number of pictures processed, so the ideal timeline graph should be a straight horizontal line. However, the collective work with memory and I/O slightly increases the execution time as the number of cores used and pictures processed increases. Statistics on usage of test beds at full workload for the most typical window size, $size_{blk} = 32$, is given in Table 2. According to these statistics, overhead costs on communications does not exceed 0.5%, which is a good result.

Table 2. Statistics on usage of test beds at full workload

Characterization	Broadwell	Skylake	KNL
Images processed	32	40	68
Execution time, sec	39.229	44.188	206.121
Energy consumed, J	9947	10002	30757
Memory usage, MB	19249.27	24178.27	43829.55
Input data, MB	1526.65	1908.31	3257.54
Output data, MB	3048.04	3810.05	6477.07
Computation, %	99.74	99.91	99.56
MPI, %	0.26	0.09	0.44

In order to ensure accurate and precise results, we measured ten training runs per input variable with freeing page cache, dentries and inodes[2] between runs.

We used the median value of all ten runs and two test data sets as the final T_{run} result.

RAPL [5] measurements for energy use were also collected to estimate the average energy consumption of the module (see Fig. 2).

As can be seen, energy consumption strongly correlates with the execution time and the size of the moving window ($size_{blk}$). Also, it is worth mentioning a slight increase in energy consumption after 16 cores for the Broadwell test bed

---

[1] The number of OpenMP threads for the hybrid version is equal to the number of hyperthreads per core, namely two OpenMP threads per core for the Broadwell test bed and four for the KNL test bed.

[2] A filesystem is represented in memory using dentries and inodes. Inodes are the objects that represent the underlying files (and also directories). A dentry is an object with a string name, a pointer to an inode, and a pointer to the parent dentry.

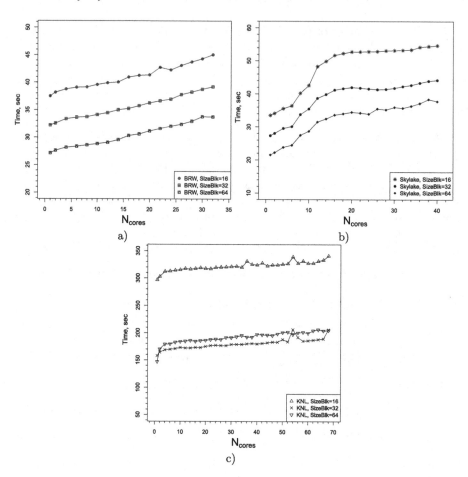

Fig. 1. Median execution times. (a) Broadwell test bed, (b) Skylake test bed, (c) KNL test bed

and after 20 cores for the Skylake test bed. To a great extent, the reason is that the first physical CPU is fully loaded to this moment, and further MPI processes are pinned to the second CPU.

Furthermore, we would like to emphasize that we use strong MPI pinning. Thus, $\langle N \rangle$ MPI ranks correspond to N cores used, $\langle N \rangle$ images processed, and $\text{N}_{\text{cores}} - \langle N \rangle$ idle cores. In this regard, there is room for time/energy consumption optimizations, based on the obtained results.

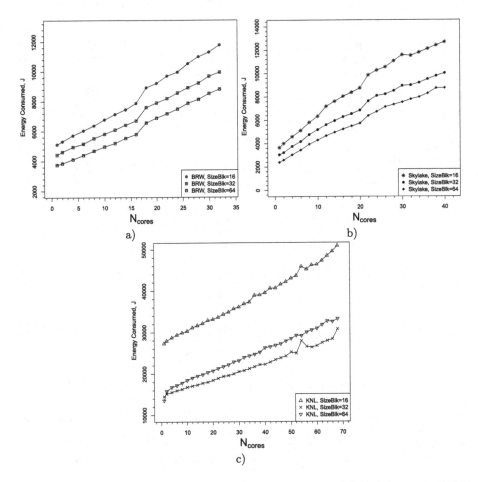

Fig. 2. Median energy consumption. (a) Broadwell test bed, (b) Skylake test bed, (c) KNL test bed

5 Analysis of the Processing Stage

According to the times measured (see Fig. 3), the `Processing` stage is the most computationally intensive in the module.

As stated above, this routine repeatedly performs computations in a $size_{blk} \times size_{blk}$ moving window.

Table 3 lists the number of iterations for the moving window sizes studied.

Each iterations includes:

- 2D direct Fourier transform for a $size_{blk} \times size_{blk}$ matrix ($O(size_{blk}^2 \cdot \log(size_{blk}))$).
- Vector logarithm calculation ($O(\log(size_{blk}/2))$).
- Solving an overdetermined real linear system ($O(size_{blk})$).

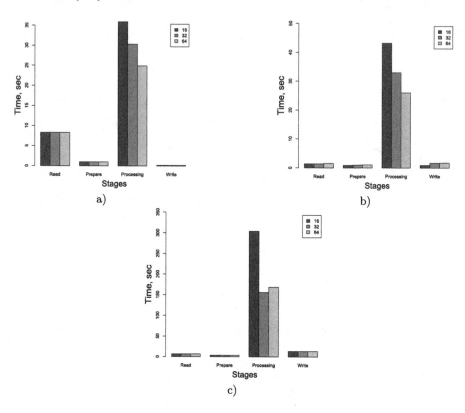

Fig. 3. Average times of stages per process. (a) Broadwell test bed, (b) Skylake test bed, (c) KNL test bed

Table 3. Number of iterations for the moving window sizes studied

Moving window size	Number of iterations
16×16	3 092 672
32×32	766 080
64×64	188 000

However, in the case of the Broadwell test bed (see Fig. 4(a)), the time required to process the data inside the moving window may be considered as a constant, since the moving window size (size_{blk}) processed at each iteration is comparatively small.

It must therefore be assumed that the number of iterations is the main contributor to the ratio of the processing stage time to size_{blk}.

For the KNL test bed (see Fig. 4(c)), a notable decline in processing time in the $\text{size}_{\text{blk}} = 32$ case is probably due to features of the cache memory subsystem implementation.

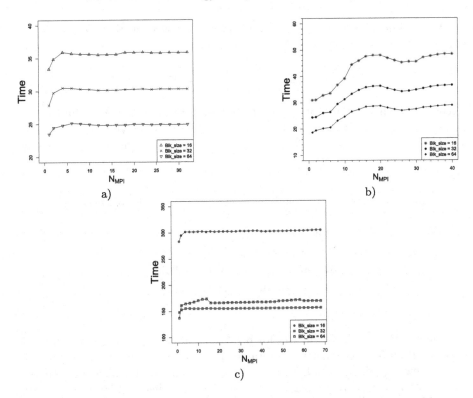

Fig. 4. Average time spent in the processing stage. (a) Broadwell (b) Skylake (c) KNL

6   Archive Processing

It is difficult to assess the comparative technical potential of each architecture using the measurement results provided in Sect. 5 because the execution time, the energy consumption and the number of processed pictures differ among the test beds.

As regards practical application of multispectral image processing routines, it may be useful to compare the results assessing test-bed performance and energy consumption in archive processing mode. Multispectral data processing algorithms are changed and upgraded systematically, so that archive data re-computations are regularly encountered in the remote sensing research area.

According to current data, one Visible Infrared Imaging Radiometer Suite (VIIRS) day/night band (DNB) image corresponds to 5 min of observation data, and therefore an observation data archive for one year contains approximately 52 560 images, as follows from Eq. (1), where $\text{Night}_{\text{AVG}} = 12\,\text{h} = 720\,\text{min}$:

$$\text{N}_{\text{pictures}} \approx (365 * \text{Night}_{\text{AVG}})/5. \tag{1}$$

An assumption about a one-year data archive could be made, as a very rough approximation, using the relation

$$\texttt{Total}_\texttt{Time}(\texttt{N}_\texttt{cores}) = \frac{\texttt{N}_\texttt{pictures}}{\texttt{N}_\texttt{cores}} \times \texttt{Time}_\texttt{AVG}(\texttt{N}_\texttt{cores}). \tag{2}$$

According to this assumption, the time required for processing the archive would amount to 17 h for the Broadwell test bed, 16.1 h for the Skylake test bed, and 42.7 h for the KNL test bed, for the standard 32×32 moving window size (see Table 4).

Table 4. Approximate time to process a one-year archive of DNB data

$\texttt{size}_\texttt{blk}$	Broadwell	Skylake	KNL
16×16	20.6 h	19.9 h	72.9 h
32×32	17.9 h	16.1 h	42.7 h
64×64	15.5 h	13.8 h	43.7 h

Energy consumption estimates for processing a one-year archive are summarized in Table 5. According to this table, the energy consumption is minimal for the new Skylake test bed. The results of this examination show that energy consumption for archive processing on the KNL test bed is 1.76 times greater than it is on the Skylake test bed.

Table 5. Approximate energy required to process a one-year archive of DNB data

$\texttt{size}_\texttt{blk}$	Broadwell	Skylake	KNL
16×16	19 340 kJ	16 651 kJ	39 214 kJ
32×32	16 425 kJ	13 142 kJ	23 156 kJ
64×64	14 543 kJ	11 478 kJ	25 564 kJ

7　Conclusions

According to our results, the architectures studied in the article are suitable for satellite image processing. Measurement studies of execution time and energy consumption indicate that the Skylake test bed shows significantly better results for execution time as well as for energy consumption in all input cases. By the way, KNL results are within an acceptable range according to archive processing requirements.

Price valuations, which include acquisition and depreciation costs, should also be taken into account. According to official data (https://ark.intel.com/), the recommended customer price for the processor used in the KNL test bed (Intel® Xeon Phi® 7250) is $2436.00 [8]. The recommended customer price for the processor used in the Broadwell test bed (Intel® Xeon® E5-2697A v4) is $2891.00 [9]. We studied a 2-socket test bed configuration, so the total acquisition value for the processors is about $5782. Finally, the recommended price for the processor used in the Skylake test bed (Intel® Xeon® Gold 6148) is $3072.00, so it is $6144 for two processors in a 2-socket configuration. Of course, this estimation does not include prices and depreciation costs for other equipment.

Moreover, the cost of developing should also be considered. Currently, the large project codes are implemented using MATLAB. So the AVX2 and AVX512 support for MATLAB codes could have a significant influence on the final decision.

8   Future Work

In the future, we plan to find mathematical models suitable to describe the workload and predict the module behavior in the cluster (multi-node) case. There are some successful examples of queue-theory applications for imaging service analysis [7], especially for the data downloading stage.

Moreover, we are looking forward to including a price estimation of our mathematical model, taking into account energy intensity per unit of output and acquisition costs.

As a first approximation, we have used a linear model to predict the execution time of a given module on a number of cores equal to N_{cores} [6]:

$$\ln(T_{run}) \approx c_0 + c_1 \times \ln(N_{cores}) + c_2 \times \ln(\texttt{size}_{blk}). \tag{3}$$

The relative error between measured time and predicted time is limited to 5% for the Broadwell test bed (see Fig. 5), so that this model could be suitable for further application.

The coefficient estimates were found using multilinear regression. The coefficients are listed below:

– **Broadwell:** $c_0 = 4.1885$, $c_1 = 0.0571$, $c_2 = -0.2201$.

Measured Time vs. Predicted Time

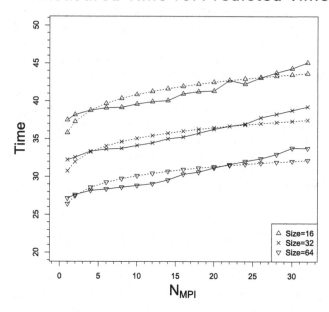

Fig. 5. Measured (*solid lines*) vs. Predicted (*dotted lines*) time on the Broadwell test bed through all available physical cores

Acknowledgments. This research was supported by a grant from the Russian Ministry of Education and Science (government contract 14.607.21.0165, unique identifier RFMEFI60716X0165).

References

1. Elvidge, C., Zhizhin, M., Baugh, K., Hsu, F.-C.: Automatic boat identification system for VIIRS low light imaging data. Remote Sens. J. **7**(3), 3020–3036 (2015). https://doi.org/10.3390/rs70303020
2. Tyutlyaeva, E., Konyukhov, S., Odintsov, I., Moskovsky, A.: The energy consumption analysis for the multispectral infrared satellite images processing algorithm. In: Voevodin, V., Sobolev, S. (eds.) RuSCDays 2017. CCIS, vol. 793, pp. 376–387. Springer, Cham (2017). https://doi.org/10.1007/978-3-319-71255-0_30
3. Lim, J.: Two-Dimensional Signal and Image Processing, p. 548. Prentice Hall, Englewood Cliffs (1990)
4. Vu, C.T., Phan, T., Chandler, D.: S3: a spectral and spatial measure of local perceived sharpness in natural images. IEEE Trans. Image Process. **21**, 934–945 (2012)
5. Rotem, E., Naveh, A., Ananthakrishnan, A., Weissmann, E., Rajwan, D.: Power-management architecture of the Intel Microarchitecture Code-Named Sandy Bridge. IEEE Micro **32**(2), 20–27 (2012). https://doi.org/10.1109/MM.2012.12

6. Barnes, B., Rountree, B., Lowenthal, D., Reeves, J., Supinski, B., Schulz, M.: A regression-based approach to scalability prediction. In: Proceedings of the 22nd Annual International Conference on Supercomputing (ICS 2008), pp. 368–377. ACM, New York (2008)
7. Chen, W., Palmer, P., Mackin, S., Crowley, G.: Queuing theory application in imaging service analysis for small Earth observation satellites. Acta Astronautica **62**(10–11), 623–631 (2008). ISSN 0094–5765. https://doi.org/10.1016/j.actaastro.2008.01.026
8. Intel® Xeon Phi™ Processor 7250 Product Specification. https://ark.intel.com/products/94035/Intel-Xeon-Phi-Processor-7250-16GB-1_40-GHz-68-core
9. Intel® Xeon® E5–2697A v4 Product Specification. https://ark.intel.com/ru/products/91768/Intel-Xeon-Processor-E5-2697A-v4-40M-Cache-2_60-GHz
10. Intel Corporation: A Guide to Vectorization with Intel® C++ Compilers (2012)

Parallel Numerical Algorithms

Fully Homomorphic Encryption for Parallel Implementation of Approximate Methods for Solving Differential Equations

Artem K. Vishnevsky[1][(✉)] and Sergey F. Krendelev[2,3]

[1] Joint Stock Company Scientific and Technical Production Enterprise
"Comprehensive Development of Technologies", Novosibirsk, Russia
vishn.artem@yandex.ru
[2] Novosibirsk State University, Novosibirsk, Russia
s.f.krendelev@gmail.com
[3] JatBrains Research, Novosibirsk, Russia

Abstract. A parallel fully homomorphic encryption for rational numbers is developed in this paper. Parallelism of processing is achieved by using methods of modular arithmetic. Encryption is constructed by mapping the field of rational numbers onto a vector space. Two operations, namely addition and multiplication, are defined. Addition and multiplication tables are constructed, which ensures that a vector space is closed under these mathematical operations. We show the implementation of protected recursive computations in rings of the form Z_M, $M = m_1 m_2 \ldots m_k$. We give a criterion of effective use of encryption for the numerical solution of the Cauchy problem. It is proved that the efficiency of encryption increases with increasing volumes and accuracy of computations.

Keywords: Fully homomorphic encryption · Parallel computations
Secure computations · Cloud computations
Chinese remainder theorem · Modular arithmetic
Differential equations · Numerical methods

1 Introduction

Cloud computing is a modern and efficient solution to the use of computing resources. High performance, parallel computing and scaling flexibility make cloud computing attractive for knowledge-intensive business areas. The problem of information security, however, remains unresolved since the model of calculations has radically changed. This means that the old methods of data protection either do not work quite right or do not work at all.

L. Sokolinsky and M. Zymbler (Eds.): PCT 2018, CCIS 910, pp. 119–134, 2018.
https://doi.org/10.1007/978-3-319-99673-8_9

How can one make cloud computing safe for users and save all its benefits? A promising direction for solving this problem is homomorphic encryption.

Homomorphic encryption is a method of secret computing in an untrusted environment. But the existing encryption methods do not allow us to take advantage of cloud computing since it is not yet clear how to use parallel computing, how to work with rational numbers in solving real applications [1–5] (solving differential equations, solving linear equations, etc.).

In this paper, a computational model based on modular arithmetic is considered for constructing parallel fully homomorphic encryption. For instance, computer systems such as K-340A and Almaz supercomputers [6–8], etc., are considered. These computers have high fault tolerance and support parallelism owing to the fact that for each modulus, calculations can be performed separately. Also, for each modulus, the table of multiplication and addition must be entered, which can significantly speed up the calculation.

The method of parallel fully homomorphic encryption for rational numbers [9] presented in the work allows us to perform secret calculations by integer methods, so that each calculation can be carried out for each modulus in parallel. The introduction of redundant moduli makes it possible to control computational errors.

1.1   Simple Description of the Idea of Constructing a Parallel Fully Homomorphic Encryption

As an example, consider the operations of multiplication and addition under a mapping of the ring of integers $\mathbb{Z}$ onto the vector space F^n.

Two integers p and q are represented by pairs of vectors:

$$p = \mathbf{ax} = \begin{bmatrix} a_1 & a_2 & a_3 \end{bmatrix} \cdot \begin{bmatrix} x_1 \\ x_2 \\ x_3 \end{bmatrix}, \quad q = \mathbf{bx} = \begin{bmatrix} b_1 & b_2 & b_3 \end{bmatrix} \cdot \begin{bmatrix} x_1 \\ x_2 \\ x_3 \end{bmatrix}.$$

The product of the numbers p and q in vector representation can be written as

$$\begin{aligned}
pq &= (\mathbf{ax})(\mathbf{bx}) \\
&= (a_1 x_1 + a_2 x_2 + a_3 x_3)(b_1 x_1 + b_2 x_2 + b_3 x_3) \\
&= a_1 x_1 b_1 x_1 + a_2 x_2 b_1 x_1 + a_3 x_3 b_1 x_1 \\
&\quad + a_1 x_1 b_2 x_2 + a_2 x_2 b_2 x_2 + a_3 x_3 b_2 x_2 \\
&\quad + a_1 x_1 b_3 x_3 + a_2 x_2 b_3 x_3 + a_3 x_3 b_3 x_3 \\
&= (\mathbf{a} \otimes \mathbf{b})(\mathbf{x} \otimes \mathbf{x}),
\end{aligned}$$

where $\otimes$ is the Kronecker product.

In general, if n is the dimension of the vectors, then the product pq takes the form

$$
\begin{aligned}
pq &= \left(\sum_{i=1}^{n} a_i x_i \right) \left(\sum_{i=1}^{n} b_i x_i \right) \\
&= \sum_{i,j=1}^{n} a_j x_j b_i x_i = \sum_{i,j=1}^{n} b_i a_j x_i x_j \qquad (1) \\
&= (\mathbf{a} \otimes \mathbf{b})(\mathbf{x} \otimes \mathbf{x}).
\end{aligned}
$$

The integers p and q are uniquely determined by the vectors $\mathbf{a}$ and $\mathbf{b}$. The vector $\mathbf{x}$ is common for both p and q, and also for their product pq. Obviously, the vector $\mathbf{x}$ must have some properties that make it possible to represent any integer using this vector. The coordinates of the vector $\mathbf{x}$ can be interpreted as the coefficients of the Diophantine equation

$$
x_1 c_1 + x_2 c_2 + \ldots + x_n c_n = u, \qquad (2)
$$

where x_i, c_i, $u \in \mathbb{Z}$. To solve Eq. (2) in integers for any $u \in \mathbb{Z}$, it is sufficient that the vector $\mathbf{x}$ has at least two mutually prime coordinates. Thus, the vector $\mathbf{x}$ must contain at least two mutually prime coordinates.

Using this example, we construct the simplest homomorphic encryption. Obviously, it is not persistent, but in this case, it is important to show the very principle of constructing an encryption, which will be considered later in the work. Let $\mathbf{x}$ be a secret key. Then the vectors $\mathbf{a}$ and $\mathbf{b}$ are public keys corresponding to the integers p and q. A homomorphism is defined for multiplication (1). Homomorphism for addition is defined as $p + q = (\mathbf{a} + \mathbf{b})\mathbf{x}$. Encryption will consist in finding the roots a_i, $b_i \in \mathbb{Z}$, $i = 1, 2, \ldots, n$ of the Diophantine equations $\sum_{i=1}^{n} a_i x_i = p$, and $\sum_{i=1}^{n} b_i x_i = q$. The disadvantage of this encryption is the exponential growth of the dimension of the vector in the case of multiplication, which makes it not applicable in practice. The construction of structural constants for tensor multiplication preserving the dimensionality of a vector completely reveals the value of the secret key $\mathbf{x}$.

Let us consider how the drawback of the encryption offered in the example can be eliminated and make it cryptographically stable.

2   Theoretical Bases for the Construction of Parallel Fully Homomorphic Encryption

Let R be a ring. It is an abelian group with respect to addition. Multiplication is not necessarily commutative, or associative, and R may even not be unitary. We will construct a homomorphic mapping. We operate according to algebraic geometry methods. For this, we choose an arbitrary set Ω and consider the set of all mappings of the set Ω into the ring R, which we denote as $\mathfrak{I}(\Omega, R)$. We assume that this set is closed under addition and multiplication. In other words, it forms a ring with respect to the pointwise multiplication of functions with values in the ring R. If $M \subset \Omega$ is a subset, it defines a mapping $i \colon M \to \Omega$ which is called embedding. The embedding mapping induces a mapping $i^* \colon \Im(\Omega, R) \to \Im(M, R)$; this mapping is a ring homomorphism. In particular, when the set M consists of a single element $m \in \Omega$, then $\Im(\Omega, R) \cong R$. This design is the basis for constructing a homomorphic encryption. Thus, in order to construct a completely homomorphic encryption for a particular ring R, it is necessary to choose a set Ω, a class of functions $\Im(\Omega, R)$, and a set M.

In concrete applications, the set of linear functions of n variables of the form

$$h(x_1, x_2, \ldots, x_n) = \alpha_1 x_1 + \alpha_2 x_2 + \ldots + \alpha_n x_n,$$

where $\alpha_i \in R$, $i = 1, 2, \ldots, n$, is considered as $\Im(M, R)$.

It is obvious that the set of such functions is closed under addition and multiplication by an element of R. However, it is not closed under multiplication. All possible multiplications are derived from the class $\Im(M, R)$. Thus, $\Im(M, R)$ is not a ring.

In order to make a ring from the set $\Im(M, R)$, we introduce a set of n^3 structure constants $\gamma_{ijk} \in R$, i, j, $k = 1, 2, \ldots, n$. This set is called the multiplication table. We define the product of two functions

$$h_1(x_1, x_2, \ldots, x_n) = \alpha_1 x_1 + \alpha_2 x_2 + \ldots + \alpha_n x_n,$$
$$h_2(x_1, x_2, \ldots, x_n) = \beta_1 x_1 + \beta_2 x_2 + \ldots + \beta_n x_n,$$

by the following rule:

$$h_1(x_1, x_2, \ldots, x_n) \otimes h_2(x_1, x_2, \ldots, x_n) = \sum_{k=1}^{n} x_k \sum_{i,j=1}^{n} \gamma_{ijk} \alpha_i \beta_j.$$

The introduction of structural constants allows us to equip the set R with the structure of a linear algebra. In this construction, this set of structural constants is absolutely arbitrary. Thus, the set $\Im(M, R)$ consisting of linear functions is, in general, not a unitary, commutative and associative ring. The set M consists of a finite number of elements and is the secret key by which the homomorphism is determined. First, we will consider a scheme of fully homomorphic encryption for integers, which will be the basis for constructing an encryption for rational numbers.

2.1 The Basic Scheme of Parallel Fully Homomorphic Encryption for Integers

Let R be an arbitrary ring, and $\Omega = R^n$ a module over R, whose elements will be called vectors. In this situation, $\Im(M, R)$ is the set of linear functions from R^n to R of the form

$$h_1(x_1, x_2, \ldots, x_n) = \alpha_1 x_1 + \alpha_2 x_2 + \ldots + \alpha_n x_n.$$

In this case, we can write $h_1(x_1, x_2, \ldots, x_n) = (\mathbf{a}, \mathbf{x})$, that is, as a scalar product where

$$\mathbf{a} = (\alpha_1, \alpha_2, \ldots, \alpha_n),$$
$$\mathbf{x} = (x_1, x_2, \ldots, x_n).$$

Thus, to define a linear function, it suffices to know the vector $\mathbf{a}$. Bearing in mind that the subsequent exposition deals only with rings of a special kind, we will replace the notation of an arbitrary ring R with F.

Suppose that a system of linear equations is given:

$$(\mathbf{a_1}, \mathbf{x}) = d_1,$$
$$(\mathbf{a_2}, \mathbf{x}) = d_2, \qquad\qquad (3)$$
$$\cdots$$
$$(\mathbf{a_k}, \mathbf{x}) = d_k,$$

where $\mathbf{a_1}, \mathbf{a_2}, \ldots, \mathbf{a_k} \in F^n$. We require that for fixed values $d_1, d_2, \ldots, d_k \in F$, Eq. (3) has a solution $\mathbf{x} \in F^n$. Suppose that it is fixed, although for us, the only important thing is that it exists. Suppose further that some $m \in F$ can be represented as

$$m = \lambda_1 d_1 + \lambda_2 d_2 + \ldots + \lambda_k d_k,$$

where $\lambda_1, \lambda_2, \ldots, \lambda_k \in F$. Then the vector $\mathbf{u} = \lambda_1 \mathbf{a_1} + \lambda_2 \mathbf{a_2} + \lambda_k \mathbf{a_k}$, according to standard linear algebra, satisfies the equation $(\mathbf{u}, \mathbf{x}) = m$. It follows that this representation is a homomorphic encryption with respect to addition. By construction, $\mathbf{x}$ is the private key, $\mathbf{u}$ is the public key corresponding to the element $m \in F$.

We shall consider multiplication. Suppose given two elements $m_1, m_2 \in F$ and two vectors $\mathbf{u}, \mathbf{v} \in F^n$ such that

$$(\mathbf{u}, \mathbf{x}) = m_1,$$
$$(\mathbf{v}, \mathbf{x}) = m_2.$$

How to find a vector $\mathbf{w} \in F^n$ such that $(\mathbf{w}, \mathbf{x}) = m_1 m_2$ and $\mathbf{w}$ depends only on the vectors $\mathbf{u}, \mathbf{v}$? From expression (1), we obtain

$$m_1 m_2 = (\mathbf{u}, \mathbf{x})(\mathbf{v}, \mathbf{x}) = \sum_{i,j=1}^{n} u_i v_j x_i x_j.$$

According to the system of Eq. (3), there exists a set of vectors $\mathbf{s_1}, \mathbf{s_2}, \ldots, \mathbf{s_n} \in F^n$ such that $(\mathbf{s_i}, \mathbf{x}) = x_i$, $i = 1, 2, \ldots, n$. If we take this set as the set of (3), then there exists a set of vectors $\lambda_{\mathbf{ij}}$, $i, j = 1, 2, \ldots, n$, such that $(\lambda_{\mathbf{ij}}, \mathbf{x}) = x_i x_j$, $i, j = 1, 2, \ldots, n$. Consequently,

$$\sum_{i,j=1}^{n} u_i v_j x_i x_j = \sum_{i,j=1}^{n} u_i v_j (\gamma_{ij}, \mathbf{x})$$

$$= \sum_{i,j=1}^{n} (u_i v_j \gamma_{ij}, \mathbf{x}) = \left(\sum_{i,j=1}^{n} u_i v_j \gamma_{ij}, \mathbf{x} \right). \tag{4}$$

Since the set of vectors γ_{ij}, $i,j = 1, 2, \ldots, n$, is fixed, a multiplication table is defined, from which it follows that $\mathbf{w} = \sum_{i,j=1}^{n} u_i v_j \gamma_{ij}$, and hence $(\mathbf{w}, \mathbf{x}) = m_1 m_2$.

This completely defines the homomorphic encryption. The set of vectors γ_{ij} is a public key. Note that when the multiplication table is available, there is no increase in the number of components of the vector under multiplication.

This variant is given in coordinate notation which is quite cumbersome, so we introduce an invariant representation.

If the vectors $\mathbf{u}, \mathbf{v}$ have, respectively, coordinates $[u_1 \, u_2 \, \ldots \, u_n]$ and $[v_1 \, v_2 \, \ldots \, v_n]$, then the vector of coordinates $u_i v_j$ is denoted by $\mathbf{u} \otimes \mathbf{v}$. In particular, the vector of coordinates $x_i x_j$ is denoted as $\mathbf{x} \otimes \mathbf{x}$. The equality $m_1 m_2 = (\mathbf{u}, \mathbf{x})(\mathbf{v}, \mathbf{x}) = \sum_{i,j=1}^{n} u_i v_j x_i x_j$ can be written as

$$m_1 m_2 = (\mathbf{u}, \mathbf{x})(\mathbf{v}, \mathbf{x}) = (\mathbf{u} \otimes \mathbf{v}, \mathbf{x} \otimes \mathbf{x}).$$

Equation (4) means that there exists a linear mapping $G: F^n \to F^n \otimes F^n$ such that $\mathbf{x} \otimes \mathbf{x} = \mathbf{G}\mathbf{x}$ ($\mathbf{G}$ is the matrix of the linear mapping G). In this case, $(\mathbf{u}, \mathbf{x})(\mathbf{v}, \mathbf{x}) = (\mathbf{u} \otimes \mathbf{v}, \mathbf{x} \otimes \mathbf{x}) = (\mathbf{u} \otimes \mathbf{v}, \mathbf{G}\mathbf{x})$. Since $(\mathbf{u} \otimes \mathbf{v}, \mathbf{G}\mathbf{x}) = (\mathbf{G}^\top(\mathbf{u} \otimes \mathbf{v}), \mathbf{x})$ ($\mathbf{G}^\top$ is the transposed matrix of $\mathbf{G}$), the multiplication is expressed in the form

$$\mathbf{w} = \mathbf{G}^\top(\mathbf{u} \otimes \mathbf{v}).$$

These notations will be used later on.

2.2   Encryption Strength

This encryption has several vulnerabilities. Let us consider them in detail.

Problem 1. Suppose that it is possible to obtain a set of correspondences, i.e. a number and the corresponding vector, which can be formally written as a set $(d_i, \mathbf{a_i})$, $d_i \in F$, $\mathbf{a_i} \in F^n$, $i = 1, 2, \ldots, r$, $r \geq n$. This means that there is a secret key $\mathbf{x} \in F^n$ such that

$$(\mathbf{a_i}, \mathbf{x}) = d_i. \tag{5}$$

If we look at (5) as a system of equations for $\mathbf{x}$ and the rank of the system turns out to be equal to n, then solving this system, we obtain a unique vector $\mathbf{x}$, therefore, there is a secret key.

In order to prevent this possibility, it is necessary to choose vectors $\mathbf{a_i}$ that can never give a complete set of linearly independent vectors. For example, this can be done by choosing $k < n$ in Eq. (3).

Problem 2. Suppose that there is a set $(d_i, \mathbf{a_i})$, $d_i \in F$, $\mathbf{a_i} \in F^n$. According to the solution of Problem 1, this set is not complete, in the sense that we will never obtain a system of equations with a rank equal to n. Assume that there is a vector $\mathbf{b}$ that does not coincide with any of the $\mathbf{a_i}$ vectors. On the other hand, the situation is possible when there are elements $\lambda_i \in F$, $i = 1, 2, \ldots, r$, such that $\lambda_1 \mathbf{a_1} + \lambda_2 \mathbf{a_2} + \ldots + \lambda_r \mathbf{a_r} = \mathbf{b}$. This means that vector $\mathbf{b}$ corresponds to the number $d = \lambda_1 d_1 + \lambda_2 d_2 + \ldots + \lambda_r d_r$.

Since it is assumed that for each specific calculation their parameters are selected, an attack with known data is not effective. Now we need to extend the range of encryption to the set of rational numbers.

2.3   Parallel Fully Homomorphic Encryption for Rational Numbers

Suppose that we are given a set of vectors $\mathbf{a_1}, \mathbf{a_2}, \ldots, \mathbf{a_k} \in F^n$, and a pair of vectors $\mathbf{x}, \mathbf{y} \in F^n$ satisfying the following conditions:

$$(\mathbf{a_1}, \mathbf{x}) = d_1, \quad (\mathbf{a_1}, \mathbf{y}) = e_1,$$
$$(\mathbf{a_2}, \mathbf{x}) = d_2, \quad (\mathbf{a_2}, \mathbf{y}) = e_2,$$
$$\cdots\cdots\cdots\cdots\cdots\cdots$$
$$(\mathbf{a_k}, \mathbf{x}) = d_k, \quad (\mathbf{a_k}, \mathbf{y}) = e_k,$$

where $d_i, e_i \in F$, $i = 1, 2, \ldots, n$.

The first question that needs to be considered is how to find a vector $\mathbf{v} \in F^n$ such that for a given pair $m_1, m_2 \in F$ the following holds:

$$(\mathbf{v}, \mathbf{x}) = m_1,$$
$$(\mathbf{v}, \mathbf{y}) = m_2.$$

Assume that $v = \lambda_1 \mathbf{a_1} + \lambda_2 \mathbf{a_2} + \ldots + \lambda_k \mathbf{a_k}$. Then

$$(\mathbf{v}, \mathbf{x}) = \lambda_1 d_1 + \lambda_2 d_2 + \ldots + \lambda_k d_k = m_1,$$
$$(\mathbf{u}, \mathbf{x}) = \lambda_1 e_1 + \lambda_2 e_2 + \ldots + \lambda_k e_k = m_2.$$

Thus, in order to obtain the desired result, it is necessary that the resulting system of equations be solvable. We assume that this equation is solvable for any right-hand side.

The subsequent implementation depends on the presentation of the data.

We will assume that $F = \mathbb{Z}$ is the set of integers. Suppose that an integer $z \in \mathbb{Z}$ is encrypted. We associate with each z a pair of numbers $q \in \mathbb{Z}$, $p \in \mathbb{N}$, such that $pz = q$. We denote this pair by (q, p).

The set of numbers satisfying the equation $pz = q$ is called the set of rational numbers. Consequently, the pair (q, p) is a representation for a rational number.

It is obvious that such a representation for a rational number is not single-valued. We choose two secret vectors $\mathbf{x}, \mathbf{y} \in \mathbb{Z}^n$ and construct $\mathbf{a} \in \mathbb{Z}^n$ such that

$$(\mathbf{a}, \mathbf{x}) = q,$$
$$(\mathbf{a}, \mathbf{y}) = p.$$

We shall consider how the addition and multiplication of numbers is defined in this representation. Consider two numbers z_1, z_2 and two pairs (q_1, p_1), (q_2, p_2) corresponding to them. According to standard arithmetic, the sum $z_1 + z_2$ corresponds to the pair $(q_1 p_2 + q_2 p_1, p_1 p_2)$, while the product $(q_1 q_2, p_1 p_2)$ corresponds to the product $z_1 z_2$.

Suppose that the secret vectors $\mathbf{x}, \mathbf{y} \in \mathbb{Z}^n$ are fixed, and two vectors $\mathbf{u}, \mathbf{v} \in \mathbb{Z}^n$ are constructed in such a way that

$$(\mathbf{u}, \mathbf{x}) = q_1, \quad (\mathbf{u}, \mathbf{y}) = p_1,$$
$$(\mathbf{v}, \mathbf{x}) = q_2, \quad (\mathbf{v}, \mathbf{y}) = p_2.$$

Multiplication. The pair $(q_1 q_2, p_1 p_2)$ corresponds to $[(\mathbf{u}, \mathbf{x})(\mathbf{v}, \mathbf{x}), (\mathbf{u}, \mathbf{y})(\mathbf{v}, \mathbf{y})]$. According to the definition of the tensor product, this pair can be rewritten in the form $[(\mathbf{u} \otimes \mathbf{v})(\mathbf{x} \otimes \mathbf{x}), (\mathbf{u} \otimes \mathbf{v})(\mathbf{y} \otimes \mathbf{y})]$. Now we can construct a matrix $\mathbf{G}$ such that $\mathbf{x} \otimes \mathbf{x} = \mathbf{G}\mathbf{x}$, $\mathbf{y} \otimes \mathbf{y} = \mathbf{G}\mathbf{y}$. Then

$$[(\mathbf{u} \otimes \mathbf{v}, \mathbf{x} \otimes \mathbf{x}), (\mathbf{u} \otimes \mathbf{v}, \mathbf{y} \otimes \mathbf{y})]$$
$$= [(\mathbf{u} \otimes \mathbf{v}, \mathbf{G}\mathbf{x}), (\mathbf{u} \otimes \mathbf{v}, \mathbf{G}\mathbf{y})]$$
$$= [(\mathbf{G}^\top (\mathbf{u} \otimes \mathbf{v}), \mathbf{x}), (\mathbf{G}^\top (\mathbf{u} \otimes \mathbf{v}), \mathbf{y})].$$

Multiplication at the vector level takes the form

$$\mathbf{w} = \mathbf{u}\mathbf{v} = \mathbf{G}^\top (\mathbf{u} \otimes \mathbf{v}).$$

The matrix $\mathbf{G}^\top$ is called the multiplication table. It is an element of the public key. By construction, $(\mathbf{w}, \mathbf{x}) = q_1 q_2$, $(\mathbf{w}, \mathbf{y}) = p_1 p_2$.

Addition. The pair $(q_1 p_2 + q_2 p_1, p_1 p_2)$ corresponds to $[(\mathbf{u}, \mathbf{x})(\mathbf{v}, \mathbf{y}) + (\mathbf{u}, \mathbf{y})(\mathbf{v}, \mathbf{x}), (\mathbf{u}, \mathbf{y})(\mathbf{v}, \mathbf{y})]$. According to the definition of the tensor product, this expression is equal to

$$[(\mathbf{u} \otimes \mathbf{v}, \mathbf{x} \otimes \mathbf{y}) + (\mathbf{u} \otimes \mathbf{v}, \mathbf{y} \otimes \mathbf{x}), (\mathbf{u} \otimes \mathbf{v}, \mathbf{y} \otimes \mathbf{y})]$$
$$= [(\mathbf{u} \otimes \mathbf{v}, \mathbf{x} \otimes \mathbf{y} + \mathbf{y} \otimes \mathbf{x}), (\mathbf{u} \otimes \mathbf{v}, \mathbf{y} \otimes \mathbf{y})].$$

Now we construct a matrix $\mathbf{H}$ such that

$$\mathbf{x} \otimes \mathbf{y} + \mathbf{y} \otimes \mathbf{x} = \mathbf{H}\mathbf{x},$$
$$\mathbf{y} \otimes \mathbf{y} = \mathbf{H}\mathbf{y}.$$

Then

$$[(\mathbf{u} \otimes \mathbf{v}, \mathbf{x} \otimes \mathbf{y} + \mathbf{y} \otimes \mathbf{x}), (\mathbf{u} \otimes \mathbf{v}, \mathbf{y} \otimes \mathbf{y})] =$$
$$[(\mathbf{H}^\top (\mathbf{u} \otimes \mathbf{v}), \mathbf{x}), (\mathbf{H}^\top (\mathbf{u} \otimes \mathbf{v}), \mathbf{y})].$$

The addition at the vector level takes the form

$$\mathbf{w} = \mathbf{u} + \mathbf{v} = \mathbf{H}^\top (\mathbf{u} \otimes \mathbf{v}).$$

By construction, $(\mathbf{w}, \mathbf{x}) = q_1 p_2 + q_2 p_1$, $(\mathbf{w}, \mathbf{y}) = p_1 p_2$. Consequently, $\mathbf{w}$ corresponds to the sum of two numbers, $z_1 + z_2$. This construction is quite similar in the case of rings of the form Z_M, $M = m_1 m_2 \dots m_k$. And calculations are carried out for each modulus separately.

Obviously, the implementation is based on the coordinate expression of all the data involved in the computation. Before considering any examples, we must decide which representation of the tensor product of the vectors $\mathbf{x}$ and $\mathbf{y}$ will be used in what follows. In these examples, we will use the left representation.

2.4   Numerical Example

We will encrypt two rational numbers $\frac{q_1}{p_1} = \frac{3}{5}$, $\frac{q_2}{p_2} = \frac{7}{10}$. The maximum calculation range for the sum of $\frac{q_1}{p_1}$ and $\frac{q_2}{p_2}$ will be 65, and for the product, 50. Then, the values of the mutually prime moduli $m_1 = 3, m_2 = 5, m_3 = 7$ satisfy the condition $3 \cdot 5 \cdot 7 > 65$.

Let us construct a cryptosystem. Generate a public key: $\mathbf{A} = \begin{bmatrix} 8 & 4 & 4 & 9 \\ 10 & 3 & 8 & 14 \\ 8 & 3 & 5 & 6 \end{bmatrix}$.

Generate private keys: $\mathbf{X} = \begin{bmatrix} 28 \\ 5 \\ 12 \\ 2 \end{bmatrix}$, $\mathbf{Y} = \begin{bmatrix} 2 \\ 7 \\ 5 \\ 3 \end{bmatrix}$, $\mathbf{D} = \begin{bmatrix} 310 \\ 419 \\ 311 \end{bmatrix}$, $\mathbf{E} = \begin{bmatrix} 91 \\ 123 \\ 80 \end{bmatrix}$. It is

not difficult to verify that $\mathbf{D} = \mathbf{AX}$, $\mathbf{E} = \mathbf{AY}$. To construct the multiplication and addition tables, we calculate the tensor products:

$$\mathbf{X} \otimes \mathbf{X} = \begin{bmatrix} 784 & 140 & 336 & 56 & 140 & 25 & 60 & 10 & 336 & 60 & 144 & 24 & 56 & 10 & 24 & 4 \end{bmatrix}^\top,$$

$$\mathbf{Y} \otimes \mathbf{Y} = \begin{bmatrix} 4 & 14 & 10 & 6 & 14 & 49 & 35 & 21 & 10 & 35 & 25 & 15 & 6 & 21 & 15 & 9 \end{bmatrix}^\top,$$

$$\mathbf{X} \otimes \mathbf{Y} + \mathbf{Y} \otimes \mathbf{X} = \begin{bmatrix} 112 & 206 & 164 & 88 & 206 & 70 & 109 & 29 & 164 & 109 & 120 & 46 & 88 & 29 & 46 & 12 \end{bmatrix}^\top.$$

Let us construct the multiplication and addition tables for moduli 3, 5, and 7. To construct the first column of the multiplication table, we solve a system of equations:

$$310\lambda_1^{(1,1)} + 419\lambda_2^{(1,1)} + 311\lambda_3^{(1,1)} = 784,$$

$$91\lambda_1^{(1,1)} + 123\lambda_2^{(1,1)} + 80\lambda_3^{(1,1)} = 4.$$

We find the solution $\lambda_1^{(1,1)} = -23569$, $\lambda_2^{(1,1)} = 17421$, $\lambda_3^{(1,1)} = 25$, which is part of the secret key. Now mask the obtained values with the public key:

$$\begin{bmatrix} g_1^{(1,1)} \\ g_2^{(1,1)} \\ g_3^{(1,1)} \\ g_4^{(1,1)} \end{bmatrix} = -23569 \begin{bmatrix} 8 \\ 4 \\ 4 \\ 9 \end{bmatrix} + 17421 \begin{bmatrix} 10 \\ 3 \\ 8 \\ 14 \end{bmatrix} + 25 \begin{bmatrix} 8 \\ 3 \\ 5 \\ 6 \end{bmatrix} = \begin{bmatrix} -14142 \\ -41938 \\ 45217 \\ 31923 \end{bmatrix}.$$

Next, we calculate the result modulo 3, 5, and 7:
$\begin{bmatrix} -14142 \\ -41938 \\ 45217 \\ 31923 \end{bmatrix} = \begin{bmatrix} 0 \\ 2 \\ 1 \\ 0 \end{bmatrix}$ (mod 3),

$\begin{bmatrix} -14142 \\ -41938 \\ 45217 \\ 31923 \end{bmatrix} = \begin{bmatrix} 3 \\ 2 \\ 2 \\ 3 \end{bmatrix}$ (mod 5), $\begin{bmatrix} -14142 \\ -41938 \\ 45217 \\ 31923 \end{bmatrix} = \begin{bmatrix} 5 \\ 6 \\ 4 \\ 3 \end{bmatrix}$ (mod 7), The remaining vec-

tor columns of the multiplication table will be calculated similarly. As a result, the multiplication tables for moduli 3, 5, and 7 are written as:

$$\mathbf{G_{m_1}} = \begin{bmatrix} 0\,2\,1\,2\,1\,1\,1\,1\,2\,2\,0\,2\,1\,2\,0\,0 \\ 2\,0\,2\,1\,2\,0\,0\,2\,0\,1\,1\,2\,0\,0\,0\,1 \\ 1\,2\,0\,2\,2\,1\,0\,1\,0\,0\,0\,0\,2\,1\,0\,1 \\ 0\,0\,2\,2\,0\,0\,1\,1\,2\,1\,2\,0\,2\,1\,0\,1 \end{bmatrix},$$

$$\mathbf{G_{m_2}} = \begin{bmatrix} 3\,2\,4\,3\,1\,0\,4\,3\,0\,1\,2\,1\,1\,4\,4\,1 \\ 2\,1\,4\,4\,0\,4\,4\,4\,0\,1\,2\,1\,2\,0\,4\,0 \\ 2\,1\,4\,2\,2\,3\,1\,4\,3\,4\,0\,1\,4\,3\,3\,4 \\ 3\,1\,3\,4\,4\,2\,3\,4\,0\,2\,4\,2\,0\,1\,3\,4 \end{bmatrix},$$

$$\mathbf{G_{m_3}} = \begin{bmatrix} 5\,0\,6\,6\,4\,1\,3\,6\,4\,2\,3\,3\,1\,3\,3\,0 \\ 6\,0\,6\,1\,5\,5\,4\,2\,0\,1\,2\,3\,0\,0\,3\,1 \\ 4\,0\,1\,5\,5\,5\,4\,2\,2\,1\,0\,6\,4\,0\,6\,0 \\ 3\,0\,0\,6\,3\,5\,3\,2\,2\,4\,4\,0\,4\,5\,0\,3 \end{bmatrix}.$$

The addition tables for moduli 3, 5, and 7 are the following:

$$\mathbf{H_{m_1}} = \begin{bmatrix} 1\,2\,0\,0\,1\,0\,2\,0\,1\,0\,1\,0\,1\,2\,1\,2 \\ 1\,2\,2\,1\,2\,1\,1\,2\,2\,1\,0\,1\,1\,2\,1\,0 \\ 1\,2\,2\,1\,2\,1\,1\,2\,2\,1\,0\,1\,1\,2\,1\,0 \\ 0\,0\,1\,1\,0\,0\,2\,2\,1\,2\,2\,1\,1\,2\,1\,0 \end{bmatrix},$$

$$\mathbf{H_{m_2}} = \begin{bmatrix} 4\,1\,4\,0\,4\,4\,1\,4\,2\,3\,4\,2\,4\,3\,2\,2 \\ 3\,0\,4\,1\,3\,3\,1\,0\,2\,3\,4\,2\,0\,4\,2\,1 \\ 0\,0\,3\,1\,2\,4\,1\,0\,0\,4\,1\,1\,2\,1\,1\,2 \\ 0\,4\,3\,3\,0\,0\,2\,1\,4\,1\,3\,4\,1\,4\,4\,1 \end{bmatrix},$$

$$\mathbf{H_{m_3}} = \begin{bmatrix} 4\,5\,4\,6\,1\,5\,6\,3\,0\,4\,0\,2\,5\,0\,2\,2 \\ 3\,6\,5\,3\,1\,1\,6\,6\,0\,0\,2\,4\,0\,4\,4\,4 \\ 1\,6\,0\,0\,1\,1\,6\,6\,2\,0\,0\,0\,4\,4\,0\,3 \\ 4\,3\,3\,5\,0\,2\,0\,2\,0\,2\,6\,6\,6\,5\,6\,6 \end{bmatrix}.$$

Now we encrypt $\frac{q_1}{p_1} = \frac{3}{5}$, $\frac{q_2}{p_2} = \frac{7}{10}$.

The encrypted values for $\frac{q_1}{p_1} = \frac{3}{5}$ modulo 3, 5, and 7 are written as: $c_{m_1}^{(1)} = [2\ 1\ 0\ 1]$, $c_{m_2}^{(1)} = [2\ 2\ 2\ 4]$, $c_{m_3}^{(1)} = [2\ 5\ 6\ 2]$. Let us check the correctness of the encryption. For this, we perform a reverse recovery using the Chinese remainder theorem: $c^{(1)} = \mathrm{CRT}_{i=1}^{3} c_{m_i}^{(1)} \pmod{105} = [2\ 82\ 27\ 79]$. Multiply the obtained value by the keys: $c^{(1)} X \pmod{105} = 3$, $c^{(1)} Y \pmod{105} = 5$.

The encrypted values for $\frac{q_2}{p_2} = \frac{7}{10}$ modulo 3, 5, and 7 are: $c_{m_1}^{(2)} = [2\ 1\ 1\ 0]$, $c_{m_2}^{(2)} = [1\ 1\ 0\ 2]$, $c_{m_3}^{(2)} = [3\ 4\ 6\ 3]$.

We shall consider how secure the computations will be. First, we consider multiplication. Secure computations will be performed in parallel for three moduli. In detail, consider multiplication modulo 3. First, multiply the encrypted values:

$$\mathbf{P}_{m_1}^* = c_{m_1}^{(1)} \otimes c_{m_1}^{(2)} = [2\ 1\ 0\ 1] \otimes [2\ 1\ 1\ 0] \pmod{3}$$
$$= [1\ 2\ 2\ 0\ 2\ 1\ 1\ 0\ 0\ 0\ 0\ 2\ 1\ 1\ 0],$$

and then multiply the result by the multiplication table: $\mathbf{G}_{m_1} \mathbf{P}_{m_1}^* = [2\ 1\ 0\ 1]$ $\pmod{3}$.

The products modulo 5, and 7 are similarly calculated: $\mathbf{P}_{m_2}^* = [1\ 1\ 2\ 2]$, $\mathbf{P}_{m_3}^* = [3\ 5\ 1\ 6]$.

Next, in accordance with the Chinese remainder theorem, we obtain the result of the protected multiplication modulo $M = 105$: $\mathbf{P}^* = \mathrm{CRT}_{i=1}^{3} \mathbf{P}_{m_i}^*$ $\pmod{105} = [101\ 61\ 57\ 97]$. Decrypt the encrypted result of the multiplication:

$$\mathbf{P}^* X \pmod{105} = \begin{bmatrix} 101 & 61 & 57 & 97 \end{bmatrix} \begin{bmatrix} 28 \\ 5 \\ 12 \\ 2 \end{bmatrix} \pmod{105} = 21 = 3 \cdot 7 = q_1 q_2 \text{ and}$$

$$\mathbf{P}^* Y \pmod{105} = \begin{bmatrix} 101 & 61 & 57 & 97 \end{bmatrix} \begin{bmatrix} 2 \\ 7 \\ 5 \\ 3 \end{bmatrix} \pmod{105} = 50 = 5 \cdot 10 = p_1 p_2.$$

Similarly to multiplication, we calculate the results of the protected addition for each of three moduli, using now the addition tables instead of the tables of multiplication: $\mathbf{S}_{m_1}^* = [0\ 1\ 2\ 0]$, $\mathbf{S}_{m_2}^* = [4\ 4\ 1\ 3]$, $\mathbf{S}_{m_3}^* = [5\ 5\ 1\ 0]$. Now, according to the Chinese remainder theorem, we obtain the result of the protected addition: $\mathbf{S}^* = \mathrm{CRT}_{i=1}^{3} \mathbf{S}_{m_i}^* \pmod{105} = [54\ 19\ 71\ 63]$. Decrypt the encrypted result of the addition: $\mathbf{S}^* X \pmod{105} = \begin{bmatrix} 54 & 19 & 71 & 63 \end{bmatrix} \begin{bmatrix} 28 \\ 5 \\ 12 \\ 2 \end{bmatrix} \pmod{105} =$

$65 = 3 \cdot 10 + 7 \cdot 5 = q_1 p_2 + q_2 p_1$ and $\mathbf{S}^* Y \pmod{105} = \begin{bmatrix} 54 & 19 & 71 & 63 \end{bmatrix} \begin{bmatrix} 2 \\ 7 \\ 5 \\ 3 \end{bmatrix}$

$\pmod{105} = 50 = 5 \cdot 10 = p_1 p_2.$

3   Parallel Implementation of Numerical Methods for Secure Computations of Differential Equations

As a particular case, we consider in this paper the numerical solution of the simplest Cauchy problem: $\frac{dy}{dx} = y$, $y(0) = 1$, $x \in [0, 1]$. We will solve the third-order Runge–Kutta method. Let us write the formula for calculating x_k:

$$x_k = 1 + kh + \frac{k(k-1)}{2!}h^2 + \frac{k(k-1)(k-2)}{3!}h^3, \tag{6}$$

where $k = 1, 2, \ldots, 100$, $h = \frac{1}{100}$. The maximum range of calculations does not exceed $2.8 \times 12 \times 10^{12}$. Calculations are performed for four moduli: $m_1 = 3931$, $m_2 = 3943$, $m_3 = 3947$, $m_4 = 3967$. So, the condition for applying the Chinese remainder theorem is met: $m_1 m_2 m_3 m_4 > 2.8 \times 12 \times 10^{12}$. We use the crypto scheme from the numerical example.

For the secure calculation of $x_1, x_2, \ldots, x_{100}$, it is necessary to encrypt the value 1 and the sampling step $h = \frac{1}{100}$; the remaining values can be obtained by performing protected calculations.

The encrypted values of 1 (we will encrypt $\frac{70}{70}$) are the following:

$$\mathbf{C_{m_1}}(1) = [1646\ 138\ 2190\ 2816],$$
$$\mathbf{C_{m_2}}(1) = [2341\ 1095\ 2615\ 2064],$$
$$\mathbf{C_{m_3}}(1) = [3851\ 1302\ 248\ 583],$$
$$\mathbf{C_{m_4}}(1) = [3227\ 1497\ 1154\ 1712].$$

The encrypted values of $h = \frac{1}{100}$ (we will encrypt $\frac{4}{400}$) are the following:

$$\mathbf{C_{m_1}}(1/100) = [172\ 3078\ 834\ 619],$$
$$\mathbf{C_{m_2}}(1/100) = [2735\ 1081\ 157\ 3412],$$
$$\mathbf{C_{m_3}}(1/100) = [2215\ 2860\ 2760\ 540],$$
$$\mathbf{C_{m_4}}(1/100) = [310\ 2501\ 1467\ 3008].$$

Below, we give some results of the calculation of (6). For $x_1 = 1.0100$:

$$\mathbf{C_{m_1}}(\mathbf{1.0100}) = [1329\ 3424\ 43\ 3421],$$
$$\mathbf{C_{m_2}}(\mathbf{1.0100}) = [3449\ 1413\ 3759\ 1757],$$
$$\mathbf{C_{m_3}}(\mathbf{1.0100}) = [1375\ 1605\ 2857\ 228],$$
$$\mathbf{C_{m_4}}(\mathbf{1.0100}) = [1706\ 3172\ 1947\ 3037].$$

For $x_2 = 1.0201$:

$$\mathbf{C_{m_1}}(\mathbf{1.0201}) = [1666\ 2170\ 78\ 2133],$$
$$\mathbf{C_{m_2}}(\mathbf{1.0201}) = [3555\ 3877\ 2848\ 84],$$
$$\mathbf{C_{m_3}}(\mathbf{1.0201}) = [1420\ 99\ 2893\ 3652],$$
$$\mathbf{C_{m_4}}(\mathbf{1.0201}) = [1566\ 1891\ 2391\ 90].$$

For $x_{100} = 2.6567$:

$$\mathbf{C_{m_1}(2.6567)} = [963\ 2486\ 3121\ 2034],$$
$$\mathbf{C_{m_2}(2.6567)} = [650\ 626\ 768\ 1244],$$
$$\mathbf{C_{m_3}(2.6567)} = [608\ 241\ 1347\ 1176],$$
$$\mathbf{C_{m_4}(2.6567)} = [2939\ 2332\ 2386\ 2121].$$

3.1 Estimation of the Effectiveness of Parallel Fully Homomorphic Encryption for Rational Numbers

In general, formula (6) takes the form:

$$x_k = 1 + kh + \frac{k(k-1)}{2!}h^2 + \ldots + \frac{k(k-1)\ldots(k-L+1)}{L!}h^L, \qquad (7)$$

Let m be the number of encrypted data, n the key dimension, h the sampling step, h^{-1} the number of recurrent calculations, and L the number of terms in (7).

The upper estimate of the homomorphic-encryption time complexity consists of the following estimates:

- the construction of a cryptosystem requires the calculation of addition and multiplication tables, for which it is necessary to compute $2n^2$ systems of equations of complexity $O(n^3)$;
- the encryption of m input data, requires the calculation of m systems of equations of complexity $O(n^3)$;
- deciphering the results for h^{-1} requires $2h^{-1}$ scalar products of vectors of dimension n.

Thus, the upper estimate of the encryption-algorithm time complexity is

$$f_1 = 2n^2n^3 + mn^3 + 2h^{-1}n. \qquad (8)$$

The upper bound of the time complexity of computing h^{-1} recursions of (7) for L in one computational flow is

$$f_2 = 2h^{-1}L. \qquad (9)$$

Thus, the effectiveness of the application of homomorphic encryption is determined by the criterion of the ratio of (8) to (9):

$$\frac{f_1}{f_2} < 1. \qquad (10)$$

Figures 1, 2 and 3 show the graphs of the dependence of the ratio (10) on the parameters n, h^{-1}, L.

From Figs. 1 and 2, we see that the encryption efficiency given by criterion (10) increases as the number of recursion computations h^{-1} increases, and it

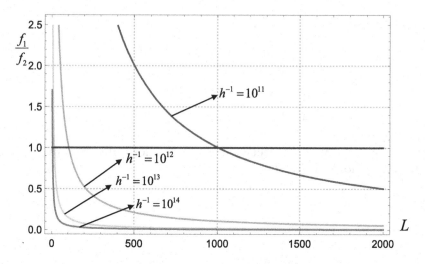

Fig. 1. Evaluating the effectiveness of encryption for $n = 10$ (*key dimension*).

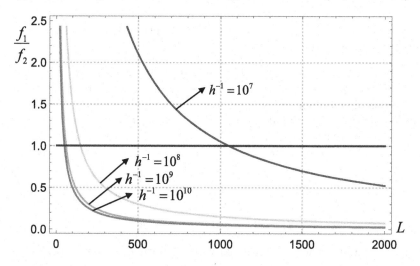

Fig. 2. Evaluating the effectiveness of encryption for $n = 100$ (*key dimension*).

decreases when the key dimension n increases. But for a certain ratio of h^{-1} and n, it satisfies the efficiency criterion (10). Figure 3 shows the dependence of criterion (10) on the number of terms in (7), there is a significant increase in efficiency when L increases. Thus, the effectiveness of encryption increases with the increase of volumes and accuracy of recurrent calculations of $\frac{dy}{dx} = y$.

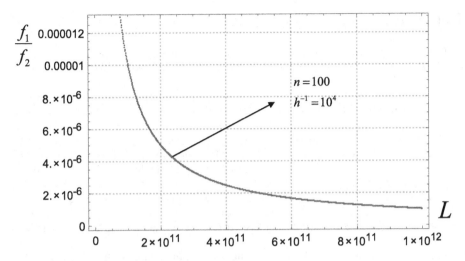

Fig. 3. Evaluating the effectiveness of encryption for $n = 100$ (*key dimension*) and fixed $h^{-1} = 10^4$ (*number of recurrent calculations*).

4   Conclusion

The paper deals with fully homomorphic encryption for rational numbers based on the methods of modular arithmetic. The operations of addition and multiplication are defined. As an example, we considered a protected calculation of a differential equation by the Runge–Kutta method of the third order. We prove the effectiveness of the use of homomorphic encryption for numerical methods for solving differential equations with high accuracy. At the same time, an increase in the accuracy of calculations associated with a decrease in the discretization step and an increase in the order of the Runge–Kutta method to values beyond the computational capacities of one calculator is successfully achieved by applying modular arithmetic methods. Processing the images of the computational problem in $Z_{m_1}, Z_{m_2}, \ldots, Z_{m_k}$ by independent computational flows makes it possible to increase the accuracy of calculations to the limit of the resource of a distributed computing environment. For example, to implement computations in the range of values from 10^{-600} to 10^{600} with moduli of dimension $\approx 10^6$ ($\approx 2^{64}$), which ensures the functioning of a 64-bit calculator without overflow, 100 grid-parallel computing flows are needed.

The strength of encryption is based on a one-time application of the key for one calculation, which excludes the development of statistics on the key by a cryptanalyst. Resistance to brute-force attacks is based on an infinite number of keys; the condition is that the size of the vector of the key must be greater than the number of secret parameters.

References

1. Gentry, C., Halevi, S.: Implementing gentry's fully-homomorphic encryption scheme. In: Paterson, K.G. (ed.) EUROCRYPT 2011. LNCS, vol. 6632, pp. 129–148. Springer, Heidelberg (2011). https://doi.org/10.1007/978-3-642-20465-4_9
2. Gentry, C., Halevi, S., Smart, N.P.: Homomorphic evaluation of the AES circuit. In: Safavi-Naini, R., Canetti, R. (eds.) CRYPTO 2012. LNCS, vol. 7417, pp. 850–867. Springer, Heidelberg (2012). https://doi.org/10.1007/978-3-642-32009-5_49
3. Brakerski, Z., Vaikuntanathan, V.: Fully homomorphic encryption from ring-LWE and security for key dependent messages. In: Rogaway, P. (ed.) CRYPTO 2011. LNCS, vol. 6841, pp. 505–524. Springer, Heidelberg (2011). https://doi.org/10.1007/978-3-642-22792-9_29
4. Brakerski, Z.: Fully homomorphic encryption without modulus switching from classical GapSVP. In: Safavi-Naini, R., Canetti, R. (eds.) CRYPTO 2012. LNCS, vol. 7417, pp. 868–886. Springer, Heidelberg (2012). https://doi.org/10.1007/978-3-642-32009-5_50
5. Lauter, K., Naehrig, M., Vaikuntanathan, V.: Can homomorphic encryption be practical? (2011). https://doi.org/10.1145/2046660.2046682, http://eprint.iacr.org/2011/405
6. Krendelev, S.F.: The soviet supercomputer K-340 and secret calculating. Ruscrypto 2015. http://www.ruscrypto.ru/resource/summary/rc2015/02_krendelev.pdf
7. Malashevich, B.M.: Unknown modular supercomputers. http://www.computer-museum.ru/books/archiv/sokcon11.pdf
8. Akushsky, I.J., Yuditsky, D.I.: Arithmetic in residual classes. Soviet radio (1968)
9. Vishnevskiy, A.K., Krendelev, S.F.: Homomorphic encryption in the ring rational numbers. Ruscrypto 2017. http://www.ruscrypto.ru/resource/summary/rc2017/02_vishnevskiy_krendelev.pdf

Static Balancing Methods in Projection-Based Mesh Generation Algorithm

Sergej K. Grigorjev and Mikhail V. Yakobovskiy[✉]

Keldysh Institute of Applied Mathematics, Russian Academy of Sciences,
Moscow, Russia
sergejgri@gmail.com, lira@imamod.ru

Abstract. The discussion is about parallel guarantied mesh-generation projection-based algorithm. Main subject of this article is load-balancing problem in distributed projection-based tetrahedral mesh generation algorithm. Algorithm is based on construction of triangle prisms, formed by orthogonal projection of base surface mesh. The advantage of using projection-based algorithm consists in guaranteed tetrahedrisation of 3-dimensional domain. Main purpose of generated meshes consists in guaranteed detection of the topology of three-dimensional domains, which can be used for mesh adaptation algorithms.

Keywords: MPI · Mesh generation · Unstructured grid
Static load balancing · Triangulation · Rational numbers

1 Introduction

Geometrical calculating meshes are used in numerical modeling of solid environment. Generating such meshes is one of the hardest computational problems due to the large amount of time that is required to generate such type of mesh. One of ways of accelerating mesh generation is using parallel mesh generation algorithms.

Creating a parallel program creates an addition amount of tasks, which must be solved for parallel algorithm to become good. There are various problems, however, in this paper we consider problem of load balancing.

The main problem of load balancing for any parallel program is distributing the load of a parallel program over multiprocessor system uniformly [3]. There are two different classes of load balancing algorithms: static and dynamic. A static load balancing algorithm does not take into account the previous state or behavior of a node while distributing the load. On the other hand, a dynamic load balancing algorithm checks the previous state of a node while distributing the load, such as CPU load, amount of memory used, delay or network load, and so on [4].

M. V. Yakobovskiy—With the support of the RFBR grant 17-07-01604 A.

© Springer Nature Switzerland AG 2018
L. Sokolinsky and M. Zymbler (Eds.): PCT 2018, CCIS 910, pp. 135–146, 2018.
https://doi.org/10.1007/978-3-319-99673-8_10

Each of these methods has their own advantages and disadvantages. Finding an optimal static load balancing is in general an NP-complete problem, unless some special cases [3]. It seems that optimal load balancing algorithm consists of good initial partitioning, which could be performed by some kind of static load balancing algorithm, and dynamic part, which redistributes load during the calculation. One of the principal costs of good initial distribution is expected to be the machine dependent cost of transferring the computational modules between processors [3].

In this particular paper we discuss static load balancing algorithm for guaranteed projection-based tetrahedral mesh generation algorithm. As this algorithm consists three major parts [1], each of them must have some kind of the algorithm. The fact that the considered algorithm tries to cover as much different surfaces as possible leads to a major problem: it is possible for one of computational modules on any of three main steps of algorithm take more time for processing, than all other modules together. This paper discusses load balancing for this type of algorithm is static load balancing algorithms. In this paper is discussed load balancing algorithms for parts of the algorithm, problems of creating balance weights for each computational module and some experiments with different size and topology objects.

2   Mesh-Generation Algorithm

For further explanation, let's point out main parts of algorithm. As input for this kind of algorithm we're using an oriented surface triangulation. For the purpose of simplicity, we suggest projection axis to be parallel to OZ. Therefore, all surface triangles are divided into 3 grand category, by the direction of z-component of their normal: TOP, WALL and BOT. Guaranteed mesh generation algorithm consists 3 main subsections: first part generates projection of BOT triangles onto TOP, triangulation of the resulting graph and creating a number of triangle prism, that covers the volume of the initial body; second - attachment of lateral faces of all faces with each other; and third - triangulating surface faces of each prism, so each prism is covered by surface triangulation. Any triangle prism is defined by six points: 3 points for each base prism triangle, even for degenerate case. Second important thing to admit is that algorithm uses rational numbers with arbitrary bit capacity for nominator and denominator in purpose of excluding any inaccuracies during the main calculation process, leaving them only to output part [1].

The first part of the algorithm can be described as follows:

1. [Cycle on i.] for $i = (1, N_t)$, for all TOP-triangles.
2. [Find nearest BOT-triangles.] For every i-th TOP-triangle, form an array of geometrically close BOT-triangles. Denote by N' the number of these triangles. Clear LAY substructure.
3. [Cycle on j.] Set $j = 1$. After the end of the cycle, go to step 8.
4. [Check hitting the projection-space.] Check the intersection of j-th triangle with the projection-space of i-th triangle. If the intersection was found, go to step 5, otherwise, go to step 7.

5. [Screen by TOP-triangles edges.] Execute screening of j-th triangle by edges of i-th triangle. Proceed to step 6.
6. [Screen by surface of BOT-triangles belonging to the projection-space.] Execute screening of the region constructed in step 5 by the surface of all triangles contained in LAY. Store the result in LAY.
7. [Termination condition for j cycle] If $j \leq N'$, then set $j = j + 1$ and go to step 4, otherwise, go to step 8.
8. [Triangulate the projection.] Execute the algorithm of 2-dimensional triangulation on data stored in LAY. As a result, LAY structure contains N'' triangles.
9. [Cycle on k.] Execute step 10 for each $k = (1, N'')$
10. [Form prisms.] For k-th triangle from LAY, reestablish its projection in the plane of i-th triangle and the corresponding BOT-triangle. Save these six points, representing two triangles as a triangle prism.

Second part of the algorithm:

1. [Cycle on i.] For $i = (1, M)$.
2. [Find nearest prisms.] For i-th prism, form an array of geometrically close prisms. Denote by M' the number of such prisms.
3. [Cycle on j.] For $j = (1, M')$. After the end of the cycle, go to step 8.
4. [Cycle on k] Execute step 5 for $k = (1, 3)$, on the sides of i-th prism.
5. [Cycle on l] Execute step 6 and, if necessary, step 7, for $l = (1, 3)$, on the vertices of the upper base of j-th prism.
6. [Check hitting the plane.] Check the hit of l-th vertex in the plane of k-th side edge of i-ith prism. The result (0 or 1) is stored in his own cell in *count* array. If hit occurs, execute step 7.
7. [Inserting vertical edges into topology.] Check the hit of vertical edge of j-th prism on the side edge of i-th prism. If the hit occus, add intersection point and parts of this edge into topology of i-th prism.
8. [Calculate number of touches.] Sum the values of *count* array in *sum* variable. If $sum = 0$, go to step 9. If $sum > 0$, go to step 10.
9. [Checking intersection of side edges of edges of bases.] Find the intersection points of vertical edges of i-th prism with edges of bases of j-th prism. Add these points to topology of i-th prism, dividing corresponding edge into parts.
10. [Building intersection of side edges.] Check hit of parts of edges of bases of j-th prism on side edge of i-th prism. Add the corresponding elements to i-th prism topology.

Next important part of the algorithm is using arbitrary bit capacity for rational numbers. It provides exclusion of any calculation inaccuracies in this particular algorithm, but leads to a few issues. First issue, that happens while using such numbers, is an exponential growth of length for nominator and denominator. During this particular algorithm, however, the maximum length of nominator and denominator during the calculation is bounded above. To prove this, let's consider every part of the algorithm.

First part includes building intersections between 2 different initial surface triangles ribs. The equation for calculation intersection point between two lines is known:

$$\frac{x - x_1}{x_2 - x_1} = \frac{x - x_3}{x_4 - x_3} \tag{1}$$

which is taken from the canonical equation of the line. So the answer is:

$$x = \frac{x_1(x_4 - x_3) - x_3(x_2 - x_1)}{(x_4 - x_3) - (x_2 - x_1)} \tag{2}$$

In the worst case, each addition and subtraction increases the length of the nominator (3) and denominator (4):

$$n_2 = 2n_1 + 1 \tag{3}$$

$$n_2 = 2n_1 \tag{4}$$

Here n_1 is a bit capacity before operation, and n_2 - after it. So, based on Eq. (2) and using (3) and (4), maximum bit capacity for any coordinate would be for the nominator (5) and the denominator (6), n_1 is the bit capacity of the initial surface point coordinate, n_2 is a bit capacity of builded point. This is the maximum theoretical length of stored coordinates for the first part of the algorithm.

$$n_2 = 6n_1 + 3 \tag{5}$$

$$n_2 = 6n_1 + 2 \tag{6}$$

For second part of the algorithm, we are building same kind of intersection points, but initial points coordinates, which are used in (2) could be from points, which are created on a first stage of the algorithm. So, in this case the result maximum bit capacity for a point coordinate nominator in the worst case would be:

$$n_2 = 36n_1 + 21 \tag{7}$$

Second issue - is very hard to handle in terms of using MPI, is described further.

The most time-consuming part of the algorithm is docking prisms with each other, as it would be shown in practical section. Therefore, the main focus of this paper would be on the static load balancing algorithm of this particular part.

Problems of its load balancing for parallel distributed realization of this algorithm are main topic of this paper.

3   Load-Balancing Problem

3.1   General Parallel Realization Problem in Case of Load Balancing

Arbitrary bit capacity, while providing exclusion of any calculation inaccuracies, is very hard to handle in terms of using MPI. It is known, that using arbitrary bit capacity for rational numbers leads to exponential growth of length for nominator and denominator. And even fixed, a priory calculable maximum length is still much larger, than any standard type. And, as long as all points coordinates is also stored in such type, it is difficult to use MPI communications with it, and it leads to an increase in transmission time. Therefore, during the realization of any balancing algorithms it is necessary to minimalize number of MPI communications.

3.2   Load Balancing for the Projection Part

As soon as algorithm is positioned as guaranteed volume coverage algorithm, it is clear that under any TOP triangle could be literally any number of triangles. Furthermore, as is shown on Figs. 1 and 2, there could be other TOP triangles underneath it. On Fig. 1 on the left is shown initial volume, which needs to be filled, and part of the surface triangulation on the right. As it was mentioned in [1] the problem of too many BOT triangles fall underneath one TOP triangles could be solved in some cases by inserting an additional 0-thickness plates between TOP and BOT triangles. For example, in Fig. 2 underneath the pointed triangles we could place an additional inner plate so under each triangle of the mesh would be close in size number of triangles. But it does not work in example, shown on Fig. 1. For this kind of surface it is impossible to find place for plate, because there are no free space left for plate where we need to insert it (see Fig. 4).

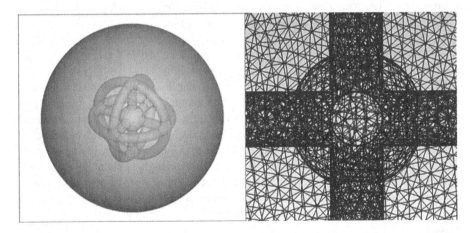

Fig. 1. Example of a surface with $N_p = 23068$ points and $N_t = 46174$ surface triangles

And, moreover, while we are creating a projection, we must check all triangles underneath considered triangle. But, during creating one projection it is unnecessary to communicate with other processes, so it is possible to create such initial triangles distribution, that at least some kind of load balancing could be achieved.

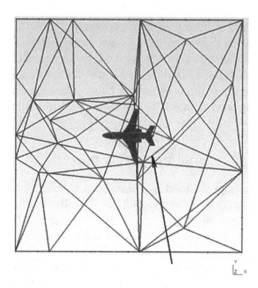

Fig. 2. Example of possible occasions for triangles mutual location

Creating a load balancing algorithm requires some kind of load weight, which could be used as weight for each element. As it is described in [1] using a 2-dimensional geometrical cache is necessary to decrease computational complexity for projection algorithm due to optimization of element search. As it is mentioned in [2], 2-dimensional cache can provide find all elements, which are close to the initial. So, as long as this type of cache could return all triangles that should be reviewed by projection algorithm, we can use number of BOT triangles, that is close to reviewed TOP triangle, as our balance weight.

But, on the other hand, such way concludes one major problem: if there are many colliding TOP and BOT triangles in the same cell of cache, One of the worst case scenarios is shown on Fig. 1 on the right. All surface of the cylinders, whose axis is parallel to the main projection axis, contains a large number of TOP and BOT triangles. That leads to one major problem: for all these TOP triangles their balance weight would be significantly higher, then for same triangles outside of the problem area. And, more than that, it is almost impossible to algorithmically find all BOT triangles, which would fall into projection onto current TOP triangle by any of their part (without, of course, creating full projection).

All this problem concludes into simple, but reasonably effective decision: for quasi-uniform surface mesh equal initial distribution leads into reasonable load balancing (the result would be mentioned below). Of course, this approach couldn't be used for other kind of meshes (for example, mesh on Fig. 2). It seems that solution for this problems, that concludes in good (by any criterion) balancing, requires and additional specific research.

3.3 Load Balancing for Attachment Part

This is the hardest part of the algorithm in terms of computational complexity and calculation time. The naïve way of balancing this part concludes into using just the same idea as in the first part. In that case, every triangle prism is represented as a triangle, which lays on OXY plane. This approach leads to two major problems. First, any prism is a 3-dimensional object. And 2 prisms could intersect with each other by a number of cases, which is shown on Fig. 3.

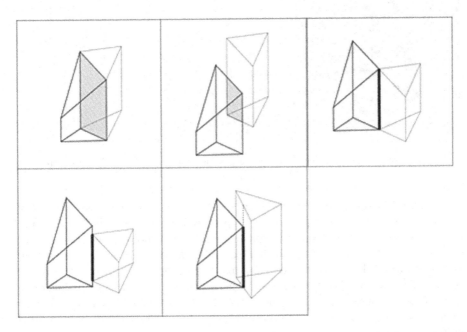

Fig. 3. Possible cases of collision of prisms, excluding the case with 0-thickness plates.

And, even more, they could be situation, when two or more prism are one above the other, without any colliding. Second, that partially follows previous paragraph, as long as docking of prisms is necessary, with every considered prism on a local process all its neighbor-prism must be stored, at minimum all 6 points, that define a prism. As we mentioned early, every point is stored as three rational numbers with arbitrary bit capacity. This concludes to extremely large memory usage, so it is necessary for the prism on each process to be as close to each other as possible.

This concludes us to main challenge: create a load balancing algorithm for prism docking, which guarantees as much domain integrity as possible and provides reasonable load balancing.

First of all, to separate whole array of prisms into any number of subdomains we need to create a graph, each vertex of this graph will represent a prism, and each edge should represent neighborhood relationship. While it is not known direct relationships between all prisms at this point of algorithm, we could use similar idea of creating of some kind of cache.

Fig. 4. Example with a large number of prisms laying in almost the same area.

On Fig. 4 is shown one of the worst case scenarios, when large amount of prisms are almost at the same place of space. On the left each triangle represents a surface for prism. On the right is shown much more closely the scale of problem in that particular case. This example leads to immediate conclusion: we cannot use 2-dimensional cache for this part of algorithm, because it would lead us directly to $O(N^2)$ operations, where N number of prisms To decrease this complexity as much as possible it is suggested to use 3-dimensional cache. There is second approach to resolving this problem, which would be described in Experiments-part.

With the same idea of caching with 2D-case, in 3D-case each cell of program cache represents a parallelepiped. For the simplicity of terminology, from here and below each cache cell would be named "cube". This approach significantly lowers the amount of prisms that have fallen into each cube of cache. The most important usage of this idea consists in using the approximate number of neighbors for each prism as balance weight of prism, and this approximate neighborship relation for constructing edges for graph. Therefore it is formed a weighted graph, which is already distributed through the all processes. It is important, that this approach is not an ideal. For example, Fig. 5, where is shown resulting surface for body from Fig. 1. These topological artifacts, so called "stars", cannot be optimized by any kind of geometrical cache.

Second trouble consists in separating this formed distributed graph on a number of domains, with minimum weight difference and least numbers of connections, which represents our demand to minimize the number of locally stored prisms. In purpose resolving this problem we use ParMETIS [5] for graph separation.

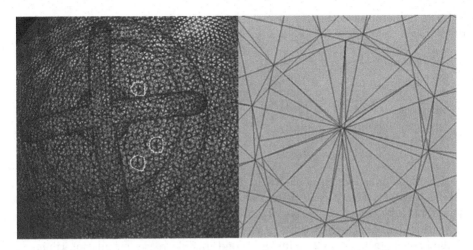

Fig. 5. Example of low efficiency of the 3D-cache.

Main problems of this algorithm consist this approach does not take in account time of inserting new vertices and edges into prisms, and does not consider any WALL triangles, that directly affect the calculation time.

3.4   Load Balancing for Triangulation of Prism Faces

It is the least expensive part of the algorithm. First thing that we have to mention that most of the prisms consists only 6 points, therefore for them it is necessary to add only 3 edge to each of them. Second thing that we should mention, that computational complexity for 2-dimensional triangulation depends mainly on number of vertices [2]. So, the main problem therefore is to decide, is it necessary to redistribute prisms after previous step of the algorithm or no.

The way of constructing previous weights for each prism is arguably inaccurate, so for each prism there would be a number of incorrect neighbors. Therefore, for the most prisms their weight is pretty much same.

This leads us to initial suggestion, that for purpose of load balancing of this stage of algorithm we could use distribution from previous stage. The advanced solution includes using number of vertices as a balance weight for this stage balancing.

4   Practical Experiments

From all of the experimental data we would consider at first all data about balancing load while docking the prisms. All experiments were done on 4 MPI-processes, on the one calculating node. All the data in Tables 1, 2 and 3 consists only time for docking of prisms.

First experiment is using surface, shown on Fig. 1. Total number of prisms in this experiment - 203 332.

Table 1. Experiment with different cache sizes, time for the second step of the algorithm. $Size_i$ – number of cells in that direction.

Dimension of cache and its size: $size_x * size_y * size_z$	Process 1, work time, minutes	Process 2, work time, minutes	Process 3, work time, minutes	Process 4, work time, minutes
2D-cache 100 * 100 * 1	248.23	239.29	224.75	248.94
3D-cache 100 * 100 * 4	188.62	235.84	230.21	212.18

The imbalance, which is shown in Table 1, is caused by the fact, that the balance weight value of each prism does not include the time of inserting new points and ribs, which are essentially created during this step of the algorithm. Increasing size of cache by 4 times caused further increase of imbalance from 10% in the first experiment to 20% in the second, but decreases calculation time.

Next idea for improving quality of balancing and further increasing load consists in rotating surface mesh around its center, so the least amount of prisms would have any additional points on their faces or edges; the result for this is shown in Table 3.

Table 2. Experiment with different cache sizes. Rotated surface.

Dimension of cache and its size: $size_x * size_y * size_z$	Process 1, work time, minutes	Process 2, work time, minutes	Process 3, work time, minutes	Process 4, work time, minutes
2D-cache 100 * 100 * 1	190.87	190.25	200.87	191.14
3D-cache 100 * 100 * 4	124.26	119.20	115.96	113.16

Simplification of prisms faces leads to significant increase of balance quality: now, first experiment provides almost only 5% of imbalance between processes, and second - near 9% of imbalance. Increasing imbalance while increasing cache size is provided by the fact that some prisms would eventually lose their balance weight, but the total number of new vertices and edges for each prism is constant, so for smaller prism there would be much less work to do, while for bigger prism there are still the same amount of work, which is represented by decrease of computational time.

The experiment shown in Table 3 shows, how initial surface triangulation of the same region will affect current algorithm (see Fig. 6). In this case the task is to fill inner part of these planes. As it is mentioned earlier, main difference is made by the amount of new vertices and edges, which are inserted into specific prism. Cache size for this experiment is 100 * 100 * 10.

Now let's consider the last stage of the algorithm.

In Table 4 is shown test of hypothesis, which was described earlier about using initial distribution for previous stage. Surprisingly, the results are much

Fig. 6. Two different surface triangulations

Table 3. Experiment with the same initial surface, N_t – number of surface triangles.

Mesh parameters, number of triangles	Process 1, work time, minutes	Process 2, work time, minutes	Process 3, work time, minutes	Process 4, work time, minutes
Left mesh, $N_t = 6115$	92.052	98.204	96.409	117.529
Right mesh, $N_t = 10997$	253.704	269.194	296.033	295.033

better than they were to be expected. The weight imbalance in all cases is near 27%. This result is arguably bad, because it is still definitely high imbalance. But, for the same time, we can admit that even without creating any specific algorithm for this stage, using just what left from previous part, we already have a distribution of data that could be used for much easier data redistribution algorithm.

Table 4. Third stage load balance using the same distribution as in the second stage.

Mesh description	Process 1, work time, minutes	Process 2, work time, minutes	Process 3, work time, minutes	Process 4, work time, minutes
First example (Fig. 1)	25.122	20.413	29.575	23.325
First example (Fig. 1) rotated	24.73	18.414	22.509	18.122
First plane (Fig. 5, left)	11.01	11.028	12.182	16.382
Second plane (Fig. 5, right)	29.852	21.071	28.441	17.489

5   Conclusion

Load balancing is hard and important task for all parallel programs. During this work created static load balancing algorithm, which is based on three-dimensional cache for generating graph and uses ParMETIS library for separating graph. Reviewed main problems of static load balancing in application to guarantied mesh generation algorithm. It is proved, that using three-dimensional cache significantly decreases total calculation time, but increases load imbalance between each processes. Founded, that using initial balancing for second stage of the algorithm provides a reasonable load balancing for the third stage.

References

1. Grigorjev, S.K., Yakobovskiy, M.V.: Practical aspects of realization of projection tetrahedral mesh generation method. In: Proceedings of an International Scientific Conference on Parallel Computational Technologies (PaVT 2016), Arkhangelsk, 28 March–1 April 2016, pp. 499–504. Publishing Center of SUSU, Chelyabinsk (2016). ISBN 978-5-696-04801-7
2. Skvorcov, A.V.: Trianguljacija Delone i ejo primenenie [Delaunay triangulation and its application], 128 p. Tomsk State University, Tomsk (2002). ISBN 5-7511-1501-5
3. Iqbal, M.A., Saltz, J.H., Bokhari, S.H.: Performance tradeoffs in static and dynamic load balancing strategies. Technical report 86–13, NASA Langley Research Center, Hampton, VA (1986)
4. Shah, N., Farik, M.: Static load balancing algorithms in cloud computing: challenges and solutions. Int. J. Sci. Technol. Res. 4(10), 365–367 (2015). ISSN 2277–8616
5. Karypis, G.: METIS and ParMETIS. In: Padua, D. (ed.) Encyclopedia of Parallel Computing, pp. 1117–1124. Springer, Boston (2011). https://doi.org/10.1007/978-0-387-09766-4. ISBN 978-0-387-09766-4

Fine-Grained Parallel Algorithms in TIM-3D Code

Andrey Alexandrovich Voropinov$^{(\boxtimes)}$ and Ivan Gennadievich Novikov

FSUE Russian Federal Nuclear Center – All-Russian Research Institute
of Experimental Physics, Sarov, Russia
{AAVoropinov,IGNovikov}@vniief.ru

Abstract. TIM-3D is a continuum-mechanics simulation code that uses arbitrary-shape unstructured polyhedral Lagrangian meshes. Parallelism in TIM-3D is provided at three levels in the mixed-memory model. The first two levels use space decomposition in the MPI-based distributed-memory model. At the first level, calculations are parallelized in task fragments (domains). At the second level, calculations within one domain are parallelized in para-domains. At the third level, iterations of calculation loops are parallelized in the OpenMP-based shared-memory model. The paper considers the fine-grained paralleling algorithms (second level). These algorithms are complementary to the OpenMP shared-memory parallelism implemented earlier. The fine-grained paralleling can be done both with overlapping in one row of para-domain interface cells and without overlapping. These approaches are compared in their parallel efficiency using one of test simulations.

Keywords: TIM-3D code · Distributed-memory parallelism · MPI
Unstructured meshes

1 Introduction

TIM-3D [1] is an unsteady continuum-mechanics simulation code that employs unstructured arbitrary-shape polyhedral Lagrangian meshes. Cells can have an arbitrary number of faces, and the faces can have an arbitrary number of nodes connecting an arbitrary number of cells and edges. Figure 1 shows some simple examples of meshes used in the code.

Parallelism in TIM-3D is provided at three levels. This approach is an extension of the three-level parallelism in TIM-2D [2] to the three-dimensional case. The first two levels use space decomposition in the MPI-based distributed-memory model. At the first level, calculations are parallelized in task fragments (domains). At the second level, calculations within one domain are parallelized in para-domains. At the third level, iterations of calculation loops are parallelized in the OpenMP-based shared-memory model. These approaches can be used both together in different combinations, and separately in one calculation.

Earlier, TIM-3D used shared-memory model parallelism [3]. Shared-memory parallelism is not sufficient, because the number of memory-sharing processor

© Springer Nature Switzerland AG 2018
L. Sokolinsky and M. Zymbler (Eds.): PCT 2018, CCIS 910, pp. 147–161, 2018.
https://doi.org/10.1007/978-3-319-99673-8_11

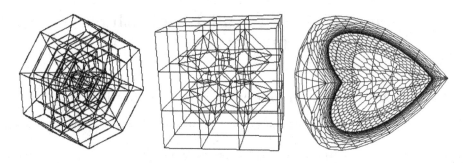

Fig. 1. Examples of polyhedral meshes used in TIM-3D

cores in up-to-date cluster computers is not very large. TIM-3D uses decomposition into domains. This involves solving a contact interaction problem between domains. Domains are calculated independently. This circumstance is used for the first level of parallelism. Its constraint is a small number of domains (usually up to 10). Thus, in order to remove the constraints on the number of computational resources engaged, an intermediate level of parallelism is required: at the sub-domain level, where the domain is divided into smaller geometric "grains". These fine-grained paralleling algorithms are described in the present paper.

The development of the parallel algorithms is based on the following principles:

– Identity of calculation outputs in any mode of calculations.
– Scalability, or possibility of running calculations on any number of cores with easy switch-over between the modes from one start to another for a single task.
– Minimum memory consumption—to run tasks that are too large for the memory available to a single core.
– Optimum utilization of computational resources. Prevention of imbalance, or in-process balancing when this occurs.

To facilitate the development of computational programs:

– Universal data representation in any mode of calculations.
– Minimum revision of codes to make them parallel; the burden of managing parallel computations lies with a set of supporting programs.

2   Data Decomposition

Efficient use of computational programs on parallel computers requires decomposition to ensure uniform distribution of work load among computer cores with as little communication as possible. Decomposition in the distributed-memory model includes distribution of data among processes (data decomposition) in such a way that the number of data transfers and the volume of communicated data between them is minimum. For distributed-memory parallelism, TIM-3D uses space decomposition. Decomposition principles for fine-grained parallelism are as follows:

- The decomposition is performed by cells (cells are the basic computing mesh elements in TIM-3D).
- All domain cells are distributed among compacts so that each cell belongs to only one compact.
- Each domain is split into compacts irrespective of other domains.

The problem of decomposition for fine-grained parallelism comes down to solving a problem of graph partitioning into subgraphs. This is accomplished by the following algorithm:

- A graph representing the mesh structure is built based on the unstructured mesh. Graph nodes correspond to mesh cells, and graph edges, to neighborhoods between cells.
- Graph nodes are assigned the weight reflecting the computational load associated with the corresponding cell. Weights of graph edges are used to introduce additional decomposition properties. For example, extending compacts along boundaries reduces the number of data transfers in contact interaction calculations.
- The problem of graph partitioning into subgraphs is solved using algorithms from the ParMeTiS or SCOTCH libraries [4–6] and our own hybrid (topological and geometric) decomposition algorithm.

Examples of resulting decompositions done by SCOTCH algorithms and by our own algorithm are shown in Fig. 2.

Fig. 2. Examples of decompositions: SCOTCH (left), hybrid (right)

3   Fine-Grained Parallelism

TIM-3D uses the staggered centering stencil. Kinematic quantities (velocities, accelerations, coordinates) are assigned to cell nodes, while thermodynamic quantities (energy, pressure, density, etc.) are assigned to cells. As a result, the key issue associated with the fine-grained parallelism is the way of calculating the mesh nodes (cell vertices) surrounding the cells belonging to different compacts (para-boundary nodes). For simplicity, let us illustrate this with the two-dimensional case shown in Fig. 3. In the figure, the white cells belong to compact 1, and the yellow ones, to compact 2.

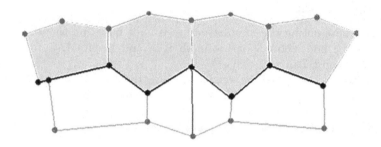

Fig. 3. A mesh fragment partitioned into compacts (Color figure online)

In accordance with the difference scheme of TIM-3D [1], a closed integration contour of cell centers and edge centers is constructed to calculate a node. For the two-dimensional case of interest, an example of an integration contour for the node V in Fig. 4 is marked by a green line. The integration contour is defined by the centers of the surrounding cells C_1, C_2, C_3, C_4, and the "centers" of the edges VV_1, VV_2, VV_3, VV_4. In conformity with the integration contour, the node V is calculated using the quantities in both the cells C_1, C_2, C_3, C_4, and the nodes V, V_1, V_2, V_3, V_4.

If the integration contour is preserved, we obtain the first type of fine-grained parallelism (with one layer of overlapping cells). In order to preserve the node integration contour on the side of the first para-domain, the cells C_2, C_3 do not need to be generated completely, i.e. no information on the nodes V_5, V_6 is required. However, if we do not attach these nodes to the first para-domain, the mesh will be incomplete, and some operations on the cells C_2, C_3, for example, definition of mesh nodes, volume calculations, or determination of the center, will be unavailable. In this case, the attached cells need to be described in the data structure in a special way and accounted for in different computing algorithms. As the highest possible transparency of parallelism for computing algorithms is one of the basic paralleling principles, it was decided to include such nodes into para-domains as attached nodes, i.e. the nodes V_3, V_5, V_6 are attached with respect to para-domain 1.

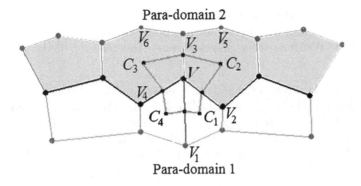

Fig. 4. Node integration contour (Color figure online)

On the other hand, the integration contour can be represented as a set of closed contours on the side of each cell (this representation is also used in the difference scheme to determine the node mass [1]). Similarly, the integration contour can also be represented as a set of closed contours on the side of each para-domain. Such a partitioning for the case under consideration is shown in Fig. 5, where the partitioning line of the integration contour is marked with red. When the integration contour is partitioned, the nodes along the para-domain interface are divided into pairs (or proportional to the number of para-domains connected at the node), for example, the node V is partitioned into V' and V''. For each node, a separate integration contour is used to determine the mass and accelerations, which are then combined to calculate the common velocity. Such an integration contour partitioning in the simulation makes it possible not to use the overlapping between para-domains. A similar approach is used for "no-slip" boundaries [7] and for fine-grained parallelism in TIM-2D [8].

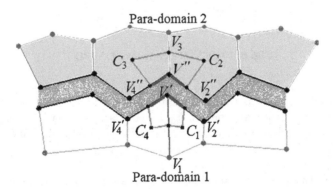

Fig. 5. Node integration contour partitioning during fine-grained paralleling without overlapping (Color figure online)

A similar volume element in the three-dimensional case is constructed as follows (see Fig. 6). Consider node i. Let the point j be the center of a cell adjacent to the node i. We draw three planes through the center j of the polyhedron (Fig. 6):

- the first passing through the node of interest i and one of the edges connected at this node and belonging to the cell having its center at j, for example, ii_1;
- the second passing through the node i and the center c_1 of one of two faces of polyhedron j to which the edge ii_1 also belongs;
- the third passing through the middle i_2 of the same edge ii_1 and the center c_1 of the face.

The triangular pyramid jii_2c_1 is part of the volume of the mass belonging to the node i. Using the same procedure, we construct another triangular pyramid with another edge belonging both to the face c_1, the node i and the cell of interest centered at j. Now we proceed to other faces belonging to the cell j and the node i at the same time. The number of such faces is equal to the number of faces of the polyhedral angle corresponding to the node i and the cell j (in most cases, they are three).

This resulting set of triangular pyramids generates the polyhedron belonging to the node on the face of the cell of interest j.

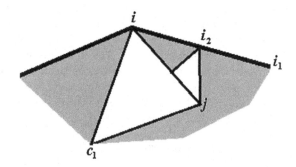

Fig. 6. Nodal cell element in TIM-3D

Interactions between para-domains are always pairwise. One can therefore speak about introducing para-boundaries. Para-boundaries include para-boundary nodes and faces separating the cells calculated in different para-domains. The para-boundaries also contain cell elements in the overlapping layer (if it is used): attached and near-boundary cells, faces, nodes.

Domain partitioning into para-domains does not change the difference scheme of TIM-3D in both fine-grained parallel modes, ensuring the identity of their results with that of the serial mode.

Finite-difference codes generally employ fine-grained parallelism with overlapping (see, e.g., [9–11]). A similar approach is to fill in missing data that are needed to calculate equations (see, e.g., [12]). The approach with node integration contour partitioning has been proposed for TIM-2D [8]. The present work considers its extension to the three-dimensional case.

If one compares the approaches, then each of them will have both advantages and disadvantages. The non-overlapping method demands less communication, because only nodal quantities are exchanged, whereas both nodal and cell-centered quantities are exchanged in the mode with overlapping. The volume of communicated data in the non-overlapping mode is also much smaller, because only data on para-boundary nodes are transferred, whereas in the mode with overlapping, this volume also includes data on the nodes of the overlapping layer. This results in higher efficiency of the non-overlapping mode. This approach is more convenient for algorithm programming, because all cell elements are computable, and the para-domain is in fact nearly identical to the mathematical domain. A constraint of the non-overlapping approach is that the difference scheme of the code should allow for integration contour partitioning. The drawback of the non-overlapping approach is that new limitations of the algorithms involving cell analysis around nodes at the para-domain interface occur. Examples of such algorithms include mesh maintenance algorithms (for example, it becomes impossible to combine directly cells from different para-domains). The computational load also increases a little because of the recovery of the common node integration contour.

The mode with overlapping is free of these limitations and drawbacks, which makes it more general. But its efficiency turns out to be a little lower because of the growing volume of exchanged data and number of data transfers.

4   Specific Features of Cell Neighborhood in the Three-Dimensional Case

In the three-dimensional case, para-domain generation has a number of specific features that distort the mesh structure in the para-domain. Such features are impossible or exceptional in the two-dimensional case, while in the three-dimensional case they are present in quite consistent decompositions.

The first class of features occurs at para-domain interfaces. They are cell neighborhoods along an edge or across a node (see Fig. 7). Such features make it difficult to describe the mesh structure [13], for example, to get a list of surrounding cells for nodes.

Features of the second class occur at outer domain boundaries. They include outbreaks of para-domains to the outer boundary with one edge or node. The simplest example of such a feature is the decomposition of a regular spherical mesh into columns or rows (see Fig. 8). The features of this class cause problems in calculations of contact interactions since they involve calculations of surface interactions, even though edges and nodes do not generate any surfaces.

These features constitute a certain challenge for the non-overlapping fine-grained parallelism because they directly influence calculations of the respective mesh nodes. These problems are not so evident in the overlapping mode since

Fig. 7. Features in para-domain cell neighborhood: neighborhood along an edge (left), neighborhood across a node (right)

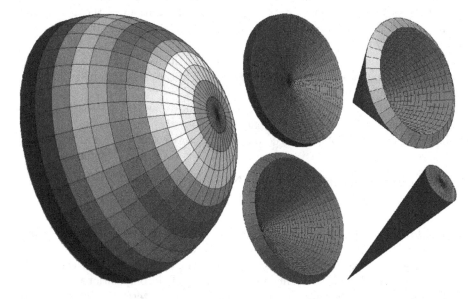

Fig. 8. Example of single-point outbreaks of para-domains onto the outer surface

the data of the feature are mostly transferred to an attached layer which is not processed. Nevertheless, these features should be kept in mind in this mode too.

To solve the above-mentioned problems, an additional object, a boundary node, has been introduced into the algorithms. This object is introduced for nodes at both outer and parallel boundaries. The following information is stored for boundary nodes:

- A full set of boundary conditions (both outer and parallel).
- A number of countable quantities essential for contact interaction calculations, such as vectors of outward normals, work, etc.

– Specifically for fine-grained parallelism, numbers of all faces connected at the boundary node are stored.

The storage of additional boundary node information makes it possible to incorporate arising features into the algorithms.

5   Fine-Grained Paralleling Algorithms

Fine-grained parallelism in para-domains includes processing of interior and near-boundary mesh elements. Cell calculations are therefore strictly associated with para-domains. However, this poses a question regarding nodes: how to calculate para-boundary nodes? As stated in Sect. 3 above, in the non-overlapping mode, para-boundary nodes are firstly calculated independently in each para-domain and then matched. This requirement is removed in the overlapping fine-grained paralleling mode, and para-boundary nodes can be calculated by any process calculating surrounding para-domains. As node calculations are rather inexpensive in TIM-3D, calculations of para-boundary nodes are backed up in this case. In TIM-3D, gas dynamic quantities are calculated in two major steps:

1. Calculations of nodal quantities, such as velocities, coordinates (calculation of the equation of motion). Node calculations are performed using data from neighbor nodes and surrounding cells from the previous time step.
2. Calculations of cell quantities, such as density, pressure, energy (calculation of the energy equation). These are done with updated node locations of the cells under consideration, based on which changes in the cell volume at an iteration time step are calculated.

The overlapping mode encounters the issue of data update in attached mesh elements (cells and nodes). This update is performed by asynchronous communication in para-boundary, near-boundary and attached elements. Calls of communication procedures are placed in such a way that the required information is updated before its use.

A flow diagram of a time step involving fine-grained paralleling with overlapping is shown in Fig. 9. The flow diagram for the non-overlapping mode remains the same, except for missing transfer of cell quantities. Likewise, once information is received, an additional integration contour recovery operation is performed.

The green boxes in Fig. 9 represent the calls of asynchronous communication procedures for nodal quantities, and the yellow boxes, for cell quantities; dashed lines are actual data streams between processes. The flow diagram shows that data transfers are combined with calculations of interior cells and nodes, which enables their parallel running. Communication involves packaging and unpackaging of quantities in the buffer array.

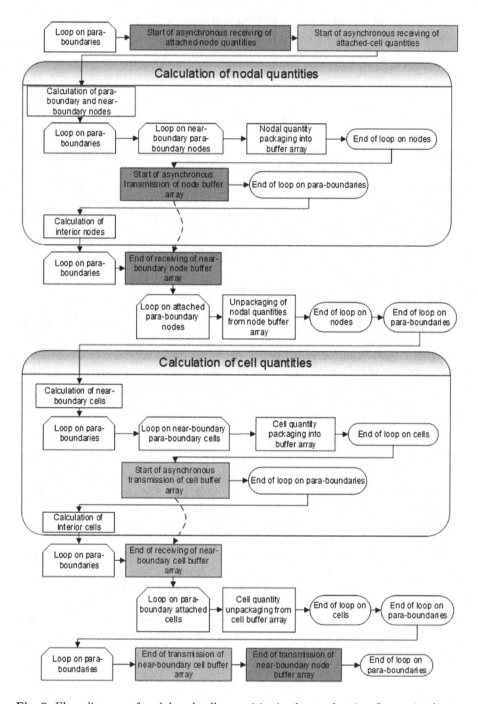

Fig. 9. Flow diagram of nodal and cell quantities in the overlapping fine-grained paralleling mode (Color figure online)

6   Distinctive Features of Non-overlapping Fine-Grained Parallelism

The approaches employed in TIM-3D paralleling algorithms are basically similar to those employed in TIM-2D [2]. In addition, most computational modules in both codes use the same programs for the two-dimensional and three-dimensional cases. The major difference in the parallelism of TIM-3D lies in the fine-grained paralleling algorithms with non-overlapping cell layers.

In TIM-2D, the integration contour partitioning scheme is used to calculate the no-slip boundary motion, when no partitioning is allowed at the domain boundary not only for the normal, but also for the tangential velocity component. In TIM-2D, the non-overlapping fine-grained parallelism is implemented by a modified algorithm for no-slip boundary calculations [7].

In the three-dimensional case, however, the approach employing the contact interaction algorithms cannot be used for a number of reasons. The main one is that the boundary interface in the two-dimensional case is a combination of two broken lines. Each line is drawn using a strictly defined series of boundary points in each domains. The three-dimensional case involves surfaces, and it is impossible in this case to set up a strict series of points and hence ensure the node-to-node point matching especially with active execution of mesh maintenance algorithms (to maintain the required shape of Lagrangian cells). In addition, contact interaction algorithms in the three-dimensional case become much more complicated themselves, primarily as a result of the transition to surface interaction. This makes the contact interaction algorithms significantly more expensive. Note that even in the two-dimensional case, calculations of boundary points are several times more expensive than calculations of interior points. It is therefore preferable to prevent the buildup of boundary points in the three-dimensional case.

For these reasons, in the three-dimensional case, for the non-overlapping fine-grained parallelism, it was decided to implement a program matching nodal quantities upon exchange of accelerations and masses for para-boundary nodes, rather than to use the contact interaction algorithms.

The node integration contour recovery algorithm is as follows:

– Prior to starting a calculation, masses of para-boundary nodes are calculated over the integration contour belonging to the para-domain of interest. That is, one para-boundary node has its own mass in different para-domains, and its total mass is the sum of its masses in the para-domains ($m_i = \sum\limits_{j=1}^{k} m_j$ is the mass of the para-boundary node i with respect to k para-domains ($k \geq 2$), m_j is the mass of the para-boundary node in the corresponding para-domain).
– The para-boundary nodes are calculated in each para-domain completely independently. Here, no forces are applied from the side of the parallel boundary (the parallel boundary serves as a free surface with pressure $P_G = 0$).
– Once all the para-boundary nodes are calculated, their resulting accelerations and masses are exchanged between the para-domains. Generally speaking,

node masses do not vary in the course of gas dynamic equation calculations, but their variations are possible as a result of the execution of mesh maintenance algorithms. In addition, more complicated algorithms may require some additional quantities, so whole sets of nodal quantities are exchanged.

– Once the exchange is over, total accelerations, velocities and positions of para-boundary nodes are calculated in each para-domain as follows:

$$\mathbf{a}_i^{n+1} = \frac{\sum\limits_{j=1}^{k} m_i \mathbf{a}_j^{n+1}}{\sum\limits_{j=1}^{k} m_j} \quad \text{– total acceleration of node } i;$$

$$\mathbf{v}_i^{n+1} = \mathbf{v}_i^n + \tau \mathbf{a}_i^{n+1} \quad \text{– total velocity of node } i;$$

$$\mathbf{r}_i^{n+1} = \mathbf{r}_i^n + \tau \mathbf{v}_i^{n+1} \quad \text{– updated node position;}$$

where τ is the time step.

Note that the additional update can lead to truncation errors due to specific features of machine arithmetics, i.e., the resulting velocity calculated by different processes can differ in its 16th digit. This error can build up with time and result in mismatch between para-boundary nodes (a gap or overlapping between para-domains). Although the mismatch is tiny, some algorithms can be sensitive even to such discrepancies. To overcome this problem, (1) the resulting velocity is rounded by assuming that near-zero accelerations and velocities are zero, and (2) the velocities and node positions are averaged at time instant n. Velocities and coordinates are calculated as follows:

$$\mathbf{v}_i^{n+1} = \frac{\sum\limits_{j=1}^{k} \mathbf{v}_j^n}{k} + \tau \mathbf{a}_i^{n+1} \quad \text{– total velocity of node } i;$$

$$\mathbf{r}_i^{n+1} = \frac{\sum\limits_{j=1}^{k} \mathbf{r}_j^n}{k} + \tau \mathbf{v}_i^{n+1} \quad \text{– updated node position.}$$

This prevents any mismatch between matched para-boundary nodes belonging to different para-domains.

7   Measurements of Parallel Efficiency

To assess the parallel efficiency, we used the functions $S_p = \frac{t_1}{t_p}$ (speedup of calculations), and $E_p = \frac{t_1}{p t_p} \cdot 100\%$ (parallel efficiency), where t_1 is the calculation time on one processor of the parallel computer (serial calculations), t_p is the calculation time on p processors.

For the test, we chose a planar-wave problem [14]. The calculation was run on an unstructured hexahedral mesh of 1 million cells. The results of time, acceleration and efficiency measurements are summarized in Table 1. We used as a basic unit the time for running the calculation of one compute node with OpenMP parallelism only. This allowed us to evaluate the efficiency of the fine-grained paralleling block in the mixed mode.

The results indicate that the efficiencies of the fine-grained parallel modes with and without overlapping are close up to 10 compute nodes, whereas, for a greater number of nodes, the efficiency of the non-overlapping parallel mode becomes higher (by 7 to 10 %). This is explained by a smaller number of data transfers and a smaller volume of communicated data.

Table 1. Measured speedup and parallel efficiency

Mode→		With overlapping			Without overlapping		
Core count	Node count	Time, s	Speedup	Efficiency	Time, s	Speedup	Efficiency
16	1	4012.10	1	100%	4012.10	1	100%
32	2	2224.96	1.80	90.16%	2268.58	1.76	88.42%
64	4	1116.10	3.59	89.86%	1144.32	3.50	87.65%
128	8	629.42	6.37	79.67%	594.82	6.74	84.31%
160	10	515.88	7.77	77.77%	495.80	8.09	80.92%
256	16	359.40	11.16	69.77%	321.14	12.49	78.08%
320	20	303.34	13.22	66.13%	266.54	15.05	75.26%
384	24	264.42	15.17	63.22%	230.37	17.41	72.56%
512	32	222.41	18.03	56.37%	186.29	21.53	67.30%
640	40	193.15	20.77	51.92%	164.17	24.43	61.09%
800	50	169.12	23.72	47.44%	143.87	27.88	55.77%
1600	100	146.71	27.34	27.34%	110.78	36.21	36.21%

8   Conclusions

The paper describes two fine-grained parallel methods used in TIM-3D code. In the first method, the whole node integration contour is preserved, and para-domains overlap in one layer of cells. The overlapping layer serves for node and cell data communication. In the second method, the node integration contour is partitioned, and para-domain interactions are calculated. These methods are close in their parallel efficiency on a small number of compute nodes (up to 10), while, on a large number of nodes, the non-overlapping method turns out to be 7 to 10 % more efficient. The higher efficiency is achieved owing to a smaller number of data transfers and a smaller volume of communicated data. Most calculations in TIM-3D are performed in the non-overlapping fine-grained parallel mode.

The algorithms developed are complementary to the OpenMP parallelism implemented earlier in TIM-3D.

References

1. Sokolov, S.S., Panov A.I., Voropinov, A.A., et al.: The code TIM for three-dimensional continuum mechanics simulations on unstructured polyhedral Lagrangian meshes. In: Voprosy atomnoi nauki i tekhniki. Ser. Matematicheskoe modelirovanie fizicheskikh protsessov, no. 3, pp. 37–52 (2005). (in Russian)
2. Voropinov, A.A., Sokolov, S.S.: The method of three-level paralleling in the code TIM-2D. In: Voprosy atomnoi nauki i tekhniki. Ser. Matematicheskoe modelirovanie fizicheskikh protsessov, no. 4, pp. 70–77 (2013). (in Russian)
3. Voropinov, A.A., Novikov, I.G., Sobolev, I.V., Sokolov, S.S.: Paralleling of the code TIM in the shared-memory model using the OpenMP interface. Vychislitel'nye metody i programmirovanie. **8**(1), 134–141 (2007). (in Russian)
4. Polovnikova, T.N., Voropinov, A.A.: Experience of using the SCOTCH and MeTiS libraries in unstructured mesh decomposition in the code TIM. In: Shagaliev, R.M. (ed.) Proceedings of the XII International Workshop on Supercomputing and Mathematical Modeling, pp. 282–288. FSUE RFNC-VNIIEF, Sarov (2011). (in Russian)
5. ParMETIS: Parallel graph partitioning and fill-reducing matrix ordering. http://glaros.dtc.umn.edu/gkhome/metis/parmetis/overview
6. Pellegrini, F.: SCOTCH: Static mapping, graph, mesh and hypergraph partitioning, and parallel and sequential sparse matrix ordering package. http://www.labri.fr/perso/pelegrin/scotch/
7. Voropinov, A.A., Novikov, I.G., Sokolov, S.S.: Calculations of contact interaction between domains in the code TIM-2D. In: Voprosy atomnoi nauki i tekhniki. Ser. Matematicheskoe modelirovanie fizicheskikh protsessov, no. 2, pp. 5–20 (2008). (in Russian)
8. Voropinov, A.A., Novikov, I.G., Sokolov, S.S.: Fine-grained paralleling methods in the code TIM-2D. In: Voprosy atomnoi nauki i tekhniki. Ser. Matematicheskoe modelirovanie fizicheskikh protsessov, no. 3, pp. 24–33 (2012). (in Russian)
9. Pronin, V.A.: Paralleling methods for two-dimensional gas dynamics simulations on unstructured meshes with variable topology in the code MEDUZA. In: Voprosy atomnoi nauki i tekhniki. Ser. Matematicheskoe modelirovanie fizicheskikh protsessov, no. 1, pp. 54–67 (2011). (in Russian)
10. Gasilov, V.A., Diachenko, S.V., Boldarev, A.S., et al.: Application package MARPLE3D for pulsed magnetically driven plasma simulations on high-performance computers, p. 20. Keldysh Institute of Applied Mathematics, Moscow (2011, preprint). (in Russian)
11. Lyapin, V.V., Korolev, R.A., Vetchinnikov, A.V.: A paralleling method with two-dimensional mesh decomposition for numerical solution of the two-dimensional heat transfer equation using the code KORONA-2D. In: Voprosy atomnoi nauki i tekhniki. Ser. Matematicheskoe modelirovanie fizicheskikh protsessov, no. 2, pp. 69–77 (2014). (in Russian)
12. Andrianov, A.N., Efimkin, K.N.: An approach to implementation of numerical methods on unstructured meshes. Vychislitel'nye metody i programmirovanie **8**, 6–17 (2007). [in Russian]
13. Voropinov, A.A., Sokolov, S.S., Panov, A.I., Novikov, I.G.: Polyhedral arbitrary-structure mesh description format in the code TIM. In: Voprosy atomnoi nauki i tekhniki. Ser. Matematicheskoe modelirovanie fizicheskikh protsessov, no. 3–4, pp. 55–63 (2007). (in Russian)

14. Bondarenko, Yu.A., Voronin, B.L., Delov, V.I., et al.: Description of a test suite for two-dimensional gas dynamic codes and programs. Part 1. Test requirements. Tests 1–7. In: Voprosy atomnoi nauki i tekhniki. Ser. Matematicheskoe modelirovanie fizicheskikh protsessov, no. 2, pp. 3–9 (1991). (in Russian)

Modified Componentwise Gradient Method for Solving Structural Magnetic Inverse Problem

Elena N. Akimova[1,2](✉)📷, Vladimir E. Misilov[1,2], and Andrey I. Tretyakov[1,2]

[1] Krasovskii Institute of Mathematics and Mechanics, Ural Branch of RAS,
Yekaterinburg, Russia
aen15@yandex.ru, out.mrscreg@gmail.com, fr1z2rt@gmail.com
[2] Yeltsin Ural Federal University, Yekaterinburg, Russia

Abstract. An original variant of the componentwise gradient method is constructed to solve a nonlinear magnetic inverse problem: using magnetic data, find a boundary surface between two layers with constant arbitrarily directed magnetizations. An efficient parallel algorithm is created and implemented on a multicore CPU and multiple GPUs to solve the problem. We study the efficiency and speedup of the parallel algorithm. We solve various model problems with synthetic magnetic data on a fine grid. A comparison of the proposed method with the conjugate gradient method shows that the new one allows for a significant reduction of computation time.

Keywords: Componentwise gradient method · Parallel algorithms
Magnetic inverse problem · Multicore CPU and multiple GPUs

1 Introduction

The solution of structural gravity problems and magnetic inverse problems has an extraordinary importance in the study of the Earth's crust structure [1–3].

This paper deals with the problem of finding an interface between layers with different magnetizations using known magnetization contrast, interface depth, and magnetic field [4,5].

The problem is described by a nonlinear integral equation of the first kind and thus is ill-posed. It is therefore necessary to use iterative regularization methods [6].

Real observations are performed on large areas. To increase the accuracy and the level of detail, it is essential to use finer grids, which leads to big data sets. The application of modern computing technologies and parallel computations makes it possible to significantly reduce computation time.

This work was partly supported by the Ural Branch of the Russian Academy of Sciences (project no. 18-1-1-8).

L. Sokolinsky and M. Zymbler (Eds.): PCT 2018, CCIS 910, pp. 162–173, 2018.
https://doi.org/10.1007/978-3-319-99673-8_12

An effective method to determine the structural boundary in the case of arbitrarily directed magnetization was constructed in [7,8] on the basis of the linearized conjugate gradient method.

A time-efficient componentwise gradient method for solving gravity inverse problems was constructed in [9]. In the present paper, we use this method to solve the magnetic inverse problem of finding a magnetization interface in the case of an arbitrarily directed magnetization. Here, we modify the method for better performance. The modification consists in offsetting the indices of the components with respect to the angle of the magnetization vector.

Moreover, we construct a parallel algorithm based on the modified componentwise method and implement this parallel algorithm using the Intel CPUs and NVIDIA Tesla GPUs of the Uran supercomputer, which is installed at the Institute of Mathematics and Mechanics of the Ural Branch of the Russian Academy of Sciences. We also investigate the efficiency and speedup of the parallel algorithm and compare it with a conjugate gradient-based algorithm in terms of iteration number and computation time.

2 Problem Statement

Let us introduce a cartesian coordinate system in which the $x0y$ plane coincides with the Earth's surface and the z axis is directed downwards, as shown in Fig. 1. Assume that the lower half-space consists of two layers with constant magnetizations J_1 and J_2, divided by the surface sought, which is described by a bounded function $\zeta = \zeta(x,y)$, and $\lim\limits_{|x|+|y|\to\infty} (h - \zeta(x,y)) = 0$ for some h.

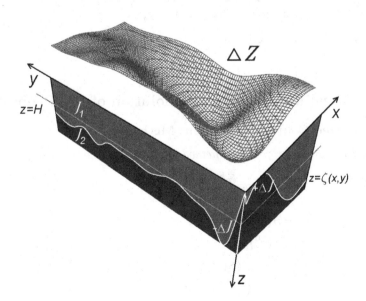

Fig. 1. Two-layer medium for the magnetic problem

The function ζ must satisfy the following equation:

$$\Delta Z(x',y',0) = \frac{1}{4\pi} \int\limits_{-\infty}^{\infty} \int\limits_{-\infty}^{\infty} \left[\frac{\Delta J_x(x-x') + \Delta J_y(y-y') - \Delta J_z h}{\left((x'-x)^2 + (y'-y)^2 + h^2\right)^{3/2}} \right.$$

$$\left. - \frac{\Delta J_x(x-x') + \Delta J_y(y-y') - \Delta J_z \zeta(x,y)}{\left((x'-x)^2 + (y'-y)^2 + \zeta^2(x,y)\right)^{3/2}} \right] dx\,dy, \qquad (1)$$

where $\Delta J_x, \Delta J_y, \Delta J_z$ are the components of the magnetization contrast $\Delta J = J_2 - J_1$, and $\Delta Z(x,y,0)$ is the vertical component of the anomalous magnetic field measured at the Earth's surface.

A preliminary processing of data with the aim of extracting the anomalous field from the measured magnetic data is performed using a technique described and implemented in [10].

Equation (1) is a nonlinear two-dimensional integral equation of the first kind.

After discretization of the region $\Pi = \{(x,y) : a \leqslant x \leqslant b, c \leqslant y \leqslant d\}$ by means of an $n = M \times N$ grid and approximation of the integral operator using quadrature rules, we obtain a vector F on the right-hand side and an approximation of the solution vector z of dimension n. Equation (1) can be thus written as

$$\Delta F_i = \frac{\Delta x \Delta y}{4\pi} \sum_{j=1..n} \left[\frac{\Delta J_x(x_i - x_j) + \Delta J_y(y_i - y_j) - \Delta J_z h}{\left((x_i - x_j)^2 + (y_i - y_j)^2 + h^2\right)^{3/2}} \right.$$

$$\left. - \frac{\Delta J_x(x_i - x_j) + \Delta J_y(y_i - y_j) - \Delta J_z z_j}{\left((x_i - x_j)^2 + (y_i - y_j)^2 + z_j^2\right)^{3/2}} \right], \qquad (2)$$

We can rewrite the equation as

$$A(z) = F. \qquad (2a)$$

3   Numerical Methods for the Solution of the Problem

3.1   Linearized Conjugate Gradient Method

The *linearized conjugate gradient method* (LCGM) has the following form [11]:

$$z^{k+1} = z^k - \psi \frac{\langle p^k, S(z^k) \rangle}{\|A'(z^k)p^k\|^2} p^k,$$

$$p^k = S(z^k) + \beta^k p^{k-1},$$

$$p^0 = S(z^0), \qquad (3)$$

$$\beta^k = \max \left\{ \frac{\langle S(z^k), \left(S(z^k) - S(z^{k-1})\right) \rangle}{\|S(z^{k-1})\|^2}, 0 \right\},$$

$$S(z) = A'(z)^T \left(A(z) - F\right),$$

where z^k is the approximation of the solution in the kth iteration, $k \in \mathbb{N}$, and ψ is a damping factor.

A parallel algorithm based on this method was developed and implemented in [8] for NVIDIA GPUs using CUDA technology.

3.2 Componentwise Gradient Method

The *componentwise gradient method* (CWM) has the following form [9]:

$$z_i^{k+1} = z_i^k - \psi \frac{A_i(z^k) - F_i}{\|\nabla A_i(z^k)\|^2} \left(\frac{\partial A_i(z^k)}{\partial z_i} \right), \tag{4}$$

where z_i is the ith component of the solution approximation, $i = 1, \ldots, n$, $k \in \mathbb{N}$, and ψ is a damping factor.

The main idea of this method is to minimize the residual $A_i(z) - F_i$ at one grid node i by changing the value z_i at this node. The idea is based on the fact that the value of a gravity or magnetic (in the case of vertically directed magnetization) field depends on $1/r^2$. Thus, the value of z_i exerts the greatest influence on the field value F_i at the node directly situated above it. In the case of an arbitrarily directed magnetization, the correlation between z_i and F_i is weaker, so this method will not be as effective as it is for vertical magnetization.

3.3 Modified Componentwise Gradient Method

Let us find the approximation of a new point j at which F_j is mostly influenced by z_i in the case of an arbitrarily directed magnetization. This point is displaced from the point i by the biases $\bar{x}$ and $\bar{y}$. To find $\bar{x}$, we need to solve the following problem:

$$\bar{x} = \arg\max_x \left[-\frac{\Delta J_x(x) - \Delta J_z h}{\left(x^2 + h^2\right)^{3/2}} \right].$$

The necessary condition for maximum is

$$\frac{d}{dx} \left[-\frac{\Delta J_x(x) - \Delta J_z h}{\left(x^2 + h^2\right)^{3/2}} \right] = 0.$$

Write the derivative:

$$-\frac{\Delta J_x(2x^2 - h^2) + 3\Delta J_z h x}{\left(x^2 + h^2\right)^{5/2}} = 0.$$

Evidently, $x \neq 0$ for the case of nonvertical magnetization and the surface lies below the Earth's level, *i.e.* $h > 0$, so that

$$\Delta J_x(2x^2 - h^2) + 3\Delta J_z h x = 0.$$

Write the roots of this equation:

$$\bar{x}_{1,2} = \frac{\left(-3\Delta J_z \pm \sqrt{8\Delta J_x^2 + 9\Delta J_z^2}\right)h}{4\Delta J_x}.$$

Assume that $\Delta J_z > 0$. Then, obviously, the relation $\mathrm{sgn}(\Delta J_x) = \mathrm{sgn}(\bar{x})$ must hold. Only the first root (the one with the plus sign) satisfies this condition. For $\Delta J_z < 0$, we have the second root (the one with the minus sign).

The $\bar{y}$ bias can be found in the same way. We can now write the *modified componentwise gradient method* (MCWM) as follows:

$$
\begin{aligned}
z_i^{k+1} = z_i^k &- \psi \frac{A_j(z^k) - F_j}{\|\nabla A_j(z^k)\|^2} \left(\frac{\partial A_j(z^k)}{\partial z_i}\right), \\
j = i + M &\frac{\left(-3\Delta J_z + \mathrm{sgn}(\Delta J_z)\sqrt{8\Delta J_y^2 + 9\Delta J_z^2}\right)h}{4\Delta J_y \Delta y} \\
&+ \frac{\left(-3\Delta J_z + \mathrm{sgn}(\Delta J_z)\sqrt{8\Delta J_x^2 + 9\Delta J_z^2}\right)h}{4\Delta J_x \Delta x},
\end{aligned}
\tag{5}
$$

where Δx and Δy are the grid element sizes.

We should also check whether the offsetted indices are out of the grid. If so, we should use the boundary values.

4   Parallel Implementation

The parallel algorithms based on the componentwise methods were implemented on a multicore CPU, using OpenMP technology, and NVIDIA M2090 GPUs, using CUDA technology.

Note that storing a Jacobian matrix for a $2^9 \times 2^9$ grid takes more than 512 GB.

The elements of the Jacobian matrix in the constructed algorithms are calculated on-the-fly, which means that the value of an element is computed when calling this element, without storing it previously in memory.

The most expensive operation is to compute the values of the integral operator and its Jacobian matrix. This operation consists of four nested loops. In the OpenMP implementation, the outer loops are parallelized using '#pragma omp parallel', whereas the inner loops are vectorized using '#pragma simd' directives. When using multiple GPUs, two outer loops are distributed to the GPUs, and two inner loops are executed on each GPU. The CPU transfers the data between the host memory and GPUs, and then calls the kernel functions.

The adjustment of the kernel execution parameters for the grid size is an important problem. In [12], we proposed an original method for automatic adjustment of parameters. This method is based on rescaling the optimal parameters found for a reference grid size.

This imposes some constraints on the input data and GPUs configuration:

- the grid size must be divisible by 128 $(128, 256, 512, 1024, \ldots)$;
- the number of GPUs must be a power of 2 $(1, 2, 4, 8, \ldots)$.

5 Numerical Experiments

The model problems consisted in finding the interface between two layers. Figure 2 shows the model surface z^* considered in all model problems.

Figures 3, 4, 5, 6 and 7 show the model magnetic fields $\Delta Z_i(x, y, 0)$. These fields were obtained by solving the direct problem for the surface with the asymptotic plane $H = 10$ km and various magnetization contrasts:

$$\Delta J_1 = (0, 0, 1)\, \text{A/m},$$
$$\Delta J_2 = (0.19, 0.19, 1)\, \text{A/m},$$
$$\Delta J_3 = (0.41, 0.41, 1)\, \text{A/m},$$
$$\Delta J_4 = (0.71, 0.71, 1)\, \text{A/m},$$
$$\Delta J_5 = (1.23, 1.23, 1)\, \text{A/m}.$$

These contrasts correspond to magnetization direction angles of $0°$, $15°$, $30°$, $45°$, and $60°$.

The problems were solved on the Uran supercomputer nodes (two eight-core Intel E5-2660 CPUs and eight NVIDIA Tesla M2090 GPUs) by the following three methods:

- linearized conjugate gradient method (LCGM) (3);
- componentwise gradient method (CWM) (4);
- modified componentwise gradient method MCWM (5).

The reconstructed interfaces are shown in Fig. 8.

The condition $\|A(z) - F\|/\|F\| < \varepsilon$, $\varepsilon = 0.011$, was taken as termination criterion for all methods. The parameter ψ was set at 0.85 in the CGM for $60°$, as well as in the CWM and MCWM for $45°$. In the CWM and MCWM for $60°$, it was set at 0.75. Everywhere else, it was set at 1.

The relative error of all solutions is $\delta = \|z - z^*\|/\|z^*\| < 0.01$.

Table 1 summarizes the numbers of iterations N and average execution times T for 10 runs on two eight-core Intel E5-2660 CPUs (16 cores) with a 512×512 grid.

Speedup and efficiency coefficients are used to analyse the scaling of parallel algorithms. The speedup is expressed as $S_m = T_1/T_m$, where T_1 is the execution time of a program running on one GPU, and T_m is the execution time for m GPUs. The efficiency is defined as $E_m = S_m/m$. The ideal values are $S_m = m$ and $E_m = 1$, but real values are lower because of the overhead.

Table 2 summarizes the average execution times for the CWM method on a 512×512 grid for various numbers of GPUs.

The experiments show that the constructed modified algorithms are very effective. New algorithms are more economical in terms of operations and time at each iteration step. For the model problems, the componentwise method has a better performance in terms of number of iterations and computation time than the conjugate gradient methods. The parallel algorithms demonstrate an excellent scaling; the efficiency is more than 100% for eight GPUs. Probably, this is due to a non-optimal automatic adjustment of the kernel execution parameters for some configurations of GPUs.

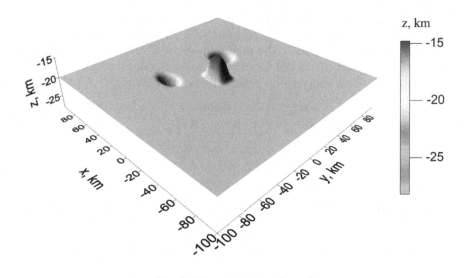

Fig. 2. The original surface z^*

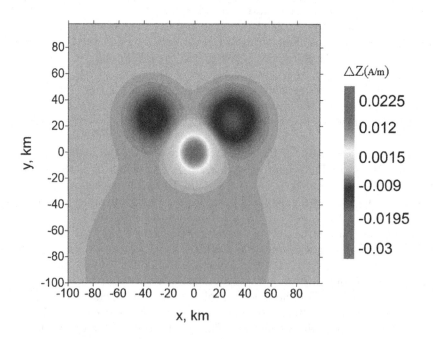

Fig. 3. Model gravitational field for an angle of $0°$

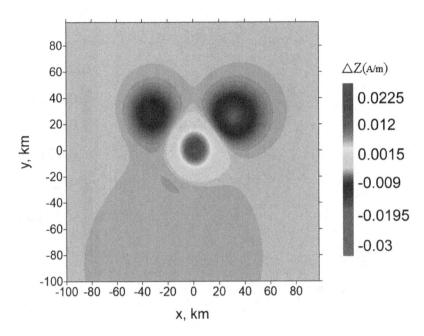

Fig. 4. Model gravitational field for an angle of 15°

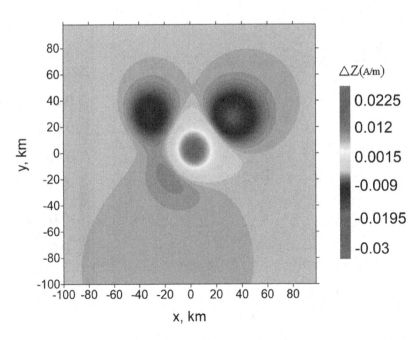

Fig. 5. Model gravitational field for an angle of 30°

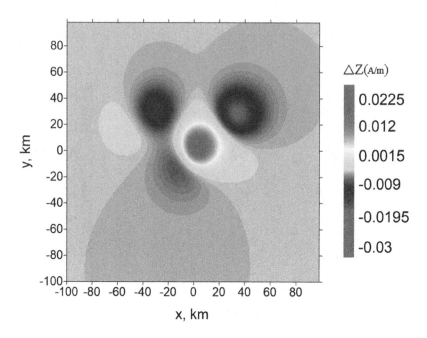

Fig. 6. Model gravitational field for an angle of 45°

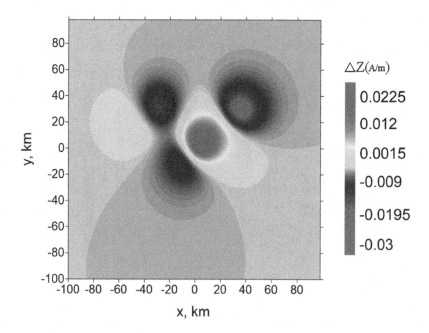

Fig. 7. Model gravitational field for an angle of 60°

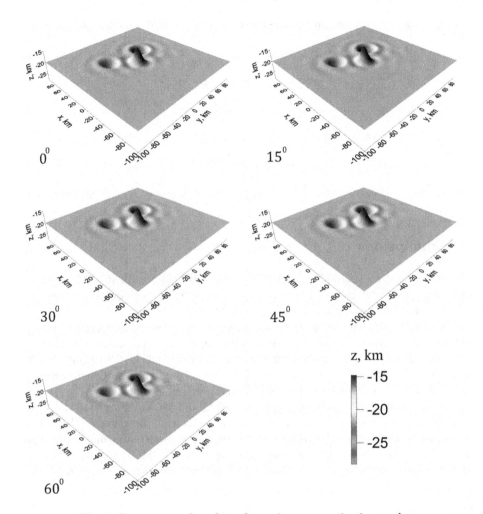

Fig. 8. Reconstructed surfaces for various magnetization angles

Table 1. Comparison of methods

Magnetization angle	CGM		CWM		MCWM	
	N	T (min)	N	T (min)	N	T (min)
0°	20	120	6	36	6	36
15°	20	120	6	36	6	36
30°	20	120	8	48	7	42
45°	25	150	10	60	9	54
60°	26	156	16	96	14	84

Table 2. Execution times (in minutes) of the parallel CWM algorithm on multiple GPUs

Magnetization angle	Number of GPUs			
	1	2	4	8
	Execution time T, minutes			
0°	7.6	2.8	1.3	0.7
15°	7.6	2.8	1.3	0.7
30°	8.3	3.3	1.5	0.9
45°	10.8	4.5	2.3	1.1
60°	17.7	7.9	4.2	2.1

6   Conclusions

We constructed an original variant of a componentwise gradient method for a structural magnetic inverse problem consisting in finding a contact surface in the case of an arbitrarily directed magnetization.

We developed parallel algorithms based on the componentwise gradient method and its modified variant. The parallel algorithms were implemented on a multicore CPU, using OpenMP technology, and on multiple GPUs, using CUDA technology. Model problems with fine grids were solved. The parallel algorithms demonstrated an excellent scaling and nearly 100% efficiency.

The componentwise gradient methods (CWM and MCWM) are very effective for solving problems with a nearly vertical magnetization direction; in this case, computation times are reduced by a factor of 2 to 4. For greater magnetization angles, the modified componentwise gradient method (MCWM) show better computation times compared to the unmodified componentwise method.

References

1. Martyshko, P.S., Byzov, D.D., Martyshko, M.P.: Solving the structural inverse problem of magnetic prospecting with respect to demagnetization for a two-layer medium model. Dokl. Earth Sci. **453**(2), 1264–1267 (2013). https://doi.org/10.1134/S1028334X1312012X
2. Akimova, E.N., Martyshko, P.S., Misilov, V.E.: Algorithms for solving the structural gravity problem in a multilayer medium. Dokl. Earth Sci. **453**(2), 1278–1281 (2013). https://doi.org/10.1134/S1028334X13120180
3. Martyshko, P.S., Pyankov, V.A., Akimova, E.N., Vasin, V.V., Misilov, V.E.: On solving a structural gravimetry problem on supercomputer "Uran" for the Bashkir Predural's area. In: GeoInformatics 2013 – 12th International Conference on Geoinformatics: Theoretical and Applied Aspects (2013)
4. Malkin, N.R.: On solution of inverse magnetic problem for one contact surface (the case of layered masses). DAN SSSR, Ser. A (9), 232–235 (1931)

5. Akimova, E.N., Martyshko, P.S., Misilov, V.E.: Parallel algorithms for solving structural inverse magnetometry problem on multicore and graphics processors. In: Proceedings of 14th International Multidisciplinary Scientific GeoConference SGEM 2014, vol. 1, no. 2, pp. 713–720 (2014)
6. Bakushinskiy, A., Goncharsky, A.: Ill-Posed Problems: Theory and Applications. Mathematics and Its Applications, 258 p., vol. 301. Springer Science & Business Media, Heidelberg (1994). https://doi.org/10.1007/978-94-011-1026-6
7. Misilov, V.E.: On solving the structural inverse magnetic problem of finding a contact surface in the case of arbitrary directed magnetization. In: 15th EAGE International Conference on Geoinformatics: Theoretical and Applied Aspects (2016)
8. Akimova, E.N., Martyshko, P.S., Misilov, V.E., Tretyakov, A.I.: On solving the inverse structural magnetic problem for large grids on GPUs. In: AIP Conference Proceedings, vol. 1863, p. 050010 (2017). https://doi.org/10.1063/1.4992207
9. Akimova, E.N., Misilov, V.E.: A fast componentwise gradient method for solving structural inverse gravity problem. In: Proceedings of 15th International Multidisciplinary Scientific GeoConference SGEM 2015, vol. 3, no. 1, pp. 775–782 (2015)
10. Martyshko, P.S., Fedorova, N.V., Akimova, E.N., Gemaidinov, D.V.: Studying the structural features of the lithospheric magnetic and gravity fields with the use of parallel algorithms. Izv. Phys. Solid Earth **50**(4), 508–513 (2014). https://doi.org/10.1134/S1069351314040090
11. Akimova, E.N., Martyshko, P.S. and Misilov, V.E.: A fast parallel gradient algorithm for solving structural inverse gravity problem. In: AIP Conference Proceedings, vol. 1648, p. 850063 (2015). https://doi.org/10.1063/1.4913118
12. Akimova, E.N., Misilov, V.E., Tretyakov, A.I.: Optimized algorithms for solving structural inverse gravimetry and magnetometry problems on GPUs. In: Sokolinsky, L., Zymbler, M. (eds.) PCT 2017. CCIS, vol. 753, pp. 144–155. Springer, Cham (2017). https://doi.org/10.1007/978-3-319-67035-5_11

Parallel Multipoint Approximation Method for Large-Scale Optimization Problems

Victor P. Gergel[1]($\boxtimes$), Konstantin A. Barkalov[1]($\boxtimes$), Evgeny A. Kozinov[1],
and Vassili V. Toropov[1,2]

[1] Lobachevsky State University of Nizhny Novgorod, Nizhny Novgorod, Russia
{victor.gergel,konstantin.barkalov,evgeny.kozinov}@itmm.unn.ru
[2] Queen Mary University of London, London, UK
v.v.toropov@qmul.ac.uk

Abstract. The paper presents a new development in the Multipoint Approximation Method (MAM) that makes it capable of handling large-scale problems. The approach relies on approximations built in the space of design variables within the iterative trust-region-based framework of MAM. With the purpose of solving high dimensionality problems in a reasonable time, a parallel variant of the Multipoint Approximation Method (PMAM) has been developed. It is supposed that the values of the objective function and those of the constraints are computed using distributed memory (on several cluster nodes), whereas the optimization module runs on a single node using shared memory. Numerical experiments have been carried out on a benchmark example of structural optimization.

Keywords: Design optimization · Multidisciplinary optimization
Multipoint approximation method · Parallel computing

1 Introduction

In the present paper, the multipoint approximation method (MAM) [1–3] and its application to large-scale optimization problems are considered. In problems with a large (in the order of hundreds) number of design variables, MAM has proved to be efficient, e.g., in turbomachinery applications [4–6]. This method is an iterative optimization technique based on mid-range approximations built in trust regions. A trust region is a subdomain of the design space in which a set of design points, produced according to a small-scale design of experiments (DoE), is evaluated. These and a subset of previously evaluated design points are used to build metamodels of the objective and constraint functions that are considered to be valid within a current trust region. The trust region will then translate

This study was supported by the Russian Science Foundation, project No. 16-11-10150.

L. Sokolinsky and M. Zymbler (Eds.): PCT 2018, CCIS 910, pp. 174–185, 2018.
https://doi.org/10.1007/978-3-319-99673-8_13

and change size as optimization progresses. The trust region strategy has gone through several stages of development to account for the presence of numerical noise in the response function values [7,8] and occasional simulation failures [9]. The mid-range approximations used in the trust regions, as originally suggested in [1] for structural optimization problems, are intrinsically linear functions (i.e. nonlinear functions that can be reduced to a linear form by a simple transformation) for individual substructures, and an assembly of them for the whole structure. This was enhanced by the use of gradient-assisted metamodels [3], the use of simplified numerical models which is also termed the multi-fidelity approach [10], and the use of analytical models derived by genetic programming [11]. One of the recent developments [12] involves the use of approximation assemblies, i.e. a two stage approximation building process that is conceptually similar to the original one used in [1] but is free from the limitation that lower level approximations are linked to individual substructures.

The Moving Least-Squares Method (MLSM) was proposed in [13] for smoothing and interpolation of scattered data and was later used in the mesh-free form of the finite element method (FEM) [14]. As suggested in [15], it can be used as a technique for metamodeling and in multidisciplinary optimization (MDO) frameworks. The MLSM is a weighted least-squares method where the weights depend on the Euclidean distance from a sample point to where the surrogate model is to be evaluated. The weight value for a certain sample point decays as the distance increases. Describing the weight decay with a Gaussian function tends to be the most useful option, even though many others have been evaluated in [16]. As demonstrated in [17], the cross-validated MLSM can be used both for design variable screening and for surrogate modeling. In order to create an efficient MDO framework for problems with disparate discipline attributes, the optimization approach of MAM was extended in [18] to the use of local DOEs and MLS approximations built in different subspaces of the total design variable space corresponding to the individual disciplines. The subspaces are finally combined into the total design variable space in which the resulting MDO problem is solved.

This paper presents a Parallel Multipoint Approximation Method that makes it capable of handling problems with numbers of design variables in the order of thousands. The parallel variant of the algorithm (with the use of shared memory) has been developed with the purpose of minimizing the work time of the part of the algorithm related to constructing the approximation and solving the approximated problem, but not related to computing the values of the objective function and constraints. The processes of computing the values of the objective function and constraints (with the use of distributed memory) are supposed to be already parallelized.

2   The Multipoint Approximation Method

It would be useful to start with a brief description of MAM. A typical formulation of a constrained optimization problem that MAM works with is as follows:

$$\min_{a_i \leq x_i \leq b_i} \quad F_0(x)$$
$$\text{s.t. } F_j(x) \leq 1, \ j = 1, \ldots, M, \tag{1}$$

where x is a vector of design variables, a and b are the lower and upper bounds for the design variables, respectively, $F_0(x)$ is the objective function, and $F_j(x)$ are the constraints. The numbers of design variables and constraints are n and M, respectively. MAM attempts to solve this problem by using approximations of the objective function and constraints in a series of trust regions. The trust region strategy seeks to zoom in on the region where the constrained minimum is achieved. It aims at finding a trust region that is sufficiently small for the approximations to be of sufficiently good quality to improve the design and contains the point of the constrained minimum as an interior point. The main loop of the MAM is organized as follows.

Algorithm (MAM).

1. Initialization: choose a starting point x^0 and initial trust region $[a^0, b^0]$ such that $x^0 \in [a^0, b^0]$.
2. On the kth iteration, the current approximation to the constrained minimum is x^k, and the current trust region is $[a^k, b^k] \subset [a^0, b^0]$.
   (a) Design of Experiments (DoE). A set of points $x_k^i \in [a^k, b^k]$ is chosen to be used for building approximations. Responses are evaluated at the DoE points and approximations are built using the obtained values. Currently, the pool of approximation methods available in MAM consists of meta-model assemblies [12] and the moving least-squares metamodels [13–16]. Other metamodel types could be used as well.
   Denote the approximate objective function and constraints by $\widetilde{F}_0^k(x)$ and $\widetilde{F}_j^k(x)$, respectively.
   (b) The original optimization problem (1) is replaced by the following:

$$\min_{a_i^k \leq x_i^k \leq b_i^k} \quad \widetilde{F}_0^k(x)$$
$$\text{s.t. } \widetilde{F}_j^k(x) \leq 1, \ j = 1, \ldots, M. \tag{2}$$

   The approximate problem (2) is solved using Sequential Quadratic Programming (SQP). The solution of this problem determines the center of the next trust region.
   (c) The size of the next trust region is determined depending on the quality of approximations on the previous iteration, on the history of points x^k, and on the size of the current trust region [7].
   (d) The termination criterion is checked (it is a part of the trust region strategy and depends on the position of the point x^{k+1} in the current trust region, the size of the current trust region and the quality of approximations). If the termination criterion is satisfied, the algorithm proceeds to step 3. Otherwise, it returns to step 2.
3. Optimization terminates. The obtained approximation to the solution of problem (1) is x^{k+1}.

The approximations $\widetilde{F}_j^k(x)$, $j = 0, \ldots, M$, are selected in such a way that their evaluation is inexpensive as compared to the evaluation of the original response functions $F_j(x)$. For example, intrinsically linear functions were successfully used for a variety of design optimization problems in [3,19]. The approximations are determined by means of the weighted least squares:

$$\min \sum_{p=1}^{P} w_{pj} \left[F_j(x_p) - \widetilde{F}_j^k(x_p, a_j) \right]^2. \tag{3}$$

In (3), minimization is carried out with respect to the tuning parameters a_j; w_{pj} are the weight coefficients, and P is the number of sampling points in Design of Experiments (DoE), which must not be less than the number of parameters in the vector a_j.

The weight coefficients w_{pj} strongly influence the difference in the quality of the approximations in different regions of the design variable space. Since in realistic constrained optimization problems the optimum point usually belongs to the boundary of the feasible region, the approximation functions should be more accurate in such domain. Thus, the information at the points located near the boundary of the feasible region is to be treated with greater weights. In a similar manner, a larger weight can be allocated to a design with a better objective function (see [3,19]).

As optimization steps are carried out, a database with response function values becomes available. In order to achieve good quality approximations in the current trust region, an appropriate selection of DoE points must be made. In this work, DoE points in each trust region are generated randomly. Generally, points located far from the current trust region would not contribute to the improvement of the quality of the resulting approximations in the trust region. For this reason, only points located in a neighborhood of the current trust region are taken into account, as depicted in Fig. 1. A box in the space of design

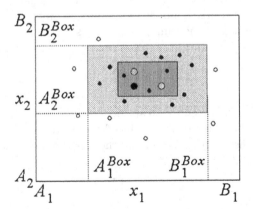

Fig. 1. Current trust region (smaller box) and its extension (larger box): points outside the larger box are not used for building the approximate functions

variables, which is approximately 1.5 to 1.8 times larger than the box representing the current trust region, was found by numerical experimentation to be a reasonable choice for the size of the neighborhood.

In this work, an approach is used that is based on the assembly of different approximate models $\{\varphi_l\}$ into one metamodel using the following form (note that the indices j and k are suppressed to simplify notation):

$$\widetilde{F}(x) = \sum_{l=1}^{NF} b_l \varphi_l(x), \tag{4}$$

where NF is the number of regressors in the model pool $\{\varphi_l\}$, and b_l are the corresponding regression coefficients. The procedure used consists of two subsequent steps. In the first step, the parameters a_l of individual functions (regressors) φ_l in (4) are determined by solving a weighted least-squares problem using a specified DoE of P points:

$$\min \sum_{p=1}^{P} w_p \left[F(x_p) - \varphi_l(x_p, a_l) \right]^2,$$

where minimization is carried out with respect to the tuning parameters a_l.

In the second step, based on the same DoE and keeping the obtained parameters a_l fixed, a vector b in (4) is estimated using the following formulation:

$$\min \sum_{p=1}^{P} w_p \left[F(x_p) - \widetilde{F}(x_p, b) \right]^2,$$

which leads to solving a linear system of NF equations with NF unknowns b_l, where NF is the number of regressors in the model pool $\{\varphi_l\}$.

The selection of the regressors $\{\varphi_l\}$ is based on the number of sampling points currently located in the trust region. In the mid-range approximation framework, inexpensive approximate models for objective and constraint functions are built using the minimum required number of sampling points. The simplest case is that of a linear function of the tuning parameters a:

$$\varphi(x) = a_0 + \sum_{i=1}^{N} a_i x_i.$$

This structure can be extended to an *intrinsically linear* function. Such functions are nonlinear but they can be reduced to linear ones by simple transformations. The most useful function among them is the multiplicative function

$$\varphi(x) = a_0 \prod_{i=1}^{N} x_i^{a_i}.$$

Intrinsically linear functions have been successfully used for a variety of design optimization problems. The advantage of these approximation functions is that a relatively small number $N+1$ (N is the number of design variables) of tuning parameters a_i is to be determined, and the corresponding least-squares problem is solved easily. This is the most important feature of such approximations as it allows applying them to large-scale optimization problems.

Other intrinsically linear functions may be considered in the model pool, e.g.,

$$\varphi(x) = a_0 + \sum_{i=1}^{N} a_i/x_i,$$

$$\varphi(x) = a_0 + \sum_{i=1}^{N} a_i x_i^2,$$

$$\varphi(x) = a_0 + \sum_{i=1}^{N} a_i/x_i^2,$$

$$\varphi(x) = a_0 + \sum_{i=1}^{N} a_i x_i^3,$$

$$\varphi(x) = a_0 + \sum_{i=1}^{N} a_i/x_i^3.$$

As more points are added to the database, the approximations may be switched to higher quality models, e.g., a rational model

$$\varphi(x) = \frac{a_1 + a_2 x_1 + a_3 x_2 + \ldots + a_{n+1} x_n}{1 + a_{n+2} x_1 + a_{n+3} x_2 + \ldots + a_{2n+1} x_n}. \tag{5}$$

The coefficients in (5) are determined using a least-squares approach which reduces to a nonlinear optimization problem with a constraint on the sign of the denominator (positive or negative). The latter is necessary in order to prevent the denominator from crossing the zero axis within a specified trust region. One may note that this formulation may yield an objective function with many local minima. Currently, this problem is resolved using optimization restarts from a specified number of initial guesses randomly generated in a trust region.

Tests results demonstrated that, although the above functions may describe the global behavior rather poorly, such approximations prove to be efficient in the mid-range approximation framework of MAM.

3   Parallel Multipoint Approximation Method

Let us consider possible methods of parallelization that could be applied to the problems considered.

First, one can parallelize the computation of the functions describing the optimized object. This way is an obvious as well as necessary one since, in industrial design optimization problems, the computation of even a single function value may take several hours. However, this method is a specific one for each particular problem. Here the computation issues are addressed at the level of the application software in which the industrial modeling is performed (e.g., Ansys, OpenFOAM, etc.).

Second, one can correct the algorithm with the purpose of parallel computing several values of the objective function and constraints at different points of the search domain. According to the MAM rules, in design of experiments, P sampling points are formed in the current trust region at each iteration. The function values at these points can be computed on different processors (nodes) in parallel. The above corresponds to the parallelization of Step 2a of the algorithm using distributed memory. The number of sampling points generated within each iteration may be set equal to $P = k \cdot NP$, where NP is the number of available processors (or nodes), and $k \geq 1$. In terms of time, the latter will be equivalent to NP function evaluations per step. This method has been implemented successfully in [20] and has demonstrated a good efficiency since, for numbers of design variables of the order of 100, time is mainly consumed by the function evaluations, whereas the work of the MAM itself (in the sequential regime) introduces a minor overhead.

However, when the number of variables becomes of the order of 1000, the work of the sequential part of MAM begins to affect the total problem solving time essentially (assuming that the time of computing the objective function values remains constant). Thus, for the problem considered in Sect. 4, the time of execution of a single iteration of the method increased by a factor of more than 4000 (from 1.5 up to 6000 s) when increasing the number of variables from 100 up to 1000. Another approach to the parallelization of the algorithm has therefore been applied within the framework of the present study: namely, the computational rules of MAM providing for the construction of the approximation, the solution of the approximated problem, and the choice of the next trust region (Steps 2b, 2c, and 2d of the algorithm) were parallelized.

In order to find the most time-consuming parts of the sequential program developed earlier, we applied the Intel VTune Amplifier XE. The analysis performed has shown that the most time-consuming operations in the execution of MAM are matrix multiplication, the solution of the SLAE when constructing the approximations, and the solution of the approximating problem by SQP. Matrix multiplication and solution of SLAE are standard operations implemented in many high-performance libraries. Here we used the corresponding parallel methods from the Intel MKL library. We parallelized the SQP method ourselves using OpenMP.

4   Numerical Example

The example considered in this study is a classical engineering optimization problem known as the scalable cantilevered beam [21]. The engineering object to be optimized is shown in Fig. 2 (taken from [21]).

The design variables are the widths b_i and heights h_i of the segments. The number of segments N can be chosen arbitrarily. The total length of the beam is

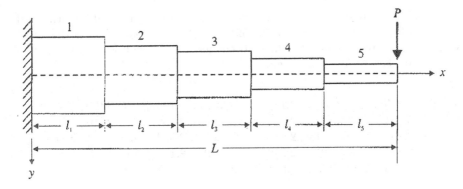

Fig. 2. The cantilevered beam

500 cm, the lengths of the segments are $l_i = 500/N$ cm. There are N geometric constraints (the aspect ratios of the blocks, i.e. heights divided by widths, should not exceed 20) and N constraints on the stress, calculated at the left end of each segment (stresses should not exceed $\bar{\sigma} = 14\,000\,\mathrm{N/cm^2}$). There is also a constraint on the displacement at the tip, which should not exceed 2.5 cm. The load is $P = 50\,000\,\mathrm{N}$; the Young's modulus is $E = 2 \cdot 10^7\,\mathrm{N/cm^2}$.

The deflection y_i at the right end of the ith segment is given by the following recursive formulas:

$$y_0 = y_0' = 0,$$

$$y_i' = \frac{P \cdot l_i}{E \cdot I_i}\left[L + \frac{l_i}{2} - \sum_{j=1}^{i} l_j\right] + y_{i-1}',$$

$$y_i = \frac{P \cdot l_i^2}{2E \cdot I_i}\left[L - \sum_{j=1}^{i} l_j + \frac{2l_i}{3}\right] + y_{i-1}'l_i + y_{i-1}.$$

The moment of inertia of the ith segment is $I_i = b_i h_i^3/12$, and the bending moment at its left end is $M_i = P[L + l_i - \sum_{j=1}^{i} l_j]$. The maximum bending stress in the ith segment is then given by the following formula:

$$\sigma_i = \frac{M_i h_i}{2I_i}$$

We should look for a design of smallest volume $V = \sum_{i=1}^{N} b_i h_i l_i$. The widths b_i vary from 1.0 to 10.0 cm and the heights h_i from 5.0 to 100.0 cm. The optimization problem is formulated as follows:

$$\min_{b,h} V(b, h)$$
$$\text{s.t. } 1.0 \leq b_i \leq 10.0,$$
$$5.0 \leq h_i \leq 100.0,$$
$$y_N \leq 2.5,$$
$$\sigma_i \leq \bar{\sigma} = 14000,$$
$$\frac{h_i}{b_i} \leq 20.$$

With $N = 50$ segments (corresponding to 100 design variables), the SQP solution of the problem is $V = 63704.598$ cm^3. The optimal values of the design variables are given below (the first 50 entries are the widths, and the last 50 entries are the heights of the segments):

$$b = [3.246, \ 3.224, \ 3.202, \ 3.179, \ 3.156, \ 3.133, \ 3.109, \ 3.085, \ 3.061, \ 3.036,$$
$$3.011, \ 2.985, \ 2.959, \ 2.933, \ 2.905, \ 2.878, \ 2.850, \ 2.821, \ 2.792, \ 2.762,$$
$$2.731, \ 2.700, \ 2.668, \ 2.635, \ 2.602, \ 2.567, \ 2.532, \ 2.495, \ 2.458, \ 2.419,$$
$$2.379, \ 2.338, \ 2.295, \ 2.250, \ 2.204, \ 2.156, \ 2.105, \ 2.052, \ 1.996, \ 1.936,$$
$$1.873, \ 1.805, \ 1.732, \ 1.651, \ 1.562, \ 1.462, \ 1.345, \ 1.196, \ 1.023, \ 1.000],$$

$$h = [64.919, \ 64.480, \ 64.033, \ 63.581, \ 63.122, \ 62.656, \ 62.184, \ 61.703, \ 61.216, \ 60.720$$
$$60.216, \ 59.703, \ 59.182, \ 58.651, \ 58.110, \ 57.558, \ 56.996, \ 56.423, \ 55.837, \ 55.239$$
$$54.628, \ 54.003, \ 53.363, \ 52.707, \ 52.035, \ 51.344, \ 50.635, \ 49.906, \ 49.155, \ 48.381$$
$$47.582, \ 46.754, \ 45.897, \ 45.008, \ 44.082, \ 43.116, \ 42.105, \ 41.041, \ 39.919, \ 38.729$$
$$37.462, \ 36.102, \ 34.630, \ 33.020, \ 31.240, \ 29.241, \ 26.910, \ 23.919, \ 20.465, \ 14.639].$$

By MAM, we obtained the solution $V = 63935.360$ cm^3, using 2201 function evaluations (as compared to almost 10 000 evaluations used by SQP). The number of points in the trust region used to build the approximations was 200. The optimal values of the design variables obtained by MAM are given below:

$$b = [3.238, \ 3.204, \ 3.202, \ 3.170, \ 3.176, \ 3.128, \ 3.108, \ 3.089, \ 3.057, \ 3.003$$
$$3.027, \ 2.986, \ 2.959, \ 2.952, \ 2.885, \ 2.855, \ 2.864, \ 2.847, \ 2.803, \ 2.762$$
$$2.737, \ 2.722, \ 2.630, \ 2.645, \ 2.590, \ 2.558, \ 2.547, \ 2.480, \ 2.693, \ 2.391$$
$$2.368, \ 2.310, \ 2.307, \ 2.227, \ 2.176, \ 2.149, \ 2.106, \ 2.016, \ 2.007, \ 1.925$$
$$1.864, \ 1.843, \ 1.758, \ 1.635, \ 1.582, \ 1.934, \ 1.332, \ 1.173, \ 1.026, \ 2.419],$$

$$h = [64.764, \ 64.083, \ 64.036, \ 63.392, \ 63.518, \ 62.566, \ 62.153, \ 61.772, \ 61.149, \ 60.054$$
$$60.534, \ 59.728, \ 59.182, \ 59.033, \ 57.695, \ 57.105, \ 57.274, \ 56.943, \ 56.057, \ 55.233$$
$$54.747, \ 54.433, \ 52.590, \ 52.902, \ 51.802, \ 51.150, \ 50.943, \ 49.597, \ 51.567, \ 47.811$$
$$47.373, \ 46.190, \ 46.140, \ 44.543, \ 43.525, \ 42.991, \ 42.119, \ 40.308, \ 40.141, \ 38.506$$
$$37.268, \ 36.839, \ 35.153, \ 32.702, \ 31.615, \ 28.304, \ 26.642, \ 23.447, \ 20.536, \ 9.417].$$

The solution obtained by MAM is very close to the reference solution obtained by SQP, except for the last design variable (the height of the last segment), which indicates that the problem is insensitive to this variable near the optimum, making it hard for metamodels to capture this dependence. Both SQP and MAM solutions are, however, feasible and differ only slightly in the value of the objective function.

Next, let us compare the work time of the sequential and parallel algorithms when solving large-scale problems. The dimensionality N of the problem being solved was varied from 100 up to 1000 design variables, which corresponds to a variation of the number of segments of the cantilevered beam from 50 to 500. The number of points in the trust region used to build the approximations was $2N$. Both algorithms were run on a single node of the cluster (the specifications of the node are listed below), the parallel algorithm employed all 16 processors cores available. Since the time of computing the objective function and the constraints in the test problem was negligible, these were computed on the same node.

Table 1. Time and speedup

N	T_{MAM}	T_{PMAM}	Speedup
100	15.4	11.9	1.3
200	167	89	1.9
400	2476	837	3.0
600	9678	2777	3.5
800	27826	6948	4.0
1000	67674	13771	4.9

Table 2. Function evaluations

N	I_{MAM}	F_{MAM}	F_{SQP}
100	10	2201	9901
200	10	4402	28143
400	11	9599	84211
600	10	13200	146046
800	10	17600	221079
1000	10	22000	305308

Table 1 reflects the work time (in seconds) of the sequential algorithm and that of the parallel one subject to the number of variables. The number of MAM iterations I_{MAM} as well as the number of function evaluations F_{MAM} required for solving the problem are presented in Table 2. For comparison purposes, the number of function value computations F_{SQP} that would be required to solve the initial (non-approximated) problem by the SQP method is also given. In all conducted experiments, the objective function values in the sequential and parallel versions of the algorithm were the same (up to computational errors) and negligibly differed from the solution obtained by SQP. The reduction of the number of the function evaluations required to solve the problem using MAM as compared to the use of SQP was demonstrated visibly.

The computational experiments were carried out on a high-performance cluster at Lobachevsky State University of Nizhny Novgorod. A cluster node includes two Intel Sandy Bridge E5-2660 2.2 GHz CPUs and 64 Gb RAM. Each CPU has 8 cores, i.e. a total of 16 physical cores were available at the node. MS Visual Studio 15 and Intel Fortran Compiler were used to implement the algorithm.

5   Conclusions

Recent developments in the Multipoint Approximation Method (MAM) made it capable of solving large-scale industrial optimization problems. The fact that MAM solves the initial problem by using approximations of the objective function and constraints is the primary distinctive feature of the method. Within the framework of the present study, we developed a parallel version of MAM oriented to the reduction of the work time of the optimization algorithm (assuming that the computation of the values of the objective function and constraints has already been parallelized). The experiments performed have demonstrated an acceptable speedup when solving large-scale problems employing 16 cores on a single cluster node. The performance was demonstrated on a benchmark example of structural optimization known as the scalable cantilevered beam.

References

1. Toropov, V.: Simulation approach to structural optimization. Struct. Optim. **1**, 37–46 (1989)
2. Toropov, V.: Multipoint approximation method in optimization problems with expensive function values. In: Sydow, A. (ed.) Proceedings of the 4th International Symposium on Systems Analysis and Simulation, pp. 207–212. Elsevier (1992)
3. Toropov, V., Filatov, A., Polynkin, A.: Multiparameter structural optimization using FEM and multipoint explicit approximations. Struct. Optim. **6**, 7–14 (1993)
4. Shahpar, S., Polynkin, A., Toropov, V.: Large scale optimization of transonic axial compressor rotor blades. In: 49th AIAA/ASME/ASCE/AHS/ASC Structures, Structural Dynamics, and Materials Conference, article no. 2008–2056. AIAA (2008)
5. Polynkin, A., Toropov, V., Shahpar, S.: Design optimization of aircraft engine components. In: Proceedings of 7th ASMO UK/ISSMO Conference on Engineering Design Optimization, Process and Product Improvement (2008)
6. Polynkin, A., Toropov, V., Shahpar, S.: Multidisciplinary optimization of turbomachinery based on metamodel built by genetic programming. In: 13th AIAA/ISSMO Multidisciplinary Analysis and Optimization Conference (2010)
7. van Keulen, F., Toropov, V., Markine, V.: Recent refinements in the multipoint approximation method in conjunction with adaptive mesh refinement. In: McCarthy, J.M. (ed.) Proceedings of ASME Design Engineering Technical Conferences and Computers in Engineering Conference, pp. 1–12. ASME, Irvine (1996)
8. Toropov, V., van Keulen, F., Markine, V., de Boer, H.: Refinements in the multipoint approximation method to reduce the effects of noisy responses. In: 6th AIAA/NASA/ISSMO Symposium Multidisciplinary Analysis and Optimization, pp. 941–951. AIAA (1996)
9. Toropov, V., Markine, V., Holden, C.: Use of mid-range approximations for optimization problems with functions of domain-dependent calculability. In: 3rd ISSMO/UBCAD/UB/AIAA World Congress of Structural and Multidisciplinary Optimization (1999)
10. Toropov, V., Markine, V.: The use of simplified numerical models as mid-range approximations. In: 6th AIAA/NASA/ISSMO Symposium on Multidisciplinary Analysis and Optimization, pp. 952–958 (1996)

11. Toropov, V., Alvarez, L.: Creation of multipoint approximations using genetic programming. In: Parmee, I.C. (ed.) Adaptive Computing in Design and Manufacture, 3rd International Conference, pp. 21–24. PEDC, Dartington (1998)

12. Polynkin, A., Toropov, V.: Mid-range metamodel assembly building based on linear regression for large scale optimization problems. Struct. Multidiscip. Optim. **45**(4), 515–527 (2012). https://doi.org/10.1007/s00158-011-0692-1

13. Lancaster, P., Salkauskas, K.: Surfaces generated by moving least squares methods. Math. Comput. **87**, 141–158 (1981)

14. Liszka, T.: An interpolation method for an irregular net of nodes. Int. J. Num. Meth. **20**, 1599–1612 (1984)

15. Choi, K., Youn, B., Yang, R.-J.: Moving least squares method for reliability-based design optimization. In: 4th World Congress of Structural and Multidisciplinary Optimization, Dalian, China (2001)

16. Toropov, V., Schramm, U., Sahai, A., Jones, R., Zeguer, T.: Design optimization and stochastic analysis based on the moving least squares method. In: Herskovits, J., Mazorche, S., Canelas, A. (eds.) 6th World Congress of Structural and Multidisciplinary Optimization, article no. 9412 (2005)

17. Polynkin, A. Toropov, V.: Recognition of design variable inter-dependencies using cross-validated moving least-squares method. In: Proceedings of the 51st AIAA/ASME/ASCE/AHS/ASC Structures, Structural Dynamics, and Materials Conference, art. no. 2010-2985. AIAA (2010)

18. Ollar, J., Toropov, V., Jones. R.: Mid-range approximations in sub-spaces for MDO problems with disparate discipline attributes. In: 15th AIAA/ISSMO Multidisciplinary Analysis and Optimization Conference, article no. 2014-2437. AIAA (2014). https://doi.org/10.2514/6.2014-2437

19. van Keulen, F., Toropov, V.: New developments in structural optimization using adaptive mesh refinement and multi-point approximations. Eng. Optim. **29**, 217–234 (1997). https://doi.org/10.1080/03052159708940994

20. Polynkin, A., Toropov, V., Shahpar, S.: Adaptive and parallel capabilities in the multipoint approximation method. In: 12th AIAA/ISSMO Multidisciplinary Analysis and Optimization Conference, art. no. 2008-5803. AIAA (2008). https://www.doi.org/10.2514/6.2008-5803

21. Vanderplaats, G.: Multidiscipline Design Optimization. Vanderplaats Research & Development Inc., Colorado Springs (2001)

High-Performance Computation of Initial Boundary Value Problems

Valery Il'in[(✉)]

Institute of Computational Mathematics and Mathematical Geophysics,
Novosibirsk State University, Novosibirsk, Russia
`ilin@sscc.ru`
`https://icmmg.nsc.ru/ru/content/employees/ilin-valeriy-pavlovich`

Abstract. This paper considers the efficient methods and high- performance parallel technologies for the numerical solution of the multidimensional initial boundary value problems, with a complicated geometry of a computational domain and contrast properties of a material on the heterogeneous multi-processor systems with distributed and hierarchical shared memory. The approximations with respect to time and space are carried out by implicit schemes on the quasi-structured grids. At each time step, the iterative algorithms are used for solving the systems of linear or nonlinear equations that, in general, are non-symmetric with a special choice of the initial guess. The scalable parallelism is provided by two-level iterative domain decomposition methods, with parameterized intersection of subdomains in the Krylov subspaces, which are accelerated by means of a coarse grid correction and polynomial or other types of preconditioning. A comparative analysis of the performance and speed up of the computational processes is presented, based on a simple model of parallel computing and data structures.

Keywords: Nonstationary boundary value problems
High-order approximations · Stability · Initial guess
Iterative processes · Domain decomposition · Scalable parallelism

1 Introduction

In this paper, we consider various numerical approaches to solving multidimensional initial boundary value problems (IBVPs) for nonstationary partial differential equations (PDEs) in complicated computational domains with real data, and offer a comparative analysis of their performance on modern heterogeneous multi-processor systems (MPS) with distributed and hierarchical shared memory. In general, we will assess the efficiency of numerical solutions of a class of mathematical problems by the volume of computational resources required to provide the accuracy needed on a particular type of MPS. Of course, such a

The work is supported by the Russian Science Foundation (grant 14-11-00485 P) and the Russian Foundation for Basic Research (grant 16-29-15122 ofi-m).

L. Sokolinsky and M. Zymbler (Eds.): PCT 2018, CCIS 910, pp. 186–199, 2018.
https://doi.org/10.1007/978-3-319-99673-8_14

statement is not quite clear, and we should refine many details in this important concept. A simple way to do this consists in measuring run time and using these measurements as a performance criterion. Other tools could be the estimation of computing time and communication time based on a certain model for the implementation of the problem on MPS. In what follows, we will use the second approach and consider simple representations for both the arithmetical execution time T_a and the communication time T_c. In total, the performance of the numerical solution of the problem will be defined by the run time $T_t = T_a + T_c$. We suppose here that the arithmetical units do not work during data transfer, although the whole picture can be more complicated.

The performance is characterized by two main aspects: mathematical efficiency of numerical methods and computational technologies for software implementation on a particular hardware architecture. The algorithmic issues depend on two main mathematical stages: discrete approximation of the original continuous problem and numerical solution of the resulting algebraic task. It is important that we do not examine model problems but problems with real data: multi-dimensional boundary value problems in computational domains with a complicated geometry, multi-connected and multi-scaled (in general) piece-wise smooth boundaries and contrast properties of a material, which provide singularities of the solution to be sought. It means that in order to ensure a high numerical resolution and accuracy of the computational model, we must use fine grids with a very small time step τ and a spatial step h. So we have, in principle, a "super task" with a very large number of degrees of freedom (d. o. f.) or a high dimension of the corresponding discrete problem. In general, the original problem can be nonlinear and multi-disciplinary or multi-physical, i.e. it is described either by a system of PDEs or by the corresponding variational relations for unknown vector functions. Also, the mathematical statement may not be a direct one with all the coefficients of the equations given and with initial and boundary conditions, it may be instead an inverse problem that includes variable parameters to be found from the condition of minimization of some given objective functional of the unknown solution. For simplicity, however, we will mainly consider direct IBVPs for a single linear scalar equation. A review of the corresponding models can be found in [11] (see also the literature cited therein). We also do not consider in detail other computational steps of the mathematical modeling (grid generation, post-processing, visualization of the results, etc.) since they are of a more general type and are almost defined by the problem specifications.

The approximation approaches are divided into temporal and spatial discretizations. If we carry out the spatial approximation at first by the finite volume method, the finite element method or any other method [2], then we will obtain a system of ordinary differential equations (ODEs). There are various explicit and implicit, multi-stage and/or multi-step algorithms of different orders [3] that may be applied to solve such a system. It is important to remark that modern computational trends give preference to methods of high order of accuracy since they make it possible to decrease the amount of data communication, which is not only a slow operation but an energy consuming process.

If we use schemes that are implicit with respect to time, thereby providing a stable procedure for numerical integration, it will be necessary to solve at each step a system of linear algebraic equations (SLAEs) of special type, with large sparse matrices. This is the most expensive computational stage as it requires a large number of arithmetical operations and a big amount of memory, and both grow nonlinearly when the number of d.o.f. increases [4]. In this case, the main tool to ensure a high performance is the scalable parallelization of domain decomposition methods (DDM), which belongs to a special field of computational algebra (see, for example, [5–7]). A detailed review of parallelization approaches for nonstationary problems is presented in [8,9]. In what follows, we will consider direct IBVPs only, whereas the ideal of engineering problems consists in solving inverse problems, which involves computing optimized parameters of the mathematical model under the condition of constrained optimization of a given objective functional. However, this is a topic that requires a special research.

The paper is structured as follows. In Sect. 2, the example of the heat transfer equation is considered regarding various aspects of temporal and spatial approximations. Section 3 deals with geometrical and algebraic issues of DDM as applied to nonstationary problems. In the last Section, we discuss an application of the given analysis for the parallel solution of practical problems.

2   Discretization Issues of Nonstationary Problems

Let us consider the initial boundary value problem (IBVP)

$$\frac{\partial u}{\partial t} + L(u) = f(\mathbf{x}, t), \quad \mathbf{x} \in \Omega \subset \mathcal{R}^d, \quad d \geq 2,$$
$$\bar{\Omega} = \Omega \cup \Gamma, \quad 0 < t \leq T_e < \infty, \quad u\big|_{t=0} = u^0(\mathbf{x}), \tag{1}$$
$$l(u)\big|_\Gamma = g(\mathbf{x}, t), \quad \mathbf{x} = (x_1, \ldots, x_d),$$

where t and $\mathbf{x}$ are, respectively, temporal and spatial variables; $u^0(\mathbf{x})$ is a given initial guess; L is some differential operator, possibly, a nonlinear one and, in general, a matrix operator. In this case, the unknown $u = (u_1, \ldots, u_{N_u})^T$ is a vector function. We call task (1) a multi-disciplinary or multi-physics problem. Here $\bar{\Omega}$ denotes a bounded d-dimensional computational domain with boundary $\Gamma = \bigcup_{k=1}^{N_\Gamma} \Gamma_k$; l is a boundary-condition operator, which can be of various types l_i (Dirichlet, Neumann or Robin) at the corresponding boundary segments Γ_i; f and g are functions that may depend on the unknown solution. We suppose that IBVP (1) describes a practical problem with real data. This means, for example, that the computational domain $\bar{\Omega}$ may have a complicated geometry, possibly, with multi-connected piecewise smooth curvilinear boundary surfaces Γ_k. As an illustration, the following linear scalar differential operator of the second order is considered in (1):

$$Lu = -\sum_{i,j=1}^{n} \frac{\partial}{\partial x_i}\left(a_{i,j}(\mathbf{x})\frac{\partial u}{\partial x_j}\right) + \sum_{i=1}^{n} b_i \frac{\partial u}{\partial x_i} + cu = f(\mathbf{x}). \tag{2}$$

The corresponding boundary conditions can be written down as

$$\alpha_k u + \beta_k \sum_{i,j=1}^{d} a_{i,j} \frac{\partial u}{\partial x_j} \cos(\mathbf{n}, x_j) = g_k, \quad |\alpha_k| + |\beta_k| \neq 0, \quad \mathbf{x} \in \Gamma_k, \tag{3}$$

where $\mathbf{n}$ denotes the outward unit normal to Γ_k.

Note that if the original system of PDEs is complex and has temporal derivatives of high order, it can always be transformed into a first order real system by including additional unknown functions. Also, formulas (1) can describe an inverse problem if it contains variable parameters $p = (p_1, \ldots, p_{N_p})^T$, which should be optimized by means of the minimization of a prescribed objective functional. For simplicity, the original IBVP is written in the classical differential form, and it can be re-described in a variational style. It is supposed that the input data ensures the smoothness of the numerical methods in all cases. One more remark: in general, some boundary segments Γ_k can move, but we will primarily consider the boundary Γ fixed.

The approximation of the original problem (1) can be made in two steps. In the first step, we generate a spatial grid Ω^h, which, in the three-dimensional case ($d = 3$), for example, a set of nodes (vertices), edges, faces (possibly, curvilinear), and finite elements or volumes. After applying the spatial approximation using the finite volume method, the finite element method, the discontinuous Galerkin method or other approaches, we obtain a system of N ordinary differential equations:

$$B\dot{u}^h + Au^h = f^h, \tag{4}$$
$$\dot{u}^h, u^h, f^h \in \mathcal{R}^N; \quad B, A \in \mathcal{R}^{N,N},$$

where $\dot{u}$ denotes the time derivative of u, and the components of the vector $f^h = \{f_l\}$ and of the matrices $B = \{b_{l,l}\}$ and $A = \{a_{l,l}\}$ may, in general, depend on the unknown solution.

In a simple case, the unknown vector $u^h = \{u^h_l\}$ consists of approximate nodal values of the original solution $(u)^h = \{u(\mathbf{x}_l)\}$ but, basically, it can include, for instance, other functionals, and some derivatives of u at different points. The vector $(u)^h$ of the discretized unknown solution satisfies the equation

$$B(\dot{u})^h + A(u)^h = (f)^h + \psi^h, \quad \psi = O(h^\gamma), \tag{5}$$

where ψ^h is the spatial approximation (truncation) error of Eq. (4), h is the maximal distance between neighboring grid nodes, and $\gamma > 0$ is the order of the approximation. The matrix A in (4) can be defined as

$$(Au^h)_l \equiv a_{l,l}u_l + \sum_{l' \in \omega_l} a_{l,l'}u_{l'} = f_l, \quad l \in \Omega^h, \tag{6}$$

where Ω^h can be considered to be a set of indices that determine the number $N = O(h^{-1})$ of all unknowns, and ω_l denotes the stencil of the lth node, i.e. the set of neighboring nodes. In other words, ω_l is the union of the column numbers of the nonzero elements in the lth row of the matrix A (the number of such values will be denoted as N_l). The total set made up by all ω_l, $l = 1, \ldots, N$, determines the portrait of the sparse matrix A ($N_l \ll N$). Note that N_l does not depend on the matrix dimension N, which can be estimated as $N \approx 10^7 \div 10^{10}$ for a large-size real problem. Moreover, for $d = 3$, we have $N_l \approx 10 \div 30$ for the first or the second order schemes, whereas $N_l > 100$ for the fourth to sixth orders of accuracy.

To solve ODEs (4), it is possible to apply various multi-stage and/or multi-step numerical integrators of different orders of accuracy with respect to the time step τ_n. For simplicity, we consider the two-step weighted scheme

$$B\frac{u^{n+1} - u^n}{\tau_n} + \theta(Au^{n+1} - f^{n+1}) = (1 - \theta)(f^n - Au^n),$$

$$\theta \in [0,1], \quad n = 0, 1, \ldots, \tag{7}$$

where n is a time-step number; $\theta = 0$ corresponds to the explicit Euler method, otherwise, we have an implicit algorithm. If $\theta = 1/2$, formula (7) corresponds to the Crank–Nicolson scheme, which has the second order approximation error $\psi^\tau = O(\tau^2)$, $\tau = \max_n\{\tau_n\}$; besides, $\psi^\tau = O(\tau)$ for $\theta \neq 1/2$. Here and in what follows, we omit the index "h" for the sake of brevity. If we denote by $(u)^n$ a vector whose components are the values of the exact solution $u(t_n, \mathbf{x}_l)$, and substitute it for u^n in (7), then we have

$$B\frac{(u)^{n+1} - (u)^n}{\tau_n} + \theta[A(u)^{n+1} - (f)^{n+1}] = (1 - \theta)[(f)^n - A(u)^n] + \psi^n, \tag{8}$$

where $\psi^n = \psi^\tau + \psi^h$ is the total, i.e. temporal and spatial, approximation error of the numerical scheme.

If relations (7) are nonlinear, we should use quasi-linearization for each n, i.e. apply the iterative process and solve SLAEs at each "nonlinear" step.

In the implicit scheme with $\theta \neq 0$, we have to solve a large algebraic system by some iterative approach, even for the original linear IBVP, since direct (noniterative) algorithms are too expensive in our case (matrices $B + \tau_n\theta A$ are supposed to be nonsingular). Finally, from (7), we do not calculate u^{n+1} but some approximate value $\tilde{u}^{n+1}$, which produces the residual vector

$$r^n = (1 - \theta)(\tilde{f}^n - A\tilde{u}^n) - B\frac{\tilde{u}^{n+1} - \tilde{u}^n}{\tau_n} + \theta(A\tilde{u}^{n+1} - \tilde{f}^{n+1}). \tag{9}$$

Now let us determine the total vector of the original solution, $z^{n+1} = (u)^{n+1} - \tilde{u}^{n+1}$. It follows from (8) and (9) that the vectors z^{n+1}, r^n and ψ^n are connected by a relation that, for the reduced original problem (the elements of the matrices A and B, as well as those of the vectors f^n are supposed to be independent of u and t), can be written down as

$$C_1 z^{n+1} = C_2 z^n + \tau_n(\psi^n - r^n),$$

$$C_1 = B + \theta A, \quad C_2 = B - (1 - \theta)A. \tag{10}$$

If

$$\|\tau_n(\psi^n - r^n)\| \le \tau\|\psi\| \tag{11}$$

for some vector norm, then we obtain from (10) the following estimate:

$$\|z^{n+1}\| \le \rho\|z^n\| + \tau\rho_1\|\psi\|,$$
$$\rho = \|C_1^{-1}C_2\|, \quad \rho = \|C_1^{-1}\|. \tag{12}$$

It follows from the considerations above that if the iterative residual r^n at each time step has the same order of accuracy as the approximation error ψ^n, then the total solution error does not change the order of accuracy. One important issue in solving a nonstationary problem consists in choosing the initial guess for the iterative solution of SLAEs at each time step. It is natural that the u^n values would be a good approximation to u^{n+1} to reduce the number of iterations, provided that the time step τ_n is sufficiently small. Another simple approach is based on the linear extrapolation with respect to time:

$$u^{n+1} = u^n + (u^n - u^{n-1})\tau_n/\tau_{n-1} + O(\tau^2). \tag{13}$$

In this case, we need to save the numerical solution for one additional time step. One of the popular methods for solving ODEs is based on the application of predictor-corrector schemes. For example, if we use in (7) the Crank–Nicolson scheme, for which $\theta = 1/2$ and $\psi^\tau = O(\tau^2)$, or any other implicit method, this involves including a preliminary predictor stage for computing an approximate value of u^{n+1} by the simple explicit formula

$$B(\hat{u}^{n+1} - u^n) = \tau_n(f^n - Au^n) \equiv \tau_n r^n, \tag{14}$$

where B is a diagonal or another easily invertible matrix, and $\hat{u}^{n+1}$ is considered to be a predicted value of u^{n+1}. It can be interpreted as a zero iteration, $u^{n+1,0} = \hat{u}^{n+1}$, and corrected by m iterations of the form

$$B(u^{n+1,s} - u^n) = \tau_n[\theta(f^{n+1} - Au^{n+1,s-1}) + (1 - \theta)(f^n - Au^n)],$$
$$s = 1, \ldots, m. \tag{15}$$

This approach is called PC^m and in practice provides an acceptable small residual

$$r^{n+1,s} = \tau_n[\theta(f^{n+1} - Au^{n+1,s-1}) + (1 - \theta)r^n] - B(u^{n+1,s} - u^n)$$

in a few iterations.

An improved idea to choose the initial guess can be proposed based on the least-squares method (LSM; see [10]). Let us save several previous time-step solutions $u^{n-1}, \ldots, u^{n-q}$, and compute the value $u^{n+1,0}$ by means of the linear combination

$$u^{n+1,0} = u^n + c_1 v_1 + \ldots + c_q v_q = u^n + Vc,$$
$$v_l = u^n - u^{n-l}, \quad l = 1, \ldots, q, \tag{16}$$
$$c = (c_1, \ldots, c_q)^T \in \mathcal{R}^q, \quad V = (v_1, \ldots, v_q) \in \mathcal{R}^{N,q}.$$

The system of Eq. (7) can be rewritten as

$$Cu^{n+1} \equiv (\tau_n^{-1}B + \theta A)u^{n+1} = g^{n+1},$$
$$g^{n+1} = [\tau_n^{-1}B + (1 - \theta)A]u^n + \theta f^{n+1} + (1 - \theta)f^n. \tag{17}$$

So it follows from relation (16) that the initial residual $r^{n+1,0} = g^{n+1} - Cu^{n+1,0}$ of system (17) satisfies the equality

$$r^{n+1,0} = r^n - CVc. \tag{18}$$

Formally, here we can set $r^{n+1,0} = 0$ and obtain overdetermined SLAE for the vector c:

$$Wc \equiv CVc = \tau^n, \quad W \in \mathcal{R}^{N,q}. \tag{19}$$

The generalized normal (with a minimal residual) solution of this system can be computed by the SVD (Singular Value Decomposition) algorithm or by the least-squares method (LSM), which gives the same result in exact arithmetics. The LSM gives the "small" symmetric system

$$Gc \equiv W^T Wc = W^T \tau^n, \quad G = V^T C^T CV \in \mathcal{R}^{q,q}, \tag{20}$$

which is nonsingular if W is a full-rank matrix. It is easy to verify that Eq. (20) implies the orthogonality property of the residual:

$$W^T r^{n+1,0} = 0. \tag{21}$$

Note that, instead of the LSM approach (20), (21), it is possible to apply the so-called deflation principle [11], which uses the following orthogonality property instead of (21):

$$V^T r^{n+1,0} = 0. \tag{22}$$

In this case, we have to solve SLAE

$$Hc \equiv V^T CVc = V^T c, \quad H \in \mathcal{R}^{q,q}, \tag{23}$$

to determine the vector c. If this vector is computed from system (20) or (23), then the initial guess $u^{n+1,0}$ for SLAEs (17) is determined from (16). For solving system (17) at each time step, it is natural to apply some preconditioned iterative method in Krylov subspaces. The stopping criterion of such iterations is

$$\|r^{n+1,m}\| = \|g^{n+1} - Cu^{n+1,m}\| \le \varepsilon \|g^{n+1}\| \tag{24}$$

for some given tolerance $\varepsilon \ll 1$. If condition (24) is satisfied, we set $u^{n+1} = u^{n+1,m}$ and go to the next time step.

3   Geometrical and Algebraic Issues of Algorithms

The general scheme of solution of nonsteady IBVPs can be described as having two main parts. The first one consists in generating an algebraic system at each time step. Usually, this stage is parallelized easily enough, with a linear speedup when the number of computer units grows. The more complicated stage includes solving the algebraic system of equations, linear or nonlinear (SLAEs or SNLAEs); such tasks require a large amount of computational resources (memory and number of arithmetic operations) as the number of d. o. f. grows.

If we have SNLAEs at each time step, the solution methods involve a two-level iterative process. At first, some type of quasi-linearization is applied, and at each "nonlinear" iteration (Newton or Jacobi type, for example), we need to solve SLAEs, usually with a large sparse ill-conditioned matrix. This second stage will be the main issue in our considerations in what follows.

The main tool to achieve scalable parallelism on modern MPS is based on a domain decomposition method that can be interpreted in an algebraic or geometrical framework. Also, domain decomposition methods can be considered at both the continuous and the discrete levels. We use the second approach and suppose that the original computational domain Ω has already been discretized into a grid computational domain Ω^h. So, in what follows, the DDM is implemented only in grid computational domains, and the upper index "h" will be omitted for brevity.

Let us decompose Ω into P subdomains (with or without overlap):

$$\Omega = \bigcup_{q=1}^{P} \Omega_q, \quad \bar{\Omega}_q = \Omega_q \cup \Gamma_q, \quad \Gamma_q = \bigcup_{q' \in \omega_q} \Gamma_{q,q'}, \quad \Gamma_{q,q'} = \Gamma_q \cap \bar{\Omega}_{q'}, \quad q' \neq q. \quad (25)$$

Here Γ_q is the boundary of Ω_q, which is composed of the segments $\Gamma_{q,q'}, q' \in \omega_q$, and $\omega_q = \{q_1, \ldots, q_{M_q}\}$ is a set of M_q contacting or conjugate subdomains. Formally, we can also denote by $\Omega_0 = R^d \setminus \Omega$ the external subdomain:

$$\bar{\Omega}_0 = \Omega_0 \cup \Gamma, \quad \Gamma_{q,0} = \Gamma_q \cap \bar{\Omega}_0 = \Gamma_q \cap \Gamma, \quad \Gamma_q = \Gamma_q^i \cup \Gamma_{q,0}, \quad (26)$$

where $\Gamma_q^i = \bigcup_{q' \neq 0} \Gamma_{q,q'}$ and $\Gamma_{q,0} = \Gamma_q^e$ stand for the internal and external parts of the boundary of Ω_q. We also define the overlap $\Delta_{q,q'} = \Omega_q \cap \Omega_{q'}$ of neighboring subdomains. If $\Gamma_{q,q'} = \Gamma_{q',q}$ and $\Delta_{q,q'} = \emptyset$, then the overlap of Ω_q and $\Omega_{q'}$ is empty. In particular, we suppose in (25) that each of the P subdomains has no intersection with Ω_0 ($\Omega_q \cap \Omega_0 = \emptyset$).

The idea of the DDM involves the definition of the sets of IBVPs that should be equivalent to the original problem (1) in all subdomains:

$$\frac{\partial u_q}{\partial t} + L u_q(\mathbf{x}) = f_q, \quad \mathbf{x} \in \Omega_q, \quad l_{q,q'}(u_q)\big|_{\Gamma_{q,q'}} = g_{q,q'} \equiv l_{q',q}(u_{q'})\big|_{\Gamma_{q',q}}, \quad (27)$$

$$q' \in \omega_q, \quad l_{q,0} u_q\big|_{\Gamma_{q,0}} = g_{q,0}, \quad q = 1, \ldots, P.$$

Interface conditions in the form of Robin boundary conditions (instead of (3), for simplicity) are imposed in each segment of the internal boundaries of the subdomains, with the operators $l_{q,q'}$ from (27):

$$\alpha_q u_q + \beta_q \frac{\partial u_q}{\partial \mathbf{n}_q}\bigg|_{\Gamma_{q,q'}} = \alpha_{q'} u_{q'} + \beta_{q'} \frac{\partial u_{q'}}{\partial \mathbf{n}_{q'}}\bigg|_{\Gamma_{q',q}},$$

$$|\alpha_q| + |\beta_q| > 0, \quad \alpha_q \cdot \beta_q \geq 0. \tag{28}$$

Here $\alpha_{q'} = \alpha_q$ and $\beta_{q'} = \beta_q$; $\mathbf{n}_q$ is the outer normal to the boundary segment $\Gamma_{q,q'}$ of the subdomain Ω_q. Strictly speaking, two pairs of different coefficients, $\alpha_q^{(1)}, \beta_q^{(1)}$ and $\alpha_q^{(2)}, \beta_q^{(2)}$, should be given for conditions of type (28) on each piece $\Gamma_{q,q'}, q' \neq 0$, of the internal boundary. For example, $\alpha_q^{(1)} = 1, \beta_q^{(1)} = 0$ and $\alpha_q^{(2)} = 0, \beta_q^{(2)} = 1$ formally correspond, respectively, to the continuity of the solution sought and its normal derivative. The additive Schwarz algorithm in DDM is based on an iterative process in which the BVPs in each subdomain Ω_q are solved simultaneously, and the right-hand sides of the boundary conditions in (27) and (28) are taken from the previous iteration.

We implement the domain decomposition in two steps. At the first one, we define subdomains Ω_q without overlap, i.e. contacting grid subdomains have no common nodes, and each node belongs to only one subdomain. Then we define the grid boundary $\Gamma_q = \Gamma_q^0$ of Ω_q, as well as the extensions of $\bar{\Omega}_q^t = \Omega_q^t \cup \Gamma_q^t$, $\Omega_q^0 = \Omega_q, t = 0, \ldots, \Delta$, layer by layer:

$$\Gamma_q \equiv \Gamma_q^0 = \left\{ l' \in \hat{\omega}_l, \ l \in \Omega_q, \ l' \notin \Omega_q, \ \Omega_q^1 = \bar{\Omega}_q^0 = \Omega_q \cup \Gamma_q^0 \right\},$$

$$\Gamma_q^t = \left\{ l' \in \hat{\omega}_l, \ l \in \Omega_q^{t-1}, \ l' \in \Omega_q^{t-1}, \ \Omega_q^t = \bar{\Omega}_q^{t-1} = \Omega_q^{t-1} \cup \Gamma_q^{t-1} \right\}. \tag{29}$$

Here Δ stands for the parameter of extension or overlap.

At each time step, the algebraic interpretation of the DDM, after the approximations of BVPs (27) and (28), is described by the block version of SLAEs (17),

$$C_{q,q} u_q + \sum_{r \in \hat{\omega}_q} C_{q,r} u_r = g_q, \quad q = 1, \ldots, P, \tag{30}$$

where indices "$n + 1$" have been omitted for brevity; $C_{q,q}$ and $u_q, f_q \in \mathcal{R}^{N_q^\Delta}$ are a block diagonal matrix and subvectors with components belonging to the corresponding subdomain Ω_q^Δ; N_q^Δ is the number of nodes in Ω_q^Δ.

The implementation of the interface conditions between adjacent subdomains can be described as follows. Let the lth node be a near-boundary one in the subdomain Ω_q. Then we write down the corresponding equation in the form

$$(D_{q,q} u)_l \equiv \left(c_{l,m} + \theta_l \sum_{m \notin \omega_q} c_{l,m} \right) u_l + \sum_{m \in \omega_q} c_{l,m} u_m =$$

$$= g_l + \sum_{m \notin \omega_q} c_{l,m} (\theta_l u_l - u_m). \tag{31}$$

Here θ_l is some parameter that corresponds to different types of boundary conditions at the boundary Γ_q, namely $\theta_l = 0$ corresponds to the Dirichlet condition,

$\theta_l = 1$ corresponds to the Neumann condition, and $\theta_l \in (0,1)$ corresponds to the Robin boundary condition.

If we denote $D = \text{block-diag}\{D_{l,l}\}$, then a simple variant of DDM is described as the Schwarz (or block Schwarz–Jacobi) iterative method

$$Du^{s+1} = (D - C)u^s + g, \quad s = 0, 1, \dots . \tag{32}$$

Improved versions of this approach are given by preconditioned algorithms in Krylov subspaces. Firstly, let us consider the advanced choice of the preconditioning matrices.

In the case of an overlapping domain decomposition, the additive Schwarz iterative algorithm is defined by the corresponding preconditioning matrix B_{AS}, which can be described as follows (see [7]). For the subdomain Ω_q^Δ with overlap parameter Δ, we define a prolongation matrix $R_{q,\Delta}^T \in \mathcal{R}^{N,N_q^\Delta}$ that extends the vectors $u_q = \{u_l, \; l \in \Omega_q^\Delta\} \in \mathcal{R}^{N_q^\Delta}$ to $\mathcal{R}^N$ according to the relations

$$(R_{q,\Delta}^T u_q)_l = \begin{cases} (u_q)_l & \text{if } l \in \Omega_q^\Delta, \\ 0 & \text{otherwise.} \end{cases}$$

The transpose of this matrix defines a restriction operator that restricts vectors in $\mathcal{R}^N$ to the subdomain Ω_q^Δ. The diagonal block of the preconditioning matrix B_{AS}, which represents the restriction of the discretized BVP to the qth subdomain, is expressed by $\hat{C}_q = R_{q,\Delta} C R_{q,\Delta}^T$. In these terms, the additive Schwarz preconditioner is defined as

$$B_{AS} = \sum_{q=1}^{P} B_{AS,q}, \quad B_{AS,q} = R_{q,\Delta}^T \hat{C}_q^{-1} R_{q,\Delta}.$$

Also, it is possible to define the so-called restricted additive Schwarz (RAS) preconditioner by considering the prolongation $R_{q,0}^T$ instead of $R_{q,\Delta}^T$, i.e.

$$B_{RAS} = \sum_{q=1}^{P} B_{RAS,q}, \quad B_{RAS,q} = R_{q,0}^T \hat{C}_q^{-1} R_{q,\Delta}.$$

Note that B_{RAS} is a nonsymmetric matrix, even if C is a symmetric one.

The third way to define the preconditioner consists in the weighted determination of the iterative values in the intersections of the subdomains. For example, if the set of node indexes $S_q^h = \bigcap_{q'} \Omega_{q'}^h$ belongs to n_q^{s+1} grid subdomains $\Omega_{q'}^h$, and we have n_q^{s+1} different values of u_l^{s+1} for $l \in S_q^h$, then it is natural to compute the real next iterative value of the subvector u_q^{n+1} by means of the least-squares condition for the corresponding residual subvector.

Another type of preconditioning matrix which is used for DDM iterations in Krylov subspaces is responsible for the coarse grid correction or aggregation approach, which is based on a low-rank approximation of the original matrix C. We define a coarse grid, or macrogrid, Ω_c and the corresponding coarse space with $N_c \ll N$ degrees of freedom, as well as some basic functions $w^k \in \mathcal{R}^N$, $k = 1, \ldots, N_c$. We suppose that the rectangular matrix $W = (w_1, \ldots, w_{N_c}) \in \mathcal{R}^{N,N_c}$ has full rank. Then we define the coarse grid preconditioner B_c as

$$B_c^{-1} = W\hat{C}^{-1}W^T, \quad \hat{C} = W^T CW \in \mathcal{R}^{N_c,N_c},$$

where the small matrix $\hat{C}$ is a low-rank approximation of C; W is called the restriction matrix, and the transposed matrix W^T is the prolongation matrix.

Let us consider now the construction of the preconditioned iterative processes in Krylov subspaces. We offer a general description of the multi-preconditioned semi-conjugate residual (MPSCR) iterative method [12]. Let $r^0 = f^0 - Cu^0$ be the initial residual of algebraic system (17), and let $B_0^{(1)}, \ldots, B_0^{(m_0)}$ be a set of some nonsingular easily invertible preconditioning matrices. Using them, we define a rectangular matrix composed of the initial direction vectors p_k^0, $k = 1, \ldots, m_0$:

$$P_0 = [p_1^0 \cdots p_{m_0}^0] \in \mathcal{R}^{N,m_0}, \quad p_l^0 = (B_0^{(l)})^{-1}r^0, \tag{33}$$

which are assumed to be linearly independent.

Successive approximations u^n and the corresponding residuals r^n will be determined with the help of the recursions

$$\begin{aligned}
u^{n+1} &= u^n + P_n\bar{\alpha}_n = u^0 + P_0\bar{\alpha}_0 + \cdots + P_n\bar{\alpha}_n, \\
r^{n+1} &= r^n - CP_n\bar{\alpha}_n = r^0 - CP_0\bar{\alpha}_0 - \cdots - CP_n\bar{\alpha}_n.
\end{aligned} \tag{34}$$

Here $\bar{\alpha}_n = (\alpha_n^1, \ldots, \alpha_n^{m_n})^T$ are m_n-dimensional vectors. The direction vectors p_l^n, $l = 1, \ldots, m_n$, which form the columns of the rectangular matrices $P_n = [P_1^n \cdots P_{m_n}^n] \in \mathcal{R}^{N,m_n}$, are defined as orthogonal vectors in the sense of satisfying the relations

$$P_n^T C^T CP_k = D_{n,k} = 0 \quad \text{for } k \neq n, \tag{35}$$

where $D_{n,n} = \text{diag}\{\rho_{n,l}\}$ is a symmetric positive definite matrix since the matrices P_k have full rank, as is supposed.

Orthogonality properties (35) provide the minimization of the residual norm $\|r^{n+1}\|_2$ in the Krylov block subspace of dimension M_n:

$$K_{M_n} = \text{Span}\{P_0, \ldots, C^{n-1}P_{n-1}\}, \quad M_n = \sum_{k=0}^{n-1} m_k \tag{36}$$

provided that we define the coefficient vectors $\bar{\alpha}_n$ and the matrices P_n by the formulas

$$\bar{\alpha}_n = \{\alpha_{n,l}\} = (D_{n,n}^{-1})^{-1}P_n^T C^T r^0, \tag{37}$$

$$P_{n+1} = Q_{n+1} - \sum_{k=0}^{n} P_k\bar{\beta}_{k,n}, \tag{38}$$

where the auxiliary matrices

$$Q_{n+1} = [q_1^{n+1} \cdots q_{m_n}^{n+1}], \quad q_l^{n+1} = (B_{n+1}^{(l)})^{-1} r^{n+1}, \quad l = 1, \ldots, m_n, \qquad (39)$$

have been introduced; $B_{n+1}^{(l)}$ are some nonsingular easily invertible preconditioning matrices, and $\bar{\beta}_{k,n}$ are coefficient vectors that are determined, after substitution of (38) into orthogonality conditions (35), by the formula

$$\bar{\beta}_{k,n} = D_{k,k}^{-1} P_k^T C^T C Q_{n+1}. \qquad (40)$$

Let us remark that a successful acceleration of various Krylov algorithms can be attained by least-squares approaches [13].

4 Parallel Implementation of the Method

The parallel implementation of the numerical approaches we have considered consists, in general, of the following main stages:

(a) at each time step the grid constructing and or reconstructing the mesh at each time step if it is necessary, i.e. if the solution changes dresfiarlly in time;

(b) computing the coefficients of a discrete algebraic system, and recomputing these coefficients if the input data of the original problem depend on time;

(c) at each time step, implementing nonlinear iterations if the coefficients of the original IBVP depend on the unknown solution;

(d) solving SLAEs by means of domain decomposition methods in Krylov subspaces;

(e) postprocessing and visualization of the numerical results obtained;

(f) solving the inverse or the optimal IBVP which includes constraint minimization of the objective parameterized functional based on the optimization methods and on solution of a set of direct problems, presented by the above stages;

(g) control of the general computational process and decision-making in the results of mathematical modeling.

The "d" stage is the most expensive in terms of the required computational resources, and it is also the most investigated in the sense of achieving scalable parallelism. The main numerical and technological tools here are based on both domain decomposition methods and hybrid programming: MPI (Message Passing Interface system), open-MP type multi-thread computing, vectorization of operations and use of special computational units, for instance, GPGPU (see [14] and references therein). The DDMs represent two-level iterative processes in the Krylov subspaces. The upper level includes the distributed version of the MPSCR method (33)–(40), for example. In the case of a symmetric matrix C, this algorithm becomes simpler and transforms into a multi-preconditioned conjugate residual (MPCR) method with short recursions. Here matrix-vector operations are parallelized easily by means of efficient functions from the SPARSE

BLAS library. To minimize inter-processor communication time, a special array buffering is implemented. The main speedup is attained by synchronously solving the auxiliary algebraic subsystems for subdomains on the corresponding processors. It is important that SLAEs can have diverse matrix structures and be solved by various direct or iterative algorithms. In a sense, we have here a heterogeneous block iterative process, and minimizing the general run-time is not simply in such cases. In this situation, the balancing domain decomposition problem is a nonstandard task that should be solved in terms of general computer resource consuming minimization.

The scalable parallelization of the other computational stages (a–c) should also be based, naturally, on the domain decomposition principle. Within the conception of the basic system of modeling (BSM; see [15]), each stage would be implemented by the corresponding BSM kernel subsystem which is interacted by means of distributed data structures.

References

1. Il'in, V.P.: Mathematical Modeling, Part I: Continuous and Discrete Models. SBRAS Publ., Novosibirsk (2017). (in Russian)
2. Il'in, V.P.: Finite Element Methods and Technologies. ICM&MG SBRAS, Novosibirsk (2007). (in Russian)
3. Il'in, V.P.: Methods of Solving the Ordinary Differential Equations. NSU Publ., Novosibirsk (2017). (in Russian)
4. Il'in, V.P.: Problems of parallel solution of large systems of linear algebraic equations. J. of Math. Sci. **216**, 795–804 (2016). https://doi.org/10.1007/s10958-016-2945-4
5. Saad, Y.: Iterative Methods for Sparse Linear Systems. PWS Publ., New York (2002). https://doi.org/10.1137/1.9780898718003
6. Il'in, V.P.: Finite Difference and Finite Volume Methods for Elliptic Equations. ICM&MG SBRAS Publisher, Novosibirsk (2001). (in Russian)
7. Dolean, V., Jolivet, P., Nataf, F.: An Introduction to Domain Decomposition Methods: Algorithms, Theory and Parallel Implementaion. SIAM, Philadelphia (2015). https://doi.org/10.1137/1.9781611974065
8. Gander, M.J., Guttel, S.: ParaExp: a parallel integrator for linear initial value problems. SIAM J. Sci. Comput. **35**, 123–142 (2013). https://doi.org/10.1137/110856137
9. Karra, S.: A hybrid Pade ADI scheme of high-order for convection-diffusion problem. Int. J. Numer. Methods Fluids **64**, 532–548 (2010). https://doi.org/10.1002/fld2160
10. Lawson, G.L., Hanson, R.J.: Solving Least Squares Problems. Prentice-Hall, Inc., Upper Saddle River (1974). https://doi.org/10.1137/1.9781611971217
11. Saad, Y., Yeung, M., Erhel, J., Guyomarc'h, F.: A deflated version of the Conjugate Gradient Algorithm. SIAM J. Sci. Comput. **21**, 1909–1926 (2000). https://doi.org/10.1137/s1064829598339761
12. Il'in, V.P.: Multi-preconditioned domain decomposition methods in the Krylov subspaces. In: Dimov, I., Faragó, I., Vulkov, L. (eds.) NAA 2016. LNCS, vol. 10187, pp. 95–106. Springer, Cham (2017). https://doi.org/10.1007/978-3-319-57099-0_9

13. Il'in, V.P.: Least squares methods in Krylov subspaces. J. Math. Sci. **224**, 900–910 (2017). https://doi.org/10.1007/s10958-017-3460-y
14. Il'in, V.: On the parallel strategies in mathematical modeling. In: Sokolinsky, L., Zymbler, M. (eds.) PCT 2017. CCIS, vol. 753, pp. 73–85. Springer, Cham (2017). https://doi.org/10.1007/978-3-319-67035-5_6
15. Gladkikh, V.S., Il'in, V.P.: Basic System of Modeling (BSM): conception, architecture and methodology. In: Conference Proceedings of "Modern Problems of Mathematical Modeling, Image Processing and Parallel Computing", pp. 151–158. RTU Publ., Rostov (2017). https://doi.org/10.23947/2587-8999-2017-2-194-200. (in Russian)

A Study of Euclidean Distance Matrix Computation on Intel Many-Core Processors

Timofey Rechkalov and Mikhail Zymbler$^{(\boxtimes)}$ (ID)

South Ural State University, Chelyabinsk, Russia
trechkalov@yandex.ru, mzym@susu.ru

Abstract. Computation of a Euclidean distance matrix (EDM) is a typical task in a wide spectrum of problems connected with data analysis. Currently, many parallel algorithms for this task have been developed for GPUs. However, these developments cannot be directly applied to the Intel Xeon Phi many-core processor. In this paper, we address the task of accelerating EDM computation on Intel Xeon Phi in the case when the input data fit into the main memory. We present a parallel algorithm based on a novel block-oriented scheme of computations that allows for the efficient utilization of Intel Xeon Phi vectorization abilities. Experimental evaluation of the algorithm on real-world and synthetic datasets shows that it is highly scalable and outruns analogues in the case of rectangular matrices with low-dimensional data points.

Keywords: Euclidean distance matrix · OpenMP · Intel Xeon Phi
Data layout · Vectorization

1 Introduction

Computation of a Euclidean distance matrix (EDM) is a typical subtask in a wide spectrum of practical and scientific problems connected with data analysis [5]. The elements of an EDM are squared Euclidean distances[1], which can be interpreted as distances between data points of a set or distances between data points belonging to two sets of data points. These two cases correspond to square and rectangular EDMs, respectively. Square EDMs are extensively exploited in audio and video information retrieval [7,19], signal processing [5], hierarchical clustering of DNA microarray data [2], and so on. Rectangular EDMs play an important role in clustering-related applications, where it is necessary to calculate distances between cluster centers and data points subject to clustering, e.g., segmentation of medical images [12,21], fuzzy clustering of DNA microarray data [4], and so on.

---

[1] Strictly speaking, an EDM should contain Euclidean distances, and not the squares thereof. However, we adhere to this ambiguous convention in order to ensure compatibility with most papers related to EDMs [5].

© Springer Nature Switzerland AG 2018
L. Sokolinsky and M. Zymbler (Eds.): PCT 2018, CCIS 910, pp. 200–215, 2018.
https://doi.org/10.1007/978-3-319-99673-8_15

In this paper, we address the computation of both square and rectangular EDMs and formally define the problem as follows. Let us consider two non-empty finite sets of n and m data points in d-dimensional Euclidean space. Now we assign the first set data points to the rows of a matrix $\mathbf{A} \in \mathbb{R}^{n \times d}$, and the second set data points to the rows of a matrix $\mathbf{B} \in \mathbb{R}^{m \times d}$. Let us denote by $a_{1,\cdot}, \ldots, a_{n,\cdot}$ and $b_{1,\cdot}, \ldots, b_{m,\cdot}$, where $a_{i,\cdot}, b_{j,\cdot} \in \mathbb{R}^d$, the rows of the matrices $\mathbf{A}$ and $\mathbf{B}$, respectively. Then the Euclidean distance matrix $\mathbf{D} \in \mathbb{R}^{n \times m}$ consists of the rows $d_{1,\cdot}, \ldots, d_{n,\cdot}$, where $d_{i,\cdot} \in \mathbb{R}^m$, $d_{i,j} = \|a_{i,\cdot} - b_{j,\cdot}\|^2$, and $\| \cdot \|$ denotes the Euclidean norm[2].

Since EDM computation has time complexity $O(nmd)$, this task is often the most time-consuming stage of an entire problem, and it is therefore considered as a subject of parallelization for different hardware architectures.

At the present time, many parallel algorithms for EDM computation have been developed for GPUs [1,2,10,13]. These developments, however, cannot be directly applied to Intel Xeon Phi many-core systems [3,18]. Intel Xeon Phi is a series of products based on Intel Many Integrated Core (MIC) architecture, which provides a large number of compute cores with a high local memory bandwidth and 512-bit wide vector processing units. Being based on the Intel x86 architecture, Intel Xeon Phi supports thread-level parallelism and the same programming tools as a regular Intel Xeon CPU, and serves as an attractive alternative to GPUs. Currently, Intel offers two generations of MIC products, namely Knights Corner (KNC) [3] and Knights Landing (KNL) [18]. The former is a coprocessor with up to 61 cores, which supports native applications as well as offloading of calculations from a host CPU. The latter provides up to 72 cores and, unlike the first, is a bootable device that runs applications only in native mode.

In this paper, we address the task of accelerating EDM computation on the Intel Xeon Phi KNL system. In what follows, we assume that all the data involved in the computation fit into the main memory. The paper makes the following contributions. We propose a parallel algorithm based on a novel block-oriented scheme of computations, which allows for the efficient utilization of Intel Xeon Phi KNL vectorization abilities, more efficient than straightforward techniques such as data alignment and auto-vectorization. The algorithm versions developed in the course of the work are experimentally evaluated on real-world and synthetic datasets, and it is shown that our approach is highly scalable and outruns analogues in the case of rectangular matrices with low-dimensional data points.

The paper is structured as follows. Section 2 discusses related works. In Sect. 3, we describe the parallel algorithm proposed for Euclidean distance matrix computation on Intel MIC systems. We give the results of the experimental evaluation of our algorithm in Sect. 4. Finally, in Sect. 5, we summarize the results obtained and propose directions for further research.

---

[2] Note that this definition also covers the case $\mathbf{A} \equiv \mathbf{B}$.

2   Related Work

Chang *et al.* [2] suggested a CUDA-based parallel algorithm for EDM computation on GPUs. This algorithm assumes that the EDM is square ($n = m$) and both n and d are multiples of 16. The number 16 comes from the algorithmic design fitting the NVIDIA GPU architecture. The algorithm basic idea can be briefly described as follows. According to the nature of CUDA, threads are organized into 16×16 two-dimensional blocks, and the blocks are then organized in an $\frac{n}{16} \times \frac{n}{16}$ two-dimensional grid. Thus, a thread orients itself through a quadruplet (b_x, b_y, t_x, t_y), where two pairs (b_x, b_y) and (t_x, t_y) are block and thread indices, respectively. In this coordinate system, a thread calculates the $d_{16 \cdot b_y + t_y, 16 \cdot b_x + t_x}$ entry of the EDM. At each iteration, all threads firstly load two 16×16 submatrices into shared memory. Each thread, after synchronization, calculates and accumulates its own partial Euclidean distance. Then the threads need to be synchronized again before proceeding to the next pair of submatrices. The authors reported on an algorithm speedup by a factor of up to 44 on NVIDIA Tesla C870 (with a peak performance of 0.5 GFLOPS) compared with the CPU implementation.

Li *et al.* [13] proposed a chunking method to compute an EDM on large datasets in a multi-GPU environment. The method supposes the implementation of a GPU algorithm that is suitable for calculating Euclidean distance submatrices. Then the authors used a MapReduce-like framework to split the computation of the final EDM into many small independent jobs which calculate partial submatrices. The framework also dynamically allocates GPU resources to those independent jobs for maximum performance. The authors reported on a speedup of the method by a factor of up to 15 on three NVIDIA Tesla 1060 (0.9 GFLOPS each).

Kim *et al.* [10] suggested a padding strategy for the algorithm given in [2], which expands the matrix of input data points by adding rows and columns of zeros, so that data of any size may be processed by a simple CUDA kernel function. These authors reported on a speedup of the algorithm by a factor of up to 47 on NVIDIA Tesla C2050 (1.03 TFLOPS) compared with the CPU implementation.

Arefin *et al.* [1] extended the approaches suggested in [2, 10, 13]. Together with the EDM, the input data points are also chunked. Since this operation is carried out by an external memory programming environment, the proposed method is comparatively slower (by a factor of up to 30) than the original one. However, this method is feasible when the input dataset is so large that it fits into neither the GPU memory nor the host memory.

Wu *et al.* [20], Lee *et al.* [12], and Jaros *et al.* [9] indirectly touched upon the problem of EDM computation on Intel MIC systems. The authors of these papers accelerated a k-means data clustering algorithm on Intel Xeon Phi and considered EDM computation as a subtask.

In [20], the authors suggested a heterogeneous approach to parallelizing a k-means algorithm in which CPU and Xeon Phi KNC are involved. According to the algorithm idea, the CPU reassigns data points to clusters and then offloads data points and cluster centroids on to the coprocessor. Thus, Xeon Phi KNC repeatedly computes an EDM for data points and centroids. To achieve a more efficient utilization of memory bandwidth and cache, the algorithm stores data as an array of structures. The authors reported that the clustering algorithm achieves a speedup by a factor of up to 24 and its scalability decreases dramatically if more than 56 threads are employed.

The authors of [9] use a relatively similar approach and offload computations to Intel Xeon Phi KNC. We include in our review the solutions given in [9, 20] regarding them as precursors of our approach, yet we avoid a comparison since those solutions employ an outdated approach and partial results on run time and speedup of the EDM computation stage cannot be extracted from the experimental results.

In [12], the authors exploit straightforward techniques such as data alignment and auto-vectorization, as depicted in Algorithm 1 (in what follows, we will refer to it as STRAIGHTFORWARD).

Algorithm 1. STRAIGHTFORWARD(IN $\mathbf{A}$, $\mathbf{B}$; OUT $\mathbf{D}$)

1: #pragma omp parallel for
2: **for** i from 1 to n **do**
3:      $sum \leftarrow 0$
4:      **for** j from 1 to m **do**
5:          __assume_aligned($a_{i,\cdot}$, 64)
6:          __assume_aligned($b_{j,\cdot}$, 64)
7:          **for** k from 1 to d **do**
8:              $sum \leftarrow sum + (a_{i,k} - b_{j,k})^2$
9:          **end for**
10:         $d_{i,j} \leftarrow sum$
11:     **end for**
12: **end for**

Here, lines 5–6 signal the C compiler that the memory space is aligned to a specific size. Otherwise, the compiler assumes that the loop accesses unaligned memory spaces, and splits the loop, even though the start addresses of the memory spaces are aligned in reality. Thus, the loop in line 7 is vectorized without loop peeling, since the start addresses of the data points involved in calculations are aligned and, from the signals received, the compiler knows that they are aligned to the vector processor unit (VPU) width (i.e. the number of floats stored in the VPU).

Next, when the loop for distance calculation is vectorized, even if the start address of the first data point is aligned to the VPU width, the start address of the second data point will not be aligned if the dimension d is not a multiple of the VPU width, and will start to cause loop peelings from then on, so the loop will therefore be vectorized inefficiently. To solve this problem, the authors pad input data points with zero elements to the nearest integer multiple of the VPU width. Since the size of each input data point is a multiple of the VPU width, the loop is vectorized without splitting and is compiled in just two vector operations.

However, in high-performance computations, data layout can significantly affect the efficiency of memory access operations [8]. In the next section, we will show an application of data layouts to EDM computation.

3   Accelerating EDM Computation with Intel Xeon Phi

Our approach is different in two ways from the STRAIGHTFORWARD algorithm. Firstly, we propose a novel scheme of computations that allows for the efficient use of Intel Xeon Phi vectorization abilities. Secondly, we exploit a sophisticated data layout to store data points in main memory. We consider these matters below, in Sects. 3.1 and 3.2, respectively.

3.1   Computational Scheme

The basic idea of our approach is to modify the computational scheme in such a way that more operations will be vectorized compared with the straightforward approach. STRAIGHTFORWARD iteratively calculates one distance value between two data points, so the inner loop (cf. Algorithm 1, line 7) is compiled in two vector operations (i.e. elementwise vector difference and multiplication).

Unlike STRAIGHTFORWARD, the method we suggest iteratively calculates several distance values between a point from the first set of data points and *block* points from the second set of data points, where *block* is a parameter of the algorithm. Algorithm 2, which we will refer to as BLOCKWISE, implements such a computational scheme.

In lines 1–7, we change the data layout of the second set of data points (we will discuss this below, in Sect. 3.2) and produce its copy for further computations. The outer loop (line 9) is parallelized. It scans the first set of data points. The loop in line 10 scans the blocks of the second set of data points. The loop in line 12 provides for calculations through the coordinates of data points within a block. The loop in line 15 calculates the distances, it is compiled in two vector operations. In lines 13 and 14, we notify the compiler about the alignment of a point from the first set and a block of points from the second set, respectively. Finally, the loop in line 20 stores distances in the resulting matrix and is compiled in one vector operation (additionally, this loop is preceded by a signal to the compiler about the alignment of the rows of the resulting matrix).

To ensure that the blocks in the matrix representing the second set of data points have the same size, the number of rows m must be a multiple of *block*. We must therefore increase m up to the nearest integer that is a multiple of *block* by padding the **B** matrix with redundant zero rows.

Moreover, in order to guarantee an efficient vectorization of operations involving the **B** matrix, the *block* parameter must be a multiple of $width_{VPU}$, where $width_{VPU}$ denotes the number of floats stored in the VPU. Also, to derive greater

Algorithm 2. BLOCKWISE(IN **A**, **B**, *layout*, *block*; OUT **D**)

```
 1: if layout is SoA then
 2:     PERMUTE(B, m, B̃)
 3: else if layout is ASA then
 4:     PERMUTE(B, block, B̃)
 5: else
 6:         ▷ Current layout is AoS, no permutation needed
 7: end if
 8: #pragma omp parallel for
 9: for i from 1 to n do
10:     for j from 1 to ⌈ m/block ⌉ do
11:         sum ← 0̄
12:         for k from 1 to d do
13:             __assume_aligned(a_{i,·}, 64)
14:             __assume_aligned(b̃_{j+k,·}, 64)
15:             for ℓ from 1 to block do
16:                 sum_ℓ ← sum_ℓ + (a_{i,k} − b̃_{j+k,ℓ})²
17:             end for
18:         end for
19:         __assume_aligned(d_{i,·}, 64)
20:         for k from 1 to block do
21:             d_{i,j·block+k} ← sum_k
22:         end for
23:     end for
24: end for
```

benefits from the vectorization of computations, the **B** matrix should be the largest of the two sets of data points considered.

We should note, however, that our approach supposes the empirical choice of the *block* parameter in accordance with the above-mentioned requirements (we discuss this below, in Sect. 4).

3.2   Application of Data Layouts

Figure 1 depicts the definitions of the basic data layouts in the C programming language [8]. The AoS (Array of Structures) layout simply stores the structures in an array; it is often referred to as a baseline implementation. In the SoA (Structure of Arrays) layout, all components are stored in separate arrays. This can lead to coalesced memory access if the access pattern supposes reading of adjoining elements. The ASA (Array of Structures of Arrays) layout partitions the data in chunks according to the *block* parameter. ASA-*block* generalizes to the other layouts, namely ASA-1 corresponds to AoS, and ASA-*m* corresponds to SoA. This sophisticated data layout allows for a reduction of the number of processor cache misses during EDM computations.

Algorithm 3 transforms a data matrix from one layout to another in parallel. For a given *block* parameter and a matrix $\mathbf{B} \in \mathbb{R}^{m \times d}$ with AoS layout, the

```
typedef struct {          typedef struct {          typedef struct {
    float  x;                 float  x [m];             float  x [ block ];
    float  y;                 float  y [m];             float  y [ block ];
    float  z;                 float  z [m];             float  z [ block ];
} AoS;                    } SoA;                    } ASA;

AoS B[m];                 SoA B;                    ASA B[⌈ m/block ⌉];
```

(a) Array   (b) Structure   (c) Array of Structures
of Structures   of Arrays   of Arrays

Fig. 1. Basic data layouts

Algorithm 3. PERMUTE(IN **B**, *block*; OUT **B̃**)

1: #pragma omp parallel for
2: **for** j from 1 to $\lceil \frac{m}{block} \rceil$ **do**
3:    **for** i from 1 to d **do**
4:        **for** k from 1 to *block* **do**
5:            $\tilde{b}_{j \cdot d + i, k} \leftarrow b_{j \cdot block + k, i}$
6:        **end for**
7:    **end for**
8: **end for**

---

algorithm produces a matrix $\tilde{\mathbf{B}} \in \mathbb{R}^{d \cdot \lceil \frac{m}{block} \rceil \times block}$ with ASA-*block* layout (or with SoA layout if $block = m$).

4   Experimental Evaluation

4.1   Background of the Experiments

Objectives. In the experiments, we studied the following aspects of our approach. We investigated its performance and scalability compared with both the STRAIGHTFORWARD algorithm of Lee *et al.* [12] and the EDM computational algorithm from Intel Math Kernel Library (MKL)[3] optimized for Intel Xeon Phi. We combined the BLOCKWISE algorithm with the AoS, SoA and ASA-512 layouts, ran all the competitors on an Intel MIC system for different datasets, measured the run time (after deduction of the I/O time required for reading input data and writing the results), and calculated their speedup and parallel efficiency.

Here we understand these characteristics of parallel-algorithm scalability in the following manner. Speedup and parallel efficiency of a parallel algorithm employing k threads are calculated, respectively, as $s(k) = \frac{t_1}{t_k}$ and $e(k) = \frac{t_1}{k \cdot t_k}$, where t_1 and t_k are the run times of the algorithm when one and k threads are employed, respectively.

---

[3] Intel Math Kernel Library 2018 Release Notes.

We compared the performance and scalability for both square and rectangular matrices; the latter were the same used by Lee *et al.*

In order to make sure that the computational scheme proposed gives benefits on vectorization for MIC systems, we compared the performances of the BLOCKWISE algorithm (we took the results for the data layout where the algorithm performed best), the STRAIGHTFORWARD algorithm, and the Intel MKL algorithm, on both Intel Xeon and Intel Xeon Phi and for the same datasets.

Also, datasets and experimental results on performance for the algorithm of Kim *et al.* [10] on NVIDIA Tesla C2050[4] were compared with the best results of BLOCKWISE on Intel Xeon Phi (the aforesaid systems have approximately the same peak performance).

Finally, we present the results of the experiments carried out to choose the number 512 as the *block* parameter value.

Datasets. In the experiments, we compared the algorithms using the datasets described in Table 1. The Census [14] and the FCS Human [6] datasets are from real-world applications. The MixSim dataset and the ADS datasets were synthesized by artificial data generators described in [15,16], respectively. The ADS (Aligned Data Set) datasets were used for the experimental evaluation of the STRAIGHTFORWARD algorithm in [12]. The PRND (Pseudo Random Numbers) datasets were used by Kim *et al.* for the experimental evaluation of their algorithm [10].

Table 1. Datasets used in experiments

Dataset	d	n	m	Type	Semantic
MixSim	5	$35 \cdot 2^{10}$	$35 \cdot 2^{10}$	Synthetic	Created by a synthetic data generator [15]
Census	67	$35 \cdot 2^{10}$	$35 \cdot 2^{10}$	Real	US Census Bureau population surveys [14]
FCS Human	423	$18 \cdot 2^{10}$	$18 \cdot 2^{10}$	Real	Aggregated human gene information [6]
ADS-16	16	10^6	10^3	Synthetic	Used in [12] for experimental evaluation
ADS-32	32				
ADS-64	64				
ADS-256	256				
PRND-50	50	$15 \cdot 10^3$	$15 \cdot 10^3$	Synthetic	Used in [10] for experimental evaluation
PRND-100	100				
PRND-150	150				
PRND-200	200				

For the experiments, we took the largest parts of the MixSim and Census datasets that fit in the main memory of the hardware the algorithms were evaluated on. In order to meet the requirements for the *block* parameter (cf. Sect. 3.1), we took from MixSim, Census and FCS Human numbers of data points that are

---

[4] NVIDIA Tesla C2050/C2070 Data sheet.

multiples of *block* = 512 (the original FCS Human dataset was padded with zero points).

To evaluate the STRAIGHTFORWARD algorithm on datasets in which the dimension is not a multiple of $width_{VPU} = 16$, we increased d up to the nearest integer multiple of 16 by padding the data points with zeros. To evaluate our approach on the datasets used by Lee *et al.* and Kim *et al.*, in which the numbers of data points are not multiples of 512, we increased n and m up to the nearest integers that are multiples of 512 by padding the datasets with zero points.

Hardware. We conducted experiments on a node of the Tornado SUSU super-computer [11] (cf. Table 2 for the specifications of both the host and the MIC system).

Table 2. Hardware specifications

Specifications	Host	MIC system
Model, Intel Xeon	X5680	Phi (KNC), SE10X
Physical cores	2×6	61
Hyperthreading factor	2	4
Logical cores	24	244
Frequency, GHz	3.33	1.1
VPU size, bit	128	512
Peak performance, TFLOPS	0.371	1.076

4.2 Results and Discussion

Scalability. Figures 2 and 3 depict the run time, speedup and parallel efficiency of the competitors on square and rectangular matrices, respectively.

Regarding the experiments on square matrices, we can see that the Intel MKL algorithm outruns the competitors, and BLOCKWISE(ASA-512) holds the second place (with roughly the same performance on the MixSim dataset with d padded to 16). At the same time, the Intel MKL algorithm shows almost the worst speedup and parallel efficiency among the competitors. All the algorithms (except Intel MKL and BLOCKWISE(SoA)) show a close-to-linear speedup and up to 80% efficiency when the number of threads matches the number of physical cores the algorithm is running on. However, when more than one thread per physical core is employed, only BLOCKWISE(ASA-512) displays the aforementioned tendency, showing a speedup by a factor of up to 200 and at least 80% efficiency, whereas the speedup of the other algorithms slows or even drops down and their parallel efficiency diminishes accordingly.

Experiments on rectangular matrices deal with larger datasets and show the following. BLOCKWISE(ASA-512) outruns the competitors on the ADS-16

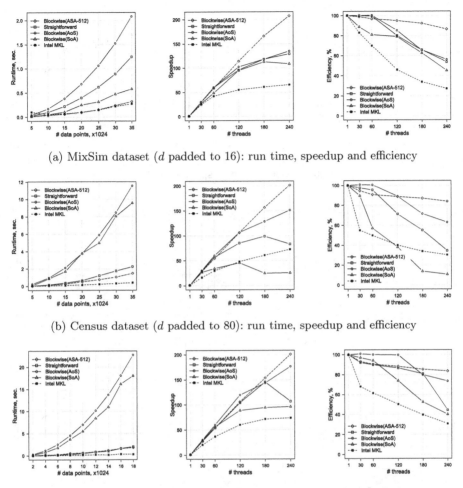

(a) MixSim dataset (d padded to 16): run time, speedup and efficiency

(b) Census dataset (d padded to 80): run time, speedup and efficiency

(c) FCS Human dataset (d padded to 432): run time, speedup and efficiency

Fig. 2. Run time and scalability on square matrices

and ADS-32 datasets, and shows roughly the same performance as the Intel MKL algorithm on the ADS-64 dataset. On the ADS-256 dataset, the Intel MKL algorithm beats the competitors. Regarding scalability, we see a similar picture as for square matrices. BLOCKWISE(ASA-512) shows a close-to-linear speedup and up to 90% parallel efficiency when the number of threads matches the number of physical cores. In the range from 60 to 240 threads, our algorithm scalability remains the best, giving a speedup by a factor of up to 160 and at least 70% efficiency. We can conclude that BLOCKWISE(ASA-512) performs its best on rectangular matrices with low-dimensional data points (approximately when $d \leq 32$).

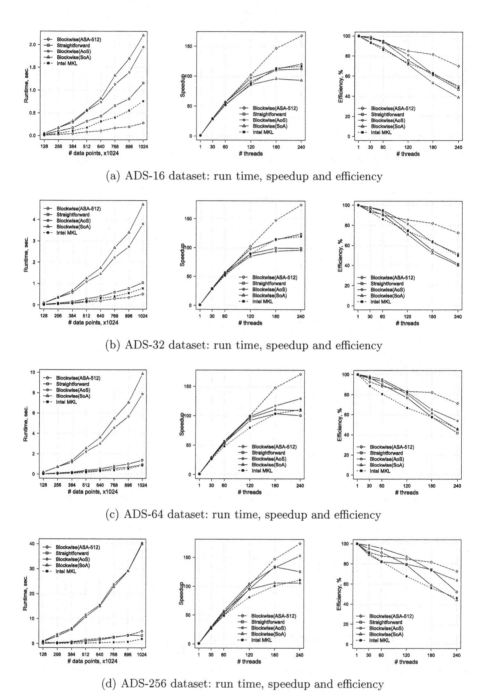

(a) ADS-16 dataset: run time, speedup and efficiency

(b) ADS-32 dataset: run time, speedup and efficiency

(c) ADS-64 dataset: run time, speedup and efficiency

(d) ADS-256 dataset: run time, speedup and efficiency

Fig. 3. Run time and scalability on rectangular matrices

Benefits of Vectorization. Table 3 shows the performance results of BLOCKWISE(ASA-512) for the Intel Xeon and Intel Xeon Phi platforms compared with STRAIGHTFORWARD. As we can see, BLOCKWISE(ASA-512) is 3.5 to 8 times faster on Intel Xeon Phi than it is on the host consisting of two Intel Xeon CPUs. The STRAIGHTFORWARD algorithm, in the same manner as BLOCKWISE(ASA-512), is faster on Intel Xeon Phi than on two Intel Xeon hosts. However, our algorithm shows a greater ratio of run times on the said platforms. Also, we should remind that the Intel MKL algorithm outruns our BLOCKWISE(ASA-512) in the case of high-dimensional data (approximately when $d > 32$) on both platforms.

Table 3. Run times on ADS datasets, s

Dataset	Intel Xeon Phi (KNC)			2×Intel Xeon CPU			Ratio of run times	
	1.076 TFLOPS			0.371 TFLOPS			2×CPU/Phi	
	Blockwise (ASA-512)	Intel MKL	Straight-forward	Blockwise (ASA-512)	Intel MKL	Straight-forward	Blockwise (ASA-512)	Straight-forward
ADS-16	0.28	0.76	1.05	1.04	3.02	1.00	3.7×	1.0×
ADS-32	0.51	0.78	1.15	1.76	3.14	1.79	3.5×	1.6×
ADS-64	0.98	0.88	1.36	3.78	3.81	4.25	3.9×	3.1×
ADS-256	3.71	1.92	3.79	30.32	5.14	31.41	8.2×	8.3×

Comparison with the GPU Solution. The performance results of our solution compared with the algorithm proposed by Kim *et al.* [10] are summarized in Table 4. We can see that BLOCKWISE(ASA-512) is up to two times faster on Intel Xeon Phi than the algorithm of Kim *et al.* is on NVIDIA Tesla C2050. However, the Intel MKL algorithm still outruns BLOCKWISE(ASA-512) on Intel Xeon Phi in the case of such small datasets.

Table 4. Run time on PRND datasets, s

Dataset	Intel Xeon Phi		2×Intel Xeon		NVIDIA Tesla
	1.076 TFLOPS		0.371 TFLOPS		1.03 TFLOPS
	Blockwise (ASA-512)	Intel MKL	Blockwise (ASA-512)	Intel MKL	Kim *et al.* [10]
PRND-50	0.19	0.07	0.35	0.74	0.82
PRND-100	0.32	0.08	0.59	0.89	1.01
PRND-150	0.45	0.10	0.78	1.01	1.21
PRND-200	0.58	0.12	1.60	1.16	1.41

Choice of the block Parameter. The preceding experimental results were obtained after an empirical research was carried out to choose the value of the *block* parameter. The value *block* = 512 was determined as follows. We ran BLOCKWISE(ASA-*block*) on Intel Xeon Phi for different values of *block* on datasets with $n = m = 2^{15}$ random data points having different dimensions: $d = 3, 5, 67$, and 129 (cf. Fig. 4). After that, we chose *block* = 512 as the value that gives the best performance for the most corresponding values of d.

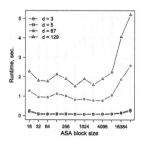

Fig. 4. Performance of BLOCKWISE(ASA-*block*) for different values of *block*

Discussion. To finish the presentation of the experimental results, we should mention both memory and run time overheads of our approach.

Memory overhead is due to the following reasons. First, for an efficient utilization of the Intel Xeon Phi vectorization abilities, our algorithm requires that the cardinality of the second set of data points be a multiple of *block*. If it is not so, then the value of m must be increased up to the nearest integer that a is multiple of *block* by padding the dataset with zero points. Thus, in the worst case, we will have $d \cdot (block - 1)$ redundant zero elements. Second, before computing an EDM, we create a copy of the matrix that represents the second set of data points and fills this copy with the elements of the original matrix permuted in a proper way. So we additionally need $d \cdot \max(n, m)$ redundant data elements (here we use the "max" function since, to derive greater benefits from the vectorization of computations, the **B** matrix should be the largest of the two sets of data points). Thus, the total memory overhead for our solution amounts to $d \cdot (block - 1 + \max(n, m))$ elements.

The STRAIGHTFORWARD algorithm, unlike our solution, requires that the dimension d be a multiple of $width_{VPU}$. If d does not meet this requirement, then it must be increased up to the nearest integer multiple of $width_{VPU}$ by padding the data points with zeros. Thus, in the worst case, it will cost $(width_{VPU} - 1) \cdot (m + n)$ redundant zero elements. Returning to the experimental results in which BLOCKWISE(ASA-512) outruns STRAIGHTFORWARD, we can conclude that, in the case of rectangular matrices with low-dimensional data points, our algorithm yields less memory overhead than STRAIGHTFORWARD.

As for the run time overhead related to the permutation of matrix elements, our experiments showed that the run time of the permutation step is negligibly small compared with the computation run time (less than one percent).

To conclude, we should also remind that the performance of the BLOCKWISE(ASA-*block*) algorithm depends on the *block* parameter, which must be determined through empirical research.

5 Conclusions

In this paper, we touched upon the problem of Euclidean distance matrix (EDM) computation, which is a typical subtask in a wide spectrum of practical and scientific problems connected with data analysis. At present, many parallel algorithms for EDM computation have been developed for GPUs. These developments, however, cannot be directly applied to modern Intel Xeon Phi many-core systems, which serve as an attractive alternative to GPUs. We addressed the task of accelerating EDM computation on the Intel Xeon Phi Knights Landing (KNL) system in the case when all data involved in the computations fit in the main memory.

We proposed a novel parallel algorithm for EDM computation, called BLOCKWISE, which is different in two ways from the approach that exploits straightforward techniques such as data alignment and auto-vectorization. Firstly, we use a block-oriented scheme of computations that allows for the efficient use of the Intel Xeon Phi vectorization abilities. Secondly, we apply a sophisticated data layout to store data points in main memory so as to reduce the number of processor cache misses during EDM computations.

We performed an experimental evaluation of the algorithm on real-world and synthetic datasets organized as square and rectangular matrices, and compared our solution with analogues. The experimental results show the following. BLOCKWISE demonstrates a close-to-linear speedup and at least 80% parallel efficiency when the number of threads matches the number of physical cores the algorithm is running on. When BLOCKWISE employs more than one thread per physical core, its speedup and parallel efficiency become sublinear but they remain the best among other competitors. Our algorithm outruns the straightforward approach and the algorithm from Intel Math Kernel Library (MKL) in the case of rectangular matrices with low-dimensional data points (approximately when $d \leq 32$). As for the case of high-dimensional data points ($d > 32$), the Intel MKL algorithm outruns the competitors on both square and rectangular matrices, while BLOCKWISE shows roughly the same performance as the straightforward approach.

Further studies of EDM computation on Intel MIC processors might elaborate on the following topics: applications of our approach to different clustering algorithms (e.g., k-means [12], PAM [17], and others), development of an analytical model that would be able to predict the performance of the BLOCKWISE algorithm and determine the value of the *block* parameter for best performance.

Acknowledgments. This work was financially supported by the Russian Foundation for Basic Research (grant No. 17-07-00463), by Act 211 of the Government of the Russian Federation (contract No. 02.A03.21.0011) and by the Ministry of Education and Science of the Russian Federation (government order 2.7905.2017/8.9).

References

1. Arefin, A.S., Riveros, C., Berretta, R., Moscato, P.: Computing large-scale distance matrices on GPU. In: The 7th International Conference on Computer Science and Education, ICCSE 2012, Melbourne, Australia, 14–17 July 2012, pp. 576–580. IEEE Computer Society (2012). https://doi.org/10.1109/ICCSE.2012.6295141

2. Chang, D., Jones, N.A., Li, D., Ouyang, M., Ragade, R.K.: Compute pairwise Euclidean distances of data points with GPUs. In: Proceedings of the IASTED International Symposium on Computational Biology and Bioinformatics, CBB'2008, Orlando, Florida, USA, 16–18 November 2008, pp. 278–283. IASTED (2008)

3. Chrysos, G.: Intel® Xeon Phi coprocessor (codename Knights Corner). In: 2012 IEEE Hot Chips 24th Symposium (HCS), Cupertino, CA, USA, 27–29 August 2012, pp. 1–31 (2012). https://doi.org/10.1109/HOTCHIPS.2012.7476487

4. Dembélé, D., Kastner, P.: Fuzzy c-means method for clustering microarray data. Bioinformatics **19**(8), 973–980 (2003)

5. Dokmanic, I., Parhizkar, R., Ranieri, J., Vetterli, M.: Euclidean distance matrices: essential theory, algorithms, and applications. IEEE Sig. Process. Mag. **32**(6), 12–30 (2015)

6. Engreitz Jr., J.M., Daigle, B.J., Marshall, J.J., Altman, R.B.: Independent component analysis: mining microarray data for fundamental human gene expression modules. J. Biomed. Inform. **43**(6), 932–944 (2010)

7. Foote, J.: An overview of audio information retrieval. Multimed. Syst. **7**(1), 2–10 (1999)

8. Hassan, Q.F.: Innovative Research and Applications in Next-Generation High Performance Computing. IGI Global, Hershey (2016). https://doi.org/10.4018/978-1-5225-0287-6

9. Jaros, M., et al.: Implementation of k-means segmentation algorithm on Intel Xeon Phi and GPU: application in medical imaging. Adv. Eng. Softw. **103**, 21–28 (2017)

10. Kim, S., Ouyang, M.: Compute distance matrices with GPU. In: Proceedings of the 3rd Annual International Conference on Advances in Distributed and Parallel Computing, ADPC'2012, Bali, Indonesia, 17–18 September 2012 (2012). https://doi.org/10.5176/2251-1652_ADPC12.07

11. Kostenetskiy, P., Safonov, A.: SUSU supercomputer resources. In: Sokolinsky, L., Starodubov, I., (eds.) PCT'2016, International Scientific Conference on Parallel Computational Technologies, Arkhangelsk, Russia, 29–31 March 2016. CEUR Workshop Proceedings, vol. 1576, pp. 561–573 (2016)

12. Lee, S., Liao, W., Agrawal, A., Hardavellas, N., Choudhary, A.N.: Evaluation of K-means data clustering algorithm on Intel Xeon Phi. In: Joshi, J., et al. (eds.) 2016 IEEE International Conference on Big Data, BigData 2016, Washington DC, USA, 5–8 December 2016, pp. 2251–2260. IEEE (2016)

13. Li, Q., Kecman, V., Salman, R.: A chunking method for Euclidean distance matrix calculation on large dataset using multi-GPU. In: Draghici, S., Khoshgoftaar, T.M., Palade, V., Pedrycz, W., Wani, M.A., Zhu, X. (eds.) The 9th International Conference on Machine Learning and Applications, ICMLA 2010, Washington, DC, USA, 12–14 December 2010, pp. 208–213. IEEE Computer Society (2010). https://doi.org/10.1109/ICMLA.2010.38

14. Meek, C., Thiesson, B., Heckerman, D.: The learning-curve sampling method applied to model-based clustering. J. Mach. Learn. Res. **2**, 397–418 (2002)

15. Melnykov, V., Chen, W.C., Maitra, R.: MixSim: an R package for simulating data to study performance of clustering algorithms. J. Stat. Softw. **51**(12), 1–25 (2012). https://doi.org/10.18637/jss.v051.i12
16. Narayanan, R., Özisikyilmaz, B., Zambreno, J., Memik, G., Choudhary, A.N.: Minebench: a benchmark suite for data mining workloads. In: Proceedings of the 2006 IEEE International Symposium on Workload Characterization, IISWC 2006, San Jose, California, USA, 25–27 October 2006, pp. 182–188. IEEE Computer Society (2006)
17. Rechkalov, T., Zymbler, M.: Accelerating medoids-based clustering with the Intel Many Integrated Core architecture. In: 9th International Conference on Application of Information and Communication Technologies, AICT 2015, 14–16 October 2015, Rostov-on-Don, Russia - Proceedings, pp. 413–417 (2015). https://doi.org/10.1109/ICAICT.2015.7338591
18. Sodani, A.: Knights Landing (KNL): 2nd generation Intel® Xeon Phi processor. In: 2015 IEEE Hot Chips 27th Symposium (HCS), Cupertino, CA, USA, 22–25 August 2015, pp. 1–24. IEEE (2015)
19. Valenzise, G., Gerosa, L., Tagliasacchi, M., Antonacci, F., Sarti, A.: Scream and gunshot detection and localization for audio-surveillance systems. In: Fourth IEEE International Conference on Advanced Video and Signal Based Surveillance, AVSS 2007, Queen Mary, University of London, London, United Kingdom, September 5–7 2007, pp. 21–26. IEEE Computer Society (2007)
20. Wu, F., Wu, Q., Tan, Y., Wei, L., Shao, L., Gao, L.: A vectorized K-means algorithm for intel many integrated core architecture. In: Wu, C., Cohen, A. (eds.) APPT 2013. LNCS, vol. 8299, pp. 277–294. Springer, Heidelberg (2013). https://doi.org/10.1007/978-3-642-45293-2_21
21. Zou, J., Chen, L., Chen, C.L.P.: Ensemble fuzzy c-means clustering algorithms based on KL-Divergence for medical image segmentation. In: Li, G., et al. (eds.) 2013 IEEE International Conference on Bioinformatics and Biomedicine, Shanghai, China, 18–21 December 2013, pp. 291–296. IEEE Computer Society (2013)

Parallel Method of Pseudoprojection
for Linear Inequalities

Irina Sokolinskaya[(✉)]

South Ural State University, 76 Lenin prospekt, Chelyabinsk 454080, Russia
Irina.Sokolinskaya@susu.ru

Abstract. This article presents a new iterative method for finding an approximate solution of a linear inequality system. This method uses the notion of pseudoprojection which is a generalization of the operation of projecting a point onto a closed convex set in Euclidean space. Pseudoprojecting is an iterative process based on Fejer approximations. The proposed pseudoprojection method is amenable to parallel implementation exploiting the subvector method, which is also presented in this article. We prove both the subvector method correctness and the convergence of the pseudoprojection method.

Keywords: Linear inequality system · Iterative method
Fejer approximations · Pseudoprojection · Parallel algorithm
Convergence

1 Introduction

In various numerical problems, we are often confronted with the task of solving a system of linear inequalities:

$$l_i(x) = \sum_{j=1}^{n} a_{ij}x_j - b_i \leqslant 0 \quad (i = 1, \ldots, m) \tag{1}$$

under the condition that system (1) is consistent. In the general case, the task of solving a system of linear inequalities is a difficult one. Thus, in practice, methods making it possible to find an approximate solution in a finite number of iterations are frequently applied. In [1,2], Motzkin and Agmon proposed a relaxation method for finding an approximate solution of a consistent system of linear inequalities. Let us consider the main idea of this relaxation method. When considering system (1), it is convenient to use a geometric language. Thus, we look upon $x = (x_1, \ldots, x_n)$ as a point in n-dimensional Euclidean space $\mathbb{R}^n$, and each inequality $l_i(x) \leqslant 0$ as a half-space P_i. The set of solutions of

I. Sokolinskaya—The study has been partially supported by the RFBR according to research project No. 17-07-00352-a and by the Government of the Russian Federation according to Act 211 (contract No. 02.A03.21.0011).

L. Sokolinsky and M. Zymbler (Eds.): PCT 2018, CCIS 910, pp. 216–231, 2018.
https://doi.org/10.1007/978-3-319-99673-8_16

system (1) therefore is the convex polytope $M = \bigcap\limits_{i=1}^{m} P_i$. Each equation $l_i(x) = 0$ defines an hyperplane H_i. In [2], the following iterative algorithm for finding an approximate solution of the system (1) is proposed. Below, λ such that $0 < \lambda < 2$ is a parameter of the algorithm. The parameter λ is called the *coefficient of relaxation*.

1. Choose an arbitrary point $x_0 \in \mathbb{R}^n$.
2. $x := x_0$.
3. If $x \in M$ then a solution is found; go to Step 8.
4. Select a half-space P_i such that $\text{dist}(x, P_i) = \max\limits_{j} \text{dist}(x, P_j)^1$.
5. Calculate the point x' which is the orthogonal projection of x onto hyperplane H_i.
6. $x := x + \lambda(x' - x)$.
7. Go to Step 3.
8. Stop.

Thus, the algorithm computes a sequence of points $x_0, x_1, \ldots, x_k, x_{k+1}, \ldots$, where $x_{k+1} = x_k + \lambda(x'_k - x_k)$, and x'_k is the orthogonal projection of the point x_k onto hyperplane H_i bounding the half-space P_i, so that

$$\text{dist}(x_k, P_i) = \max\limits_{j} \text{dist}(x_k, P_j).$$

There are two alternatives: (1) the process terminates after K steps with the point $x_k \in M$; (2) the process continues indefinitely, producing an infinite sequence $\{x_k\}$. In [1], Agmon showed that if $0 < \lambda < 2$ and the sequence $\{x_k\}$ is infinite, then x_k converges, as $k \to \infty$, to a point on the boundary of the polytope M. In this case, we can use the condition $\text{dist}(x_k, P_i) < \varepsilon$ as a stopping criterion. Here, $\varepsilon > 0$ is an arbitrarily small positive quantity. After stopping, the last point x_k is taken as an approximate solution of system (1).

The Motzkin–Agmon method has been extended in a number of works. In [3], a generalized relaxation method was proposed and investigated, based on the introduction of so-called *subcavities*. In certain cases, this generalized method provides faster convergence in comparison with the Motzkin–Agmon method. In [4,5], an extension of the relaxation method was considered for finding the common point of convex sets. In [6], the relaxation method is extended for solving systems of non-linear inequalities. In [7], a new parameter "cone angle" is introduced and the convergence and finiteness of the relaxation method for different values of this parameter are investigated. In [8,9], an extension of the relaxation method for systems with an infinite number of linear inequalities in a finite-dimensional space was proposed and investigated. In [10], the *underrelaxation* method with $0 < \lambda < 1$ is studied, and new bounds on convergence are obtained when the linear inequalities are processed in a cyclical order. In [11,12], a combined relaxation method for non-linear convex variational inequalities is described and studied.

---

[1] Here $\text{dist}(x, P) = \inf \{\|x - y\| : y \in P\}$.

In 1922, Leopold Fejer introduced the following definition of the closeness of points to a closed set M in the Euclidean space $\mathbb{R}^n$ (see [13]). If x and x' are points of $\mathbb{R}^n$ such that

$$\|x - y\| > \|x' - y\| \tag{2}$$

for every $y \in M$, then we say that x' is *point-wise* closer than x to the set M. If x is such that there is no point x' which is point-wise closer than x to M, then x is called the closest point to the set M. Fejer pointed out that the set of closest points to M is identical to the convex hull of the set M. Using this observation, Eremin in [14,15] introduced and investigated Fejer mappings, making it possible to construct iterative methods for solving problems of various types: systems of convex inequalities and problems of convex programming, ill-posed problems of mathematical physics in the presence of additional functional constraints, and others. The notion of pseudoprojecting a point onto a convex bounded set was introduced in [16]. The pseudoprojection operation is an extension of the projection operation using Fejer mappings. Based on the pseudoprojection operation, the authors of [16] developed a pseudoprojection method for solving linear inequality systems. This method is an extension of the relaxation method proposed by Motzkin and Agmon. Based on the pseudoprojection method, a set of parallel methods for solving large-scale non-stationary linear programming problems was developed and investigated in [16–19].

An iterative method for solving systems of linear inequalities based on determining the centroid is proposed in [20]. Each inequality defines a half-space of feasible points. The method starts with an arbitrary point in $\mathbb{R}^n$ as an initial approximation, and then calculates at each step the centroid of a subsystem of masses placed at the reflections of the previous iterate with respect to the bounding hyperplanes of only the violated half-spaces defined by the system of inequalities. This centroid is taken as the new iterate. In [21], a similar method is presented. In this method, each iterate lies in the half line determined by the previous one and a convex combination of its orthogonal projections on all the half spaces defined by the inequalities. The authors of [22] describe another iterative method for solving a system of linear inequalities in which each step consists of finding the orthogonal projection of the current point onto a hyperplane corresponding to a surrogate constraint constructed through a positive combination of a group of violated constraints. Note that the last three methods can be efficiently parallelized.

The present article is devoted to the development and investigation of a parallel pseudoprojection method to find an approximate solution of a system of linear inequalities. The method starts with an arbitrary point in $\mathbb{R}^n$ as an initial approximation, and then calculates a pseudoprojection of this point onto a convex polytope defined as the set of feasible solutions of linear inequality system (1). The subvector method is used to parallelize the Fejer process. The main idea of this method is that the vector determining the current approximation is divided into subvectors. For each subvector, a certain number of Fejer iterations is performed in parallel. Then the modified subvectors are combined into a single vector. The calculations are repeated until the required precision of approximation is obtained.

The rest of the paper is organized as follows. The formal definitions of Fejer mapping, Fejer process, as well as that of the pseudoprojection operation are given in Sect. 2. Section 3 is devoted to describing the algorithm for constructing a pseudoprojection onto a convex closed set. Section 4 describes the method of subvectors used for parallelization of the pseudoprojection algorithm. In Sect. 5, we prove the convergence theorem for the pseudoprojection calculation algorithm. The results obtained are summarized in Sect. 6, and further research directions are outlined herein.

2 Fejer Mappings and the Pseudoprojection Operation

Let us consider a consistent system of m linear inequalities,

$$Ax \leqslant b, \tag{3}$$

given in the n-dimensional Euclidean space $\mathbb{R}^n$ and written in matrix form. The matrix A has dimension $m \times n$. Let M be a polytope defined as the set of feasible solutions of linear inequality system (3). Such a polytope is always a closed convex set. A single-valued mapping $\psi \colon \mathbb{R}^n \to \mathbb{R}^n$ is said to be fejerian relatively to a set M (or briefly, M-fejerian) if

$$\psi(y) = y, \forall y \in M; \qquad \|\psi(x) - y\| < \|x - y\|, \ \forall y \in M, \ \forall x \notin M. \tag{4}$$

Let a_i be an i-th row of the matrix A $(i = 1, \ldots, m)$. Let us denote by $\langle a_i, x \rangle$ the dot product of vectors a_i and x. It is known [15, 23] that the mapping

$$\varphi(x) = x - \frac{\lambda}{m} \sum_{i=1}^{m} \frac{\max\left\{\langle a_i, x \rangle - b_i, 0\right\}}{\|a_i\|^2} \cdot a_i \tag{5}$$

is a continuous single-valued M-fejerian mapping for the relaxation coefficient $0 < \lambda < 2$. We will use the notation

$$\varphi^s(x) = \underbrace{\varphi \ldots \varphi(x)}_{s}.$$

The *Fejer process* generated by the mapping φ for an arbitrary initial approximation $x_0 \in \mathbb{R}^n$ is the sequence $\{\varphi^s(x_0)\}_{s=0}^{+\infty}$. It is known [15] that the Fejer process converges to a point belonging to the polytope M:

$$\{\varphi^s(x_0)\}_{s=0}^{+\infty} \to \bar{x} \in M. \tag{6}$$

Let us denote this concisely as $\lim_{s \to \infty} \varphi^s(x_0) = \bar{x}$. Let the φ-*projection* (*pseudo-projection*) of a point $x \in \mathbb{R}^n$ on the polytope M be understood as the mapping $\pi_M^\varphi(x) = \lim_{s \to \infty} \varphi^s(x)$.

3   Parallel Algorithm for Constructing a Pseudoprojection

Let us introduce the following notation. Given an arbitrary linear subspace $\mathbb{P} \subset \mathbb{R}^n$, let us denote by $\pi_{\mathbb{P}}(x)$ the orthogonal projection of $x \in \mathbb{R}^n$ onto the linear subspace $\mathbb{P}$. Everywhere below, a linear subspace will be called simply a subspace. Denote by $\rho(\mathbb{P}, x) = \min\limits_{p \in \mathbb{P}} \|p - x\|$ the distance between the point x and the subspace $\mathbb{P}$. Let the linear manifold $\mathbb{L}$ be constructed from subspace $\mathbb{P}$ by translating it by a vector z: $\mathbb{L} = \mathbb{P} + z$. Denote by $\pi_{\mathbb{L}}(x)$ the orthogonal projection of $x \in \mathbb{R}^n$ onto the linear manifold $\mathbb{L}$:

$$\pi_{\mathbb{L}}(x) = \pi_{\mathbb{P}}(x) + z. \tag{7}$$

Let $\varphi \in \{\mathbb{R}^n \to \mathbb{R}^n\}$ be a single-valued continuous M-fejerian mapping, where M is a convex closed set. Let us define a decomposition of the space $\mathbb{R}^n$ into a direct sum of orthogonal subspaces: $\mathbb{R}^n = \mathbb{P}_1 \oplus \ldots \oplus \mathbb{P}_r$, where $\mathbb{P}_i \perp \mathbb{P}_j$ for $i \neq j$. Let us construct a linear manifold $\mathbb{L}_i$ for each subspace

$$\mathbb{P}_i \ (i = 1, \ldots, r)$$

in the following way. Suppose that $\bar{x}^i \in \text{Arg} \min\limits_{x \in M} \rho(\mathbb{P}_i, x)$. Define $\bar{z}^i = \pi_{\mathbb{P}_i^\perp}(\bar{x}^i) \in \mathbb{P}_i^\perp$. Here, $\mathbb{P}_i^\perp$ denotes the orthogonal complement to the subspace $\mathbb{P}_i$. Let us construct the linear manifold $\mathbb{L}_i$ by translating $\mathbb{P}_i$ by a vector $\bar{z}^i$:

$$\mathbb{L}_i = \mathbb{P}_i + \bar{z}^i. \tag{8}$$

For each $i \in \{1, \ldots, r\}$, define the mapping $\varphi_i \in \{\mathbb{R}^n \to \mathbb{L}_i\}$ as

$$\varphi_i(x) = \pi_{\mathbb{L}_i}(\varphi(\pi_{\mathbb{L}_i}(x))). \tag{9}$$

Assume that s is a positive integer and ε is a positive real number. The following algorithm calculates the pseudoprojection of the point $\mathbf{0} \in \mathbb{R}^n$ ($\mathbf{0}$ is the zero vector) onto the polytope M.

Algorithm $\mathfrak{S}$:

1. $k := 0$; $x_0 = \mathbf{0} \in \mathbb{R}^n$.
2. $x_{k+1} := \sum\limits_{i=1}^{r} \left(\varphi_i{}^s \left(\pi_{\mathbb{L}_i}(x_k) \right) - \bar{z}^i \right).$
3. If $\|x_{k+1} - x_k\| < \varepsilon \vee d_M(x_{k+1}) < \varepsilon$ then go to 6.
4. $k := k + 1$.
5. Go to 2.
6. Stop.

The performance of algorithm $\mathfrak{S}$ for $n = 2$ and $s = 2$ is shown in Fig. 1. To apply the algorithm $\mathfrak{S}$ to an arbitrary initial point $x_0 \in \mathbb{R}^n$, you must transfer the origin to the point x_0. In Step 3, the algorithm computes the residual function

$$d_M = \sum\limits_{j=1}^{m} \max \{\langle a_j, x \rangle - b_j, 0\}. \tag{10}$$

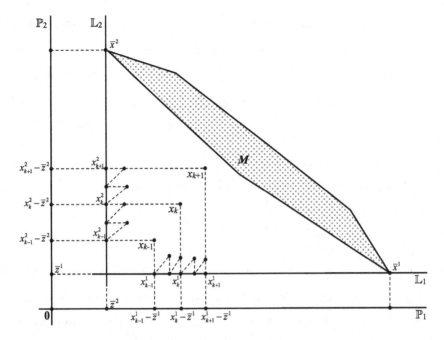

Fig. 1. The work of algorithm $\mathfrak{S}$: $x_k^1 = \pi_{\mathbb{L}_1}(x_k)$, $x_{k+1}^1 = \varphi_1^{\,2}(x_k^1)$; $x_k^2 = \pi_{\mathbb{L}_2}(x_k)$, $x_{k+1}^2 = \varphi_2^{\,2}(x_k^2)$.

This function determines the degree of closeness of the point x_{k+1} to the polytope M. We will show later on that the positive integer s is an important parameter influencing the potential scalability of algorithm $\mathfrak{S}$. By increasing s, we increase the resource of parallelism inherent in algorithm $\mathfrak{S}$. However, if one takes too large a value for the parameter s, then the sequence $\{x_k\}$ may converge to a point that does not belong to the polytope M. In this case, the iterative process will stop when the condition $\|x_{k+1} - x_k\| < \varepsilon$, included in the stopping criterion, is satisfied. If this happens, one needs to decrease the value of s and repeat the computational process. It is obvious that the most compute-intensive step of algorithm $\mathfrak{S}$ is Step 2, in which the Fejer process is implemented. To parallelize this step, the subvector method discussed below can be applied. The main idea of the method is as follows. For each subspace, a simultaneous Fejer process is performed. After every s steps, the results obtained on the subspaces are combined into one vector which is taken as the next approximation. If the stopping criterion checked in Step 3 is satisfied, then the last approximation is accepted as pseudoprojection. Otherwise, calculations continue.

4   Subvector Method

Let us consider the *subvector method*, applied to parallelize Step 2 of the $\mathfrak{S}$ algorithm. Let $r \in \mathbb{N}$ be such that $r \leqslant n$, where n is the space dimension. For simplicity, we always assume that r is a multiple of n: $n = r \cdot l$. Assume that

$$\{e_1, \ldots, e_n\} \tag{11}$$

is an orthonormal basis of the space $\mathbb{R}^n$. Let us define the linear subspaces as

$$\mathbb{P}_i = \mathrm{Lin}(\{e_{1+(i-1)l}, \ldots, e_{l+(i-1)l}\}) \tag{12}$$

for $i = 1, \ldots, r$. In Eq. (12), Lin denotes the linear hull of vectors. It is obvious that $\mathbb{P}_i \perp \mathbb{P}_j$ for $i \neq j$, and $\mathbb{P}_1 \oplus \ldots \oplus \mathbb{P}_r = \mathbb{R}^n$. Let $\bar{x}^i \in \underset{x \in M}{\mathrm{Arg\,min}}\, \rho(\mathbb{P}_i, x)$. Denote $\bar{z}^i = \pi_{\mathbb{P}_i^\perp}(\bar{x}^i) \in \mathbb{P}_i^\perp$ $(i = 1, \ldots, r)$. For $i = 1, \ldots, r$, define the mapping $\tau_i \in \{\mathbb{R}^n \to \mathbb{R}^l\}$ as follows. Let $(x_1, \ldots, x_n)$ be the coordinates of a vector $x \in \mathbb{R}^n$ in the orthonormal basis (11). Then,

$$\tau_i(x) = (x_{1+(i-1)l}, \ldots, x_{l+(i-1)l}). \tag{13}$$

Denote by $\bar{\tau}_i \colon \mathbb{P}_i \to \mathbb{R}^l$ the restriction of the mapping τ_i to subspace $\mathbb{P}_i \subset \mathbb{R}^n$. In the basis (11), an arbitrary vector has the following coordinates:

$$x = (0, \ldots, 0, x_{1+(i-1)l}, \ldots, x_{l+(i-1)l}, 0, \ldots, 0).$$

By comparing this with (13), we see that $\bar{\tau}_i$ is a one-to-one correspondence. Hence, $\bar{\tau}_i$ has the inverse mapping $\bar{\tau}_i^{-1}$. In the context of Eqs. (5) and (8), let us define the mapping $\varphi_i \in \{\mathbb{R}^n \to \mathbb{L}_i\}$ as

$$\varphi_i(x) = \bar{\tau}_i^{-1}\left(\tau_i(x) - \frac{\lambda}{m}\sum_{j=1}^{m}\frac{\max\left\{\langle \tau_i(a_j), \tau_i(x)\rangle + \langle a_j, \bar{z}^i\rangle - b_j, 0\right\}}{\|a_j\|^2} \cdot \tau_i(a_j)\right). \tag{14}$$

The following theorem shows that we can use Eq. (14) to calculate $\varphi_i(x)$ in Step 2 of algorithm $\mathfrak{S}$.

Theorem 1. *The mapping φ_i $(i = 1, \ldots, r)$ defined by Eq. (14) satisfies Eq. (9).*

Proof. From Eq. (5), we obtain

$$\pi_{\mathbb{L}_i}\left(\varphi\left(\pi_{\mathbb{L}_i}(x)\right)\right) = \pi_{\mathbb{L}_i}\left(\pi_{\mathbb{L}_i}(x) - \frac{\lambda}{m}\sum_{j=1}^{m}\frac{\max\left\{\langle a_j, \pi_{\mathbb{L}_i}(x)\rangle - b_j, 0\right\}}{\|a_j\|^2} \cdot a_j\right).$$

By expanding the parentheses, we obtain

$$\pi_{\mathbb{L}_i}\left(\varphi\left(\pi_{\mathbb{L}_i}(x)\right)\right) = \pi_{\mathbb{L}_i}(x) - \pi_{\mathbb{L}_i}\left(\frac{\lambda}{m}\sum_{j=1}^{m}\frac{\max\left\{\langle a_j, \pi_{\mathbb{L}_i}(x)\rangle - b_j, 0\right\}}{\|a_j\|^2} \cdot a_j\right).$$

According to Eq. (8), we have

$$\pi_{\mathbb{L}_i}\left(\varphi\left(\pi_{\mathbb{L}_i}(x)\right)\right)$$
$$= \pi_{\mathbb{P}_i}(x) + \bar{z}_i - \left(\pi_{\mathbb{P}_i}\left(\frac{\lambda}{m}\sum_{j=1}^{m}\frac{\max\left\{\langle a_j, \pi_{\mathbb{P}_i}(x) + \bar{z}_i\rangle - b_j, 0\right\}}{\|a_j\|^2} \cdot a_j\right) + \bar{z}_i\right).$$

By expanding the parentheses and eliminating $\bar{z}_i$, we transform the equation above to the form

$$\pi_{\mathbb{L}_i}\left(\varphi\left(\pi_{\mathbb{L}_i}\left(x\right)\right)\right) = \pi_{\mathbb{P}_i}\left(x\right) - \pi_{\mathbb{P}_i}\left(\frac{\lambda}{m}\sum_{j=1}^{m}\frac{\max\left\{\left\langle a_j, \pi_{\mathbb{P}_i}\left(x\right)+\bar{z}\right\rangle - b_j, 0\right\}}{\|a_j\|^2}\cdot a_j\right).$$

By distributivity of the dot product over the addition, this is equivalent to the equation

$$\pi_{\mathbb{L}_i}\left(\varphi\left(\pi_{\mathbb{L}_i}\left(x\right)\right)\right)$$
$$= \pi_{\mathbb{P}_i}\left(x\right) - \pi_{\mathbb{P}_i}\left(\frac{\lambda}{m}\sum_{j=1}^{m}\frac{\max\left\{\left\langle a_j, \pi_{\mathbb{P}_i}\left(x\right)\right\rangle + \left\langle a_j, \bar{z}\right\rangle - b_j, 0\right\}}{\|a_j\|^2}\cdot a_j\right).$$

Transform now the right side of the last equation as follows:

$$\pi_{\mathbb{L}_i}\left(\varphi\left(\pi_{\mathbb{L}_i}\left(x\right)\right)\right)$$
$$= \bar{\tau}_i^{-1}\left(\bar{\tau}_i\left(\pi_{\mathbb{P}_i}\left(x\right) - \pi_{\mathbb{P}_i}\left(\frac{\lambda}{m}\sum_{j=1}^{m}\frac{\max\left\{\left\langle a_j, \pi_{\mathbb{P}_i}\left(x\right)\right\rangle + \left\langle a_j, \bar{z}\right\rangle - b_j, 0\right\}}{\|a_j\|^2}\cdot a_j\right)\right)\right).$$

Since the mapping $\bar{\tau}_i$ is linear, this implies that

$$\pi_{\mathbb{L}_i}\left(\varphi\left(\pi_{\mathbb{L}_i}\left(x\right)\right)\right)$$
$$= \bar{\tau}_i^{-1}\left(\bar{\tau}_i\left(\pi_{\mathbb{P}_i}\left(x\right)\right) - \bar{\tau}_i\left(\pi_{\mathbb{P}_i}\left(\frac{\lambda}{m}\sum_{j=1}^{m}\frac{\max\left\{\left\langle a_j, \pi_{\mathbb{P}_i}\left(x\right)\right\rangle + \left\langle a_j, \bar{z}\right\rangle - b_j, 0\right\}}{\|a_j\|^2}\cdot a_j\right)\right)\right).$$

By comparing the subscripts in (12) and (13), we find that $\bar{\tau}_i\left(\pi_{\mathbb{P}_i}\left(x\right)\right) = \tau_i\left(x\right)$. Applying this to the right side of the preceding equation, we obtain

$$\pi_{\mathbb{L}_i}\left(\varphi\left(\pi_{\mathbb{L}_i}\left(x\right)\right)\right)$$
$$= \bar{\tau}_i^{-1}\left(\tau_i\left(x\right) - \tau_i\left(\frac{\lambda}{m}\sum_{j=1}^{m}\frac{\max\left\{\left\langle a_j, \pi_{\mathbb{P}_i}\left(x\right)\right\rangle + \left\langle a_j, \bar{z}\right\rangle - b_j, 0\right\}}{\|a_j\|^2}\cdot a_j\right)\right).$$

The mapping $\bar{\tau}_i$ is linear, therefore this means that

$$\pi_{\mathbb{L}_i}\left(\varphi\left(\pi_{\mathbb{L}_i}\left(x\right)\right)\right) =$$
$$\bar{\tau}_i^{-1}\left(\tau_i\left(x\right) - \frac{\lambda}{m}\sum_{j=1}^{m}\frac{\max\left\{\left\langle a_j, \pi_{\mathbb{P}_i}\left(x\right)\right\rangle + \left\langle a_j, \bar{z}\right\rangle - b_j, 0\right\}}{\|a_j\|^2}\cdot \tau_i\left(a_j\right)\right).$$

Let us compare again the subscripts in (12) and (13), we obtain

$$\pi_{\mathbb{L}_i}\left(\varphi\left(\pi_{\mathbb{L}_i}\left(x\right)\right)\right) = \bar{\tau}_i^{-1}\left(\tau_i\left(x\right) - \frac{\lambda}{m}\sum_{j=1}^{m}\frac{\max\left\{\left\langle \tau_i(a_j), \tau_i(x)\right\rangle + \left\langle a_j, \bar{z}\right\rangle - b_j, 0\right\}}{\|a_j\|^2}\cdot \tau_i\left(a_j\right)\right).$$

Finally, compare the last equation and Eq. (14), and we obtain that

$$\varphi_i\left(x\right) = \pi_{\mathbb{L}_i}\left(\varphi\left(\pi_{\mathbb{L}_i}\left(x\right)\right)\right).$$

Q.E.D.

5   Convergence Theorem

We will prove now the convergence theorem for algorithm $\mathfrak{G}$. For this we will need the two lemmas given below. The first lemma shows that each mapping $\varphi_i \in \{\mathbb{L}_i \rightarrow \mathbb{L}_i\}$ constructed by algorithm $\mathfrak{G}$ is Fejerian for the set $\mathbb{L}_i \cap M$.

Lemma 1. *Consider a convex closed set $M \subset \mathbb{R}^n$ and a single-valued M-fejerian mapping $\varphi \in \{\mathbb{R}^n \rightarrow \mathbb{R}^n\}$. Let $\mathbb{P}$ be a proper linear subspace of the space $\mathbb{R}^n$, and suppose that $\mathbb{T} = \mathbb{P}^\perp$ is the orthogonal complement of the subspace $\mathbb{P}$. Assume that*

$$\bar{x} \in \operatorname*{Arg\,min}_{x \in M} \rho(\mathbb{P}, x).$$

Write $\bar{x}$ as a sum of orthogonal vectors taken from the subspaces $\mathbb{P}$ and $\mathbb{T}$:

$$\bar{x} = \pi_{\mathbb{P}}(\bar{x}) + \pi_{\mathbb{T}}(\bar{x}).$$

Denote $\bar{z} = \pi_{\mathbb{T}}(\bar{x}) \in \mathbb{T}$. Construct the linear manifold $\mathbb{L}$ as a translation of $\mathbb{P}$ by the vector $\bar{z}$:

$$\mathbb{L} = \mathbb{P} + \bar{z}.$$

Define the mapping $\varphi_{\mathbb{L}} \in \{\mathbb{L} \rightarrow \mathbb{L}\}$ as

$$\varphi_{\mathbb{L}}(x) = \pi_{\mathbb{L}}(\varphi(\pi_{\mathbb{L}}(x))). \tag{15}$$

Take

$$M_{\mathbb{L}} = \mathbb{L} \cap M. \tag{16}$$

Then, the mapping $\varphi_{\mathbb{L}}$ is $M_{\mathbb{L}}$-fejerian.

Proof. Let us start by showing that

$$\varphi_{\mathbb{L}}(y) = y, \quad \forall y \in M_{\mathbb{L}}. \tag{17}$$

Let $y \in M_{\mathbb{L}}$. Then, by (16), $y \in M$. Since the mapping φ is M-fejerian, then $\varphi(y) = y$. Taking into account that $y \in \mathbb{L}$, we see that

$$\varphi_{\mathbb{L}}(y) = \pi_{\mathbb{L}}(\varphi(\pi_{\mathbb{L}}(y))) = \pi_{\mathbb{L}}(\varphi(y)) = \pi_{\mathbb{L}}(y) = y,$$

and so Eq. (17) holds.

Let us show now that

$$\|\varphi_{\mathbb{L}}(x) - y\| < \|x - y\|, \quad \forall y \in M_{\mathbb{L}}, \forall x \notin M_{\mathbb{L}}. \tag{18}$$

Assume that

$$y \in M_{\mathbb{L}}, \ x \in \mathbb{L}, \ x \notin M_{\mathbb{L}}.$$

By (16), it follows that $x \notin M$ in this case. Since the mapping φ is M-fejerian, then

$$\|\varphi(x) - y\| < \|x - y\|. \tag{19}$$

Construct the decomposition of $\varphi(x)$ and y as a sum of two orthogonal vectors belonging to $\mathbb{P}$ and $\mathbb{T}$:

$$\varphi(x) = \pi_{\mathbb{P}}(\varphi(x)) + \pi_{\mathbb{T}}(\varphi(x)), \tag{20}$$

$$y = \pi_{\mathbb{P}}(y) + \bar{z}. \tag{21}$$

We now substitute these decompositions into (19) and obtain

$$\|\pi_{\mathbb{P}}(\varphi(x)) + \pi_{\mathbb{T}}(\varphi(x)) - (\pi_{\mathbb{P}}(y) + \bar{z})\| < \|x - y\|, \tag{22}$$

which, after rearrangement, yields

$$\|(\pi_{\mathbb{P}}(\varphi(x)) - \pi_{\mathbb{P}}(y)) + (\pi_{\mathbb{T}}(\varphi(x)) - \bar{z})\| < \|x - y\|. \tag{23}$$

Note that $(\pi_{\mathbb{P}}(\varphi(x)) - \pi_{\mathbb{P}}(y)) \in \mathbb{P}$ and $(\pi_{\mathbb{T}}(\varphi(x)) - \bar{z}) \in \mathbb{T}$ are mutually orthogonal vectors. As we know, the square of the norm of a sum of orthogonal vectors is equal to the sum of the squares of their norms, so it follows from (23) that

$$\|\pi_{\mathbb{P}}(\varphi(x)) - \pi_{\mathbb{P}}(y)\|^2 + \|\pi_{\mathbb{T}}(\varphi(x)) - \bar{z}\|^2 < \|x - y\|^2. \tag{24}$$

The left side of the inequality is a sum of two non-negative terms. This means that, if we remove the second one, we obtain a valid inequality:

$$\|\pi_{\mathbb{P}}(\varphi(x)) - \pi_{\mathbb{P}}(y)\|^2 < \|x - y\|^2,$$

from which, in turn, we get

$$\|\pi_{\mathbb{P}}(\varphi(x)) - \pi_{\mathbb{P}}(y)\| < \|x - y\|. \tag{25}$$

By construction of $\mathbb{L}$, we have $\pi_{\mathbb{P}}(\varphi(x)) = \pi_{\mathbb{L}}(\varphi(x)) - \bar{z}$. Substitute this expression into (25), and we obtain

$$\|\pi_{\mathbb{L}}(\varphi(x)) - \bar{z} - \pi_{\mathbb{P}}(y)\| < \|x - y\|,$$

which is equivalent to

$$\|\pi_{\mathbb{L}}(\varphi(x)) - (\pi_{\mathbb{P}}(y) + \bar{z})\| < \|x - y\|. \tag{26}$$

But $x \in L$, so we may conclude that $x = \pi_{\mathbb{L}}(x)$. If we substitute the expression $\pi_{\mathbb{L}}(x)$ for x into the left side of (26), then we obtain

$$\|\pi_{\mathbb{L}}(\varphi(\pi_{\mathbb{L}}(x))) - (\pi_{\mathbb{P}}(y) + \bar{z})\| < \|x - y\|.$$

Taking into account (15) and (21), this implies

$$\|\varphi_{\mathbb{L}}(x) - y\| < \|x - y\|,$$

i.e., inequality (18) holds. Q.E.D.

To prove the convergence theorem, we need an additional lemma.

Lemma 2. *Let $\{x_k\}$ be the sequence of points produced by algorithm $\mathfrak{S}$ in Step 2:*

$$x_{k+1} := \sum_{i=1}^{r} \left(\varphi_i{}^s \left(x_k \right) - \bar{z}^i \right); \; k = 0, 1, \ldots \tag{27}$$

Under the conditions of algorithm $\mathfrak{S}$, let us define

$$M_{\mathbb{L}_i} = \mathbb{L}_i \cap M \quad (i = 1, \ldots, r).$$

If

$$x_0^i = \pi_{\mathbb{L}_i}(x_0), \; x_{k+1}^i = \varphi_i(x_k^i), \tag{28}$$

then

$$\varphi_i{}^s \left(x_k \right) = x_{s \cdot (k+1)}^i, \quad \forall k \in \mathbb{Z}_{\geqslant 0}. \tag{29}$$

Proof. Our proof will be by induction on k. Let $k = 0$. By Eq. (9), we have

$$\varphi_i{}^s \left(x_0 \right) = \varphi_i^{s-1} \left(\pi_{\mathbb{L}_i} \left(\varphi \left(\pi_{\mathbb{L}_i} \left(x_0 \right) \right) \right) \right).$$

According to the first equation in (28), we obtain

$$\varphi_i{}^s \left(x_0 \right) = \varphi_i^{s-1} \left(\pi_{\mathbb{L}_i} \left(\varphi \left(x_0^i \right) \right) \right). \tag{30}$$

But $x_0^i \in \mathbb{L}_i$, so we may write $x_0^i = \pi_{\mathbb{L}_i}(x_0^i)$. Now we substitute the expression $\pi_{\mathbb{L}_i}(x_0^i)$ for x_0^i into (30), and obtain

$$\varphi_i{}^s \left(x_0 \right) = \varphi_i^{s-1} \left(\pi_{\mathbb{L}_i} \left(\varphi \left(\pi_{\mathbb{L}_i} \left(x_0^i \right) \right) \right) \right).$$

Taking (9) into account, we get

$$\varphi_i{}^s \left(x_0 \right) = \varphi_i{}^s \left(x_0^i \right).$$

By applying the second equation from (28), we obtain

$$\varphi_i{}^s \left(x_0 \right) = \varphi_i^{s-1} \left(x_1^i \right).$$

If we repeat this substitution another $(s - 1)$ times, we obtain

$$\varphi_i{}^s \left(x_0 \right) = x_s^i,$$

that is, the induction basis holds. Now let $k > 0$, and consider the trivial equation

$$\varphi_i{}^s \left(x_k \right) = \varphi_i{}^s \left(x_k \right). \tag{31}$$

According to (27), $x_k := \sum\limits_{j=1}^{r} \left(\varphi_j^s \left(x_{k-1} \right) - \bar{z}^j \right)$. Substitute this expression into the right side of (31) to obtain

$$\varphi_i{}^s \left(x_k \right) = \varphi_i{}^s \left(\sum_{j=1}^{r} \left(\varphi_j^s \left(x_{k-1} \right) - \bar{z}^j \right) \right).$$

From this, by the induction hypothesis, it follows that

$$\varphi_i{}^s\left(x_k\right) = \varphi_i{}^s\left(\sum_{j=1}^{r}\left(x_{s\cdot k}^j - \bar{z}^j\right)\right),$$

which is equivalent to

$$\varphi_i{}^s\left(x_k\right) = \varphi_i^{s-1}\left(\varphi_i\left(\sum_{j=1}^{r}\left(x_{s\cdot k}^j - \bar{z}^j\right)\right)\right).$$

Using now (9), we obtain

$$\varphi_i{}^s\left(x_k\right) = \varphi_i^{s-1}\left(\pi_{\mathbb{L}_i}\left(\varphi\left(\pi_{\mathbb{L}_i}\left(\sum_{j=1}^{r}\left(x_{s\cdot k}^j - \bar{z}^j\right)\right)\right)\right)\right).$$

According to (7) and (8), this implies

$$\varphi_i{}^s\left(x_k\right) = \varphi_i^{s-1}\left(\pi_{\mathbb{L}_i}\left(\varphi\left(\pi_{\mathbb{P}_i}\left(\sum_{j=1}^{r}\left(x_{s\cdot k}^j - \bar{z}^j\right)\right) + \bar{z}^i\right)\right)\right).$$

Remember that $x_{s\cdot k}^j - \bar{z}^j \in \mathbb{P}_j$ $(j = 1,\ldots,r)$ and $\mathbb{P}_i \perp \mathbb{P}_j$ for $i \neq j$. Then the last implies

$$\varphi_i{}^s\left(x_k\right) = \varphi_i^{s-1}\left(\pi_{\mathbb{L}_i}\left(\varphi\left(x_{s\cdot k}^i - \bar{z}^i + \bar{z}^i\right)\right)\right),$$

i.e.

$$\varphi_i{}^s\left(x_k\right) = \varphi_i^{s-1}\left(\pi_{\mathbb{L}_i}\left(\varphi\left(x_{s\cdot k}^i\right)\right)\right). \tag{32}$$

Since $x_{s\cdot k}^i \in \mathbb{L}_i$, we have $x_{s\cdot k}^i = \pi_{\mathbb{L}_i}\left(x_{s\cdot k}^i\right)$. Substitute the expression $\pi_{\mathbb{L}_i}\left(x_{s\cdot k}^i\right)$ for $x_{s\cdot k}^i$ into (32) and obtain

$$\varphi_i{}^s\left(x_k\right) = \varphi_i^{s-1}\left(\pi_{\mathbb{L}_i}\left(\varphi\left(\pi_{\mathbb{L}_i}\left(x_{s\cdot k}^i\right)\right)\right)\right).$$

Together with Eq. (9), this implies

$$\varphi_i{}^s\left(x_k\right) = \varphi_i^{s-1}\varphi_i\left(x_{s\cdot k}^i\right),$$

which is equivalent to

$$\varphi_i{}^s\left(x_k\right) = \varphi_i{}^s\left(x_{s\cdot k}^i\right).$$

By applying the second equation from (28), we obtain

$$\varphi_i{}^s\left(x_k\right) = \varphi_i^{s-1}\left(x_{s\cdot k+1}^i\right).$$

Repeating this substitution another $(s-1)$ times, we finally arrive at

$$\varphi_i{}^s\left(x_k\right) = x_{s\cdot k+s}^i,$$

i.e.

$$\varphi_i{}^s\left(x_k\right) = x_{s(k+1)}^i.$$

Q.E.D.

Now we are ready to prove the *convergence theorem* for algorithm $\mathfrak{S}$.

Theorem 2. *Let $\{x_k\}$ be the sequence of points produced by algorithm $\mathfrak{S}$ in Step 2:*

$$x_{k+1} := \sum_{i=1}^{r} \left(\varphi_i^{\,s}(x_k) - \bar{z}^i\right); \; k = 0, 1, \ldots \tag{33}$$

Then

$$\{x_k\}_{k=0}^{+\infty} \to \bar{x} \in M.$$

Proof. Under the conditions of algorithm $\mathfrak{S}$, let us define

$$M_{\mathbb{L}_i} = \mathbb{L}_i \cap M \quad (i = 1, \ldots, r).$$

According to Lemma 1, the mapping φ_i is $M_{\mathbb{L}_i}$-fejerian. Take

$$x_0^i = \pi_{\mathbb{L}_i}(x_0), \quad x_{k+1}^i = \varphi_i(x_k^i). \tag{34}$$

The continuity of the mappings $\pi_{\mathbb{L}_i}$ and φ (see [23]) implies the continuity of the mappings φ_i. Hence, by Lemma 39.1 in [23], we may affirm that

$$\{x_k^i\}_{k=0}^{+\infty} \to \bar{x}^i \in M_{\mathbb{L}_i}, \forall i \in \{1, \ldots, r\}. \tag{35}$$

Now let us define

$$\bar{x} = \sum_{i=1}^{r} \left(\bar{x}^i - \bar{z}^i\right). \tag{36}$$

Fix an arbitrary real number $\varepsilon > 0$. By (35), there exists a number K_i such that

$$\left\|x_k^i - \bar{x}^i\right\| < \frac{\varepsilon}{\sqrt{r}}, \forall k > K_i. \tag{37}$$

Let $K = \max\limits_{1 \leqslant i \leqslant r} K_i$. We will show that the inequality $\|x_k - \bar{x}\| < \varepsilon$ holds for any $k > K$. Fix an arbitrary $k > K$. By (33), we may write

$$\|x_k - \bar{x}\| = \left\|\left(\sum_{i=1}^{r} \left(\varphi_i^{\,s}(x_{k-1}) - \bar{z}^i\right)\right) - \bar{x}\right\|,$$

and by (36), we obtain

$$\|x_k - \bar{x}\| = \left\|\sum_{i=1}^{r} \left(\varphi_i^{\,s}(x_{k-1}) - \bar{z}^i\right) - \sum_{i=1}^{r} \left(\bar{x}^i - \bar{z}^i\right)\right\|.$$

After rearrangement, this yields

$$\|x_k - \bar{x}\| = \left\|\sum_{i=1}^{r} \left(\varphi_i^{\,s}(x_{k-1}) - \bar{z}^i - \bar{x}^i + \bar{z}^i\right)\right\|,$$

which is equivalent to

$$\|x_k - \bar{x}\| = \left\| \sum_{i=1}^{r} \left(\varphi_i^{\,s} \left(x_{k-1} \right) - \bar{x}^i \right) \right\|. \tag{38}$$

According to Lemma 2, this implies

$$\|x_k - \bar{x}\| = \left\| \sum_{i=1}^{r} \left(x_{s\cdot k}^i - \bar{x}^i \right) \right\|. \tag{39}$$

Remember now that $x_{s\cdot k}^i = \pi_{\mathbb{P}_i} \left(x_{s\cdot k}^i \right) + \bar{z}^i$ and $\bar{x}^i = \pi_{\mathbb{P}_i} \left(\bar{x}^i \right) + \bar{z}^i$, and substitute these expressions into (39):

$$\|x_k - \bar{x}\| = \left\| \sum_{i=1}^{r} \left(\pi_{\mathbb{P}_i} \left(x_{s\cdot k}^i \right) + \bar{z}^i - \pi_{\mathbb{P}_i} \left(\bar{x}^i \right) - \bar{z}^i \right) \right\|,$$

i.e.

$$\|x_k - \bar{x}\| = \left\| \sum_{i=1}^{r} \left(\pi_{\mathbb{P}_i} \left(x_{s\cdot k}^i \right) - \pi_{\mathbb{P}_i} \left(\bar{x}^i \right) \right) \right\|. \tag{40}$$

Note that both vectors under the summation sign in (40) are mutually orthogonal. The square of the norm of a sum of orthogonal vectors is equal to the sum of the squares of their norms. It thus follows from (40) that

$$\|x_k - \bar{x}\|^2 = \sum_{i=1}^{r} \left\| \pi_{\mathbb{P}_i} \left(x_{s\cdot k}^i \right) - \pi_{\mathbb{P}_i} \left(\bar{x}^i \right) \right\|^2.$$

This is equivalent to

$$\|x_k - \bar{x}\|^2 = \sum_{i=1}^{r} \left\| \pi_{\mathbb{P}_i} \left(x_{s\cdot k}^i \right) + \bar{z}^i - \pi_{\mathbb{P}_i} \left(\bar{x}^i \right) - \bar{z}^i \right\|^2.$$

Since $x_{s\cdot k}^i = \pi_{\mathbb{P}_i} \left(x_{s\cdot k}^i \right) + \bar{z}^i$ and $\bar{x}^i = \pi_{\mathbb{P}_i} \left(\bar{x}^i \right) + \bar{z}^i$, we obtain

$$\|x_k - \bar{x}\|^2 = \sum_{i=1}^{r} \left\| x_{s\cdot k}^i - \bar{x}^i \right\|^2.$$

In view of (37), this implies

$$\|x_k - \bar{x}\|^2 < \sum_{i=1}^{r} \left(\frac{\varepsilon}{\sqrt{r}} \right)^2 = r \left(\frac{\varepsilon}{\sqrt{r}} \right)^2 = \varepsilon^2,$$

i.e.

$$\|x_k - \bar{x}\| < \varepsilon.$$

Q.E.D.

6   Conclusion

A new iterative method for solving linear inequality systems is proposed in the article. This method is based on the operation of pseudoprojecting a point onto a polytope which is defined as the set of feasible solutions of a linear inequality system in Euclidean space. The pseudoprojection operation is an extension of the projection operation. It exploits Fejer iterative processes developed by Eremin in [14,15,23]. For an effective parallelization of the pseudoprojection algorithm, we suggest here the subvector method. Also, we proved the convergence theorem for the pseudoprojection algorithm. The algorithm that computes the pseudoprojection was implemented in C++ using the OpenMP parallel programming library. Computational experiments have confirmed the effectiveness of the proposed method of parallelization for computer systems using multi-core accelerators Intel Xeon Phi [16]. As future research, we intend to do the following: implement the pseudoprojection algorithm in C++ language using the MPI library and the BSF algorithmic skeleton [24]; perform an analytical and experimental evaluation of the scalability of this parallel program on cluster computing systems; compare the proposed algorithm with other parallel iterative algorithms by performing computational experiments on cluster computing systems.

References

1. Agmon, S.: The relaxation method for linear inequalities. Can. J. Math. **6**, 382–392 (1954). https://doi.org/10.4153/CJM-1954-037-2
2. Motzkin, T.S., Schoenberg, I.J.: The relaxation method for linear inequalities. Can. J. Math. **6**, 393–404 (1954). https://doi.org/10.4153/CJM-1954-038-x
3. Merzlyakov, Y.I.: On a relaxation method of solving systems of linear inequalities. USSR Comput. Math. Math. Phys. **2**, 504–510 (1963). https://doi.org/10.1016/0041-5553(63)90463-4
4. Bregman, L.M.: The relaxation method of finding the common point of convex sets and its application to the solution of problems in convex programming. USSR Comput. Math. Math. Phys. **7**, 200–217 (1967). https://doi.org/10.1016/0041-5553(67)90040-7
5. Gubin, L.G., Polyak, B.T., Raik, E.V.: The method of projections for finding the common point of convex sets. USSR Comput. Math. Math. Phys. **7**, 1–24 (1967). https://doi.org/10.1016/0041-5553(67)90113-9
6. Germanov, M.A., Spiridonov, V.S.: On a method of solving systems of non-linear inequalities. USSR Comput. Math. Math. Phys. **6**, 194–196 (1966). https://doi.org/10.1016/0041-5553(66)90066-8
7. Goffin, J.L.: The relaxation method for solving systems of linear inequalities. Math. Oper. Res. **5**, 388–414 (1980). https://doi.org/10.1287/moor.5.3.388
8. González-Gutiérrez, E., Todorov, M.I.: A relaxation method for solving systems with infinitely many linear inequalities. Optim. Lett. **6**, 291–298 (2012). https://doi.org/10.1007/s11590-010-0244-4
9. González-Gutiérrez, E., Hernández Rebollar, L., Todorov, M.I.: Relaxation methods for solving linear inequality systems: converging results. TOP **20**, 426–436 (2012). https://doi.org/10.1007/s11750-011-0234-4

10. Mandel, J.: Convergence of the cyclical relaxation method for linear inequalities. Math. Program. **30**, 218–228 (1984). https://doi.org/10.1007/BF02591886
11. Konnov, I.: Combined Relaxation Methods for Variational Inequalities. LNE, vol. 495. Springer, Heidelberg (2001). https://doi.org/10.1007/978-3-642-56886-2
12. Konnov, I.V.: A modified combined relaxation method for non-linear convex variational inequalities. Optimization **64**, 753–760 (2015). https://doi.org/10.1080/02331934.2013.820298
13. Fejér, L.: Über die Lage der Nullstellen von Polynomen, die aus Minimumforderungen gewisser Art entspringen. In: Hilbert, D. (ed.) Festschrift, pp. 41–48. Springer, Heidelberg (1982). https://doi.org/10.1007/978-3-642-61810-9_6
14. Eremin, I.I.: Methods of Fejer's approximations in convex programming. Math. Notes Acad. Sci. USSR **3**, 139–149 (1968). https://doi.org/10.1007/BF01094336
15. Vasin, V.V., Eremin, I.I.: Operators and Iterative Processes of Fejér Type. Theory and Applications. Walter de Gruyter, Berlin, New York (2009)
16. Sokolinskaya, I., Sokolinsky, L.: Revised pursuit algorithm for solving non-stationary linear programming problems on modern computing clusters with many-core accelerators. In: Voevodin, V., Sobolev, S. (eds.) RuSCDays 2016. CCIS, vol. 687, pp. 212–223. Springer, Cham (2016). https://doi.org/10.1007/978-3-319-55669-7_17
17. Sokolinskaya, I.M.: Scalable algorithm for non-stationary linear programming problems solving. In: 2017 2nd International Ural Conference on Measurements (UralCon), pp. 49–53 (2017). https://doi.org/10.1109/URALCON.2017.8120685
18. Sokolinskaya, I., Sokolinsky, L.B.: Scalability evaluation of NSLP algorithm for solving non-stationary linear programming problems on cluster computing systems. In: Voevodin, V., Sobolev, S. (eds.) RuSCDays 2017. CCIS, vol. 793, pp. 40–53. Springer, Cham (2017). https://doi.org/10.1007/978-3-319-71255-0_4
19. Sokolinskaya, I., Sokolinsky, L.B.: On the solution of linear programming problems in the age of big data. In: Sokolinsky, L., Zymbler, M. (eds.) PCT 2017. CCIS, vol. 753, pp. 86–100. Springer, Cham (2017). https://doi.org/10.1007/978-3-319-67035-5_7
20. Censor, Y., Elfving, T.: New methods for linear inequalities. Linear Algebra Appl. **42**, 199–211 (1982). https://doi.org/10.1016/0024-3795(82)90149-5
21. De Pierro, A.R., Iusem, A.N.: A simultaneous projections method for linear inequalities. Linear Algebra Appl. **64**, 243–253 (1985). https://doi.org/10.1016/0024-3795(85)90280-0
22. Yang, K., Murty, K.G.: New iterative methods for linear inequalities. J. Optim. Theory Appl. **72**, 163–185 (1992). https://doi.org/10.1007/BF00939954
23. Eremin, I.I.: Teoriya lineynoy optimizatsii [The theory of linear optimization]. Publishing House "Yekaterinburg", Ekaterinburg (1999). (in Russian)
24. Sokolinsky, L.B.: Analytical estimation of the scalability of iterative numerical algorithms on distributed memory multiprocessors. Lobachevskii J. Math. **39**, 571–575 (2018). https://doi.org/10.1134/S1995080218040121

Supercomputer Simulation

GPU Acceleration of Bubble-Particle Dynamics Simulation

Ilnur A. Zarafutdinov[1(✉)], Yulia A. Pityuk[1], Azamat R. Gainetdinov[1], Nail A. Gumerov[1,2], Olga A. Abramova[1], and Iskander Sh. Akhatov[3]

[1] Center for Micro and Nanoscale Dynamics of Dispersed Systems,
Bashkir State University, Ufa, Russia
`ilnurzaraf2@gmail.com`, `pityukyulia@gmail.com`
[2] Institute for Advanced Computer Studies, University of Maryland,
College Park, USA
[3] Skolkovo Institute of Science and Engineering (Skoltech), Moscow, Russia

Abstract. Clusters containing bubbles and solid particles are used in many fields of industry. For instance, the study of bubble-particle interaction can be useful for surface cleaning in microelectronics and froth flotation in oil distillation.

The present work discusses the joint 3D dynamics of bubbles and solid spherical particles in the presence of an acoustic field in an unbounded ideal incompressible liquid. To solve the problem, we chose the boundary element method (BEM) which requires only the discretization of the boundary of the computational domain. However, the application of the conventional BEM for the direct simulation of large particle-bubble systems is normally limited by memory, computational complexity, and speed. To perform such simulations, we propose a numerical approach based on the GPU acceleration of the BEM code.

The method includes the solution of linear algebraic equations with a dense matrix of special type, using for this the generalized minimal residual method with calculation of the matrix-vector product on graphics processors using CUDA technology. We also discuss the performance of the method and run test computations of bubble-particle clusters.

Keywords: Bubbles · Particles · Acoustic field · Potential flow
Boundary element method · Graphics processors · CUDA technology

1 Introduction

Clusters containing bubbles and solid particles have wide applications in many industries. For example, the study of the interaction between bubbles and solid particles can be used in biotechnology and medicine for sterilization of medical instruments at low temperatures, tartar removal, lithotripsy, drug delivery to diseased organs, surgical operations on complex organs, as well as for microsurface cleaning (microchips, silicon substrate) in microelectronics and semiconductor

© Springer Nature Switzerland AG 2018
L. Sokolinsky and M. Zymbler (Eds.): PCT 2018, CCIS 910, pp. 235–250, 2018.
https://doi.org/10.1007/978-3-319-99673-8_17

industry. Froth flotation is used for water purification from organic substances and solid sediment, separation of mixtures, accelerating sedimentation in chemical, petrochemical, food and other industries. Thus, the study of bubble-particle clusters is relevant from a fundamental point of view, and has wide applications.

Recent theoretical, numerical and experimental investigations are mainly devoted to the analysis of the behavior of the bubble near a solid surface [1,2] and near a solid sphere [3,4]. There currently exist several models of the interaction between bubbles and solid objects. In [5], Lagrange equations of the fifth order of accuracy were used for the description of the dynamics of a cluster containing spherical bubbles and spherical elastic or solid particles in a potential flow of liquid in the presence of an acoustic field or shock wave.

A large number of works are devoted to the bubble-particle interaction in flotation processes. Derjaguin and Dukhin [6] studied the interaction of bubbles and particles by introducing a three-zone model for particles of small and medium sizes. In zone 1, when the particle is far from the bubble, hydrodynamic forces prevail. The particle moves towards the bubble under the action of inertial and gravitational forces. In zone 2, when the particle approaches the bubble, a tangential flow appears near the bubble surface. In zone 3, when the particle is located very close to the bubble, surface forces, such as Van der Waals and electrostatic forces, should be considered. Accounting for forces in zones 2 and 3 is a highly difficult task. For this reason, a model of interaction between bubbles and particles only for zone 1 was considered in [7]. It is shown that the concentration of bubbles and particles depends on the collision intensity, adhesion, and bubble stability. A more detailed overview of the bubble-particle interaction in zone 1 can be found in [8].

The adhesion process of particle-bubble interaction was experimentally studied in [9,10] by measuring a particle trajectory on the surface of a large air bubble. Basarova et al. [11] carried out similar experiments to measure the trajectory of bubbles on the surface of a large stationary particle. In recent years, researchers have begun to use the methods of computational fluid dynamics for efficient simulation of collisions of bubbles and particles in a flotation cell [12], and in a turbulent environment [13]. Despite the fact that these models have practical significance for industry, many physical phenomena, such as the motion of bubbles and particles, interfacial forces and film drainage process, have not been fully understood. Furthermore, the dynamics of the cluster containing bubbles and solid particles, especially in the three-dimensional case, has been poorly studied. In most theories associated with the dynamics of bubbles and particles, three-dimensional effects are neglected. Moreover, a qualitative study of the behavior of bubbles and particles in the presence of an acoustic field is not carried out, for an accurate description of the process of interaction between particles and bubbles at a micro-level and small time. They are rather focused on a quantitative and approximate description of the problem.

The present study is devoted to the development of mathematical models and the implementation of the corresponding program codes based on efficient methods and algorithms for three-dimensional simulation of bubble-particle interaction in an acoustic field. The numerical approach is based on the boundary element method (BEM), which requires only the surface discretization, thereby allowing to reduce the effective dimension of the problem by one. Thus, larger size problems can be solved using the BEM. Moreover, using known boundary values, we can calculate unknown values at any point of the computational domain. We used the BEM for potential flows [14]. Despite the BEM advantages, large-scale three-dimensional problems remain computationally very complex and resource-intensive. So the most important aspect is the development and application of methods to accelerate resource-intensive computing. For many years, progress in studying the behavior of large disperse systems was mainly empirical, with just several examples of large-scale direct simulations (e.g., for bubble dynamics [15]). Nowadays, modern computational methods and powerful computer resources allow to implement codes for fast large-scale microfluid dynamics simulations. A striking example of powerful hardware resource is that of a highly parallel, multi-threaded, multi-core graphics processing unit (GPU) having a large processing power and large memory bandwidth. Using GPUs not only to display graphics but also to speed up problems unrelated to graphics became a revolution in the last decade. GPUs could overtake central processing units (CPUs) owing to the large number of vertex and fragment processors.

In the present study, to accelerate computations of bubble-particle dynamics, a GPU is used since its architecture is best suited to conduct calculations with parallelization for a dense matrix of special type. We chose for this purpose a GPU manufactured by NVIDIA owing to its prevalence and accessibility, as well as to the availability of a user-friendly technology for CUDA programming. The CUDA programming language is an extension of the C/C++ language [16]. Multi-core processors work in groups: a group of threads simultaneously performs the same instructions for a set of data, and at the same time, can work with thousands of threads in parallel [17]. So the parallel algorithms should be developed taking into account the specific features of multi-core processors. In previous works, the authors successfully applied GPUs to accelerate BEM simulations of droplet [19,20] and bubble [21] dynamics. The developed approach is validated for bubble-particle interaction, including the spherical case of a bubble in an acoustic field. After that, simulations are extended to study the dynamics of small and large clusters with different structures including deformable bubbles and spherical solid particles.

2   Problem Statement

2.1   Mathematical Model

Consider the motion of a bubble (corresponding to the index "b") of volume V_b bounded by a surface S_b, and a solid spherical particle (corresponding to the index "p") of volume V_p bounded by a surface S_p, both in an unlimited inviscid incompressible liquid. The motion of a liquid in a gravity field of acceleration $\mathbf{g}$ is described by the Euler equations:

$$\rho_l \frac{d\mathbf{v}}{dt} = -\nabla p + \rho_l \mathbf{g}, \quad \nabla \cdot \mathbf{v} = 0, \quad \frac{d}{dt} = \frac{\partial}{\partial t} + \mathbf{v} \cdot \nabla, \tag{1}$$

where p, $\mathbf{v}$ and ρ_l are, respectively, the pressure, the velocity and the density of the liquid. Next, we assume that the liquid is at rest at infinity:

$$\mathbf{v}\big|_{|r|\to\infty} \to 0. \tag{2}$$

Moreover, we assume that the flow is potential, that is, $\mathbf{v} = \nabla\phi$, where ϕ is the velocity potential, which satisfies the Laplace equation $\nabla^2\phi = 0$ and can be expressed by a Cauchy–Lagrange integral,

$$\frac{\partial\phi}{\partial t} + \frac{1}{2}|\nabla\phi|^2 = \frac{p_\infty - p_g + 2\gamma k}{\rho_l} + \mathbf{g}\cdot\mathbf{x}, \quad \mathbf{x} \in S_b, \tag{3}$$

where $\mathbf{x}$ is the space coordinate of the point to which this equation is applied, p_∞ is the liquid pressure far from the bubble and the particle, γ is the surface tension, and k is the mean surface curvature.

The gas pressure p_g is determined according to some polytropic process:

$$p_g(t) = p_{g0}\left(\frac{V_{b0}}{V_b}\right)^k, \quad p_{g0} = p_0 + \frac{2\gamma}{a_0}, \tag{4}$$

where κ is the polytropic exponent, subscript "0" refers to the initial value at $t = 0$, V is the bubble volume, and a is the equivalent bubble radius.

The liquid pressure p_∞ changes according to the acoustic field:

$$p_\infty(t) = p_0 + p_a(t), \quad p_a(t) = P_a\sin(\omega t + \varphi), \tag{5}$$

where p_0 is the liquid pressure in the absence of the acoustic field, p_a is the acoustic pressure; P_a, ω and φ are, respectively, the amplitude, the frequency and the phase shift of the acoustic field.

The motion of the nodes on the surface of the bubble and the particle is described by the kinematic equation

$$\mathbf{n}_b\cdot\mathbf{v}\big|_{\mathbf{x}=\mathbf{x}_b} = \mathbf{n}_b\cdot\frac{d\mathbf{x}_b}{dt}, \quad \mathbf{n}_p\cdot\mathbf{v}\big|_{\mathbf{x}=\mathbf{x}_p} = \mathbf{n}_p\cdot\frac{d\mathbf{x}_p}{dt}, \tag{6}$$

or relative to the velocity potential:

$$\frac{\partial\phi}{\partial n}\bigg|_{\mathbf{x}=\mathbf{x}_b} = \mathbf{n}_b\cdot\frac{d\mathbf{x}_b}{dt}, \quad \frac{\partial\phi}{\partial n}\bigg|_{\mathbf{x}=\mathbf{x}_p} = \mathbf{n}_p\cdot\frac{d\mathbf{x}_p}{dt}, \tag{7}$$

where $\mathbf{n}$ is the normal to the surface pointing towards the liquid.

In the absence of a particle, the problem for the bubble can be integrated using relation (6). Indeed, this relation shows that the pressure can be computed if the position of the bubble surface is known. If the potential on the surface is also known, then one can solve the Dirichlet problem for the Laplace equation and determine the normal derivative, and according to Eq. (7), the normal velocity of the points on the surface, which can be integrated and the bubble surface position can be updated. The value of the potential then can be updated using the Cauchy–Lagrange integral (3), and so one can solve the problem for the next instance of time. The velocity of the bubble surface motion can be found from the first condition (6) (more detailed information can be found in [21]).

Unfortunately, this scheme does not work for the rigid body since the pressure here is not determined by the position of the surface alone, as it is in the case of the bubble. The Cauchy–Lagrange integral (3) can be used to determine the surface pressure, and then the force on the rigid body could be computed using Newton's second law. Thus, the force acting on a particle has two components, the first due to gravity, and the second due to the liquid:

$$\mathbf{F}_p = m_p\mathbf{g} - \int_{S_p} p\mathbf{n}\,dS. \qquad (8)$$

The mathematical model can be easily expanded to the case of multiple bubble-particle clusters.

2.2 Boundary-Integral Formulation

To simulate a cluster containing only bubbles, the mathematical model is reduced to the solution of the Laplace equation, for which the boundary-integral equations (for the velocity potential ϕ, where $\phi|_{|\mathbf{y}|\to\infty} = 0$) can be written as follows [21]:

$$L\left[q\right]\left(\mathbf{y}\right) - M\left[\phi\right]\left(\mathbf{y}\right) = \begin{cases} -\phi\left(\mathbf{y}\right), & \mathbf{y} \notin S_b, \quad \mathbf{y} \notin V_b, \\ -\dfrac{1}{2}\phi\left(\mathbf{y}\right), & \mathbf{y} \in S_b, \\ 0, & \mathbf{y} \in V_b. \end{cases} \qquad (9)$$

Here the BEM uses a formulation in terms of boundary integral equations (BIE) whose solution with boundary conditions gives $\phi(\mathbf{x})$ and $q(\mathbf{x}) = \partial\phi(\mathbf{x})/\partial n(\mathbf{x})$ on the boundary and subsequently determines $\phi(\mathbf{y})$ for the external and boundary domain point $\mathbf{y}$; $L[q]$ and $M[\phi]$ are potentials of simple and double layers.

In the presence of rigid particles in a cluster, the following modifications of the boundary-integral formulation (9) is used. The momentum conservation equation describing the particle dynamics can be written as

$$\frac{d\mathbf{V}_p}{dt} = \left(1 - \frac{\rho_l}{\rho_p}\right)\mathbf{g} + \frac{\rho_l}{m_p}\left[\frac{d}{dt}\int_{S_p}\phi\mathbf{n}\,dS + \int_S\left(\frac{1}{2}|\nabla\phi|^2\mathbf{n} - \frac{\partial\phi}{\partial n}\nabla\phi\right)dS\right] \qquad (10)$$

or

$$\frac{d\mathbf{U}_p}{dt} = \left(1 - \frac{\rho_l}{\rho_p}\right)\mathbf{g} + \frac{\rho_l}{m_p}\int_S\left(\frac{1}{2}|\nabla\phi|^2\mathbf{n} - \frac{\partial\phi}{\partial n}\nabla\phi\right)dS, \qquad (11)$$

where $\mathbf{U}_p = \mathbf{V}_p - \dfrac{\rho_l}{m_l}\displaystyle\int_{S_p}\phi\mathbf{n}\,dS.$

Taking into account (10), relation (7) can be reduced to the form

$$q(\mathbf{x}) = \mathbf{n}(\mathbf{x})\frac{\rho}{m_p}\int_{S_p}\phi(\mathbf{z})\mathbf{n}(\mathbf{z})dS(\mathbf{z}) + F(\mathbf{x}), \quad \mathbf{x} \in S_p, \quad F(\mathbf{x}) = \mathbf{U}_p\mathbf{n}(\mathbf{x}). \qquad (12)$$

Equation (12) makes it possible to eliminate the normal derivative on the particle surface and results in the following boundary integral equation

$$\frac{1}{2}\phi(\mathbf{y}) = M_b[\phi] - L_b[q] + M_p[\phi] - Q_p[\phi] - H_p[\mathbf{y}], \quad \mathbf{y} \in S = S_p \cup S_b, \quad (13)$$

where

$$Q_p[\phi](\mathbf{y}) = \int_{S_p} \Theta(\mathbf{y}, \mathbf{x}) \phi(\mathbf{x}) \, dS(\mathbf{x}), \quad \Theta(\mathbf{y}, \mathbf{x}) = \frac{\rho}{m_p} \mathbf{N}_p(\mathbf{y}) \cdot \mathbf{n}(\mathbf{x}),$$

$$H_p(\mathbf{y}) = L_p[F](\mathbf{y}) = \int_{S_p} F(\mathbf{x}) G(\mathbf{y}, \mathbf{x}) \, dS(\mathbf{x}) = \mathbf{U}_p \cdot \mathbf{N}_p(\mathbf{y}), \quad (14)$$

$$\mathbf{N}_p(\mathbf{y}) = \mathbf{L}_p[\mathbf{n}](\mathbf{y}) = \int_{S_p} \mathbf{n}(\mathbf{x}) G(\mathbf{y}, \mathbf{x}) \, dS(\mathbf{x}),$$

$$L_i[q](\mathbf{y}) = \int_{S_i} q(\mathbf{x}) G(\mathbf{y}, \mathbf{x}) \, dS(\mathbf{x}),$$

$$M_i[\phi](\mathbf{y}) = \int_{S_i} \phi(\mathbf{x}) \frac{\partial G(\mathbf{y}, \mathbf{x})}{\partial n(\mathbf{x})} \, dS(\mathbf{x}), \quad i = b, p. \quad (15)$$

Here $G(\mathbf{y}, \mathbf{x})$ and $\partial G(\mathbf{y}, \mathbf{x})/\partial n(\mathbf{x})$ are the Green's functions for the Laplace equation and its normal derivative:

$$G(\mathbf{y}, \mathbf{x}) = \frac{1}{4\pi r}, \quad \frac{\partial G(\mathbf{y}, \mathbf{x})}{\partial n(\mathbf{x})} = \frac{\mathbf{n} \cdot \mathbf{r}}{4\pi r^3}, \quad \mathbf{r} = \mathbf{y} - \mathbf{x}, \quad r = |\mathbf{y} - \mathbf{x}|. \quad (16)$$

If the potential and its normal derivative are known at time t on the surface $S(t) = S_b(t) \cup S_p(t)$, then the surface can be propagated to time $t + \triangle t$, as well as the values of the potential on the bubble surface, $\phi_b(t + \triangle t, \mathbf{x}_b(t + \triangle t))$. Then the unknowns at $t + \triangle t$ in integral equation (13) are the normal derivative on the bubble surface, which we denote by q_b, and the potential on the particle surface, which we denote by ϕ_p. This integral equation can be split into two parts corresponding to evaluation points $\mathbf{y}$ on the bubble surface and on the particle surface, respectively. We have then

$$L_b[q_b](\mathbf{y}) - J_p[\phi_p](\mathbf{y}) = \left(M_b - \frac{1}{2}I \right) [\phi_b](\mathbf{y}) - H_p(\mathbf{y}), \quad \mathbf{y} \in S_b,$$

$$L_b[q_b](\mathbf{y}) - \left(J_p - \frac{1}{2}I \right) [\phi_p](\mathbf{y}) = M_p[\phi_b](\mathbf{y}) - H_p(\mathbf{y}), \quad \mathbf{y} \in S_p, \quad (17)$$

where $J_p[\phi_p](\mathbf{y}) = M_p[\phi_p](\mathbf{y}) - Q_p[\phi_p](\mathbf{y})$ is a singular boundary operator combining the double layer M_p and the regular operator Q_p.

3   The Algorithm

The program code is developed on the basis of the above-mentioned mathematical model and the corresponding boundary-integral formulation for joint

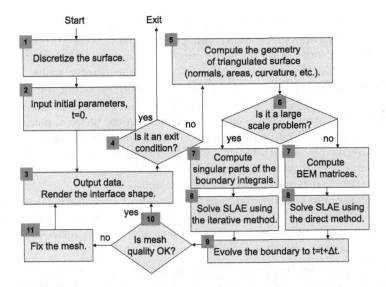

Fig. 1. Issues related to the algorithm

bubble-particle dynamics. The steps of algorithm using the BEM are shown in Fig. 1. The numbers indicate the basic program modules needed for the correct operation of the software. The purpose of each module is described below.

1. Discretize the surface. The surface is approximated by a mesh with triangular elements. We used a Delaunay triangulation in which the valence of the vertices is not less than 5 and the triangles satisfy the quality criterion $q > 0.6$, where

$$q = \frac{4S\sqrt{3}}{l_1^2 + l_2^2 + l_3^2}. \tag{18}$$

Here l_1, l_2 and l_3 are the length of the triangle sides, and S is its area. The accuracy of the approximation depends on the surface density of the mesh, which is characterized by the number of vertices N of the triangular elements. In the case of smooth surfaces, the computational nodes are chosen in the vertices. It is important that if the bubble shape deviates from that of a sphere, the surface topology still remains unchanged. In the case of several bubbles and particles, the surface of each object is discretized. A discretization of the spherical object used for calculations with $N = 642$ is shown in Fig. 2(a).

2. Input initial parameters. The determination of the initial physical and computational parameters of the problem is an important point. The initial parameters for the bubble are the initial position and the potential in the nodes on the bubble surface. The initial parameters for the spherical particle are the initial position and the velocity of the mass center. The physical parameters are the cluster structure (the number of particles and bubbles, their sizes) and its physical characteristics (particle density, gas pressure, polytropic exponent), the liquid parameters (density, pressure), and the acoustic field parameters (amplitude, frequency, phase shift). All physical parameters are expressed in a dimensionless form. The reference length is chosen as the initial bubble radius: $L^* = a_{b0}$; the pressure scale: $P^* = p_0$; the density scale: $\rho^* = \rho_l$; the reference velocity: $U^* = \sqrt{p_0/\rho_l}$. Then the corresponding reference time is $T^* = L^*/U^* = a_{b0}/\sqrt{p_0/\rho_l}$. Dimensionless parameters and variables will be decorated with a prime. Note that when determining the physical parameters of a problem, it is important to pay attention to the fact that the physical process must be described by a mathematical model (in this case, the model of potential flow). Computational tests are performed for $a'_{b0} = 1$, $a'_{p0} = 1$, $\rho'_l = 1$, $\rho'_p = 2$, $p'_0 = 1$, $\omega'/(2\pi) = 0.2$, $\gamma' = 0.073$, $\kappa' = 1.4$, $g' = 0$.

The computational parameters are the number of computational nodes, the time step (in this case, $\triangle t' = 0.01$, which satisfies Courant criterion), the maximum number of time steps, the visualization and storage methods, and the schemes for solving the ordinary differential equation (ODE).

3. Output data, render the interface shape. At each time step, the system dynamics is visualized and the results are saved using a chosen method (plots, screenshots), and data is saved (coordinates and velocity potentials of the vertices of the bubble, coordinate and velocity of the center of the particle, the bubble volume).

4. Is it an exit condition? The software execution is terminated when a certain maximum time value is reached or when the system reaches a state which is not taken into account in the model (e.g., topology change).

5. Compute the geometry of the triangulated surface. To calculate the boundary integrals by quadrature formulas, it is necessary to calculate geometric characteristics, such as the area of triangular elements, the area of the surface part related to a certain node (median splitting is used), the normal in the computational nodes of the surface. To calculate the mean curvature of the surface, the fitted paraboloid method is used on each mesh node [22].

6. Is it a large-scale problem? Depending on the problem size, we choose an approach to the solution. If the number of unknown systems exceeds a certain value determined by the memory and performance characteristics of the workstation, then we use the generalized minimal residual method (GMRES) with calculation of the matrix-vector product on graphics processors. Otherwise, the BEM matrix is calculated directly and stored in the system memory, and the system of linear algebraic equations (SLAE) is solved by direct methods. More detailed information can be found in Sect. 4.

7. Compute the BEM matrices and the singular parts of the boundary integrals. Using the vertex collocation method and quadrature formulas, each boundary integral from Eqs. (14) and (15) can be represented as a matrix of the appropriate size (depending on the number of computational nodes). Singular elements of matrices are determined on the basis of integral identities and test solutions. In case of a large-scale problem, it is necessary to storage just the singular parts of the BEM matrices.

8. Solve a linear algebraic system. Equations (17) can be discretized, and as a result, a SLAE is obtained for the unknown values q_b and ϕ_p in the nodes on the surface of the bubble and the particle, respectively.

9. Evolve the boundary. To describe the system dynamics, the following time marching is used. Time marching is the determination of all needed variables at time step $t + \triangle t$ as all those variables are known at time step t. The variables can be divided into two groups: those whose values at the next time step are determined immediately using a system of differential equations (principal or primary variables), and those whose values are computed from the values of the primary variables and needed for computation of the right-hand sides of the differential equations for primary variables (auxiliary or secondary variables). Of course, some physically important quantities can be also computed and output.

The primary variables for the present problem are the coordinates of the bubble surface $\mathbf{x}_b$ ($3N_b$ cartesian coordinates for N_b points sampling the bubble surface), the potential ϕ_b on the bubble surface (N_b values at the bubble surface sampling points), the coordinates of the center of mass of the particle $\mathbf{R}_b$ (3 cartesian coordinates), the modified velocity of the particle center $\mathbf{U}_p$ (3 cartesian coordinates). This amounts to $4N_b \times M_b + 6M_p$ primary variables, where M is the number of corresponding objects. The ODE describing the bubble-particle dynamics are

$$
\begin{aligned}
\frac{d\mathbf{x}_b}{dt} &= \mathbf{v}_b, \\
\frac{d\phi_b}{dt} &= \frac{1}{2}|\mathbf{v}_b|^2 + \frac{p_\infty - p_\mathrm{g} + 2\gamma k}{\rho_l} + \mathbf{g} \cdot \mathbf{x}_b, \\
\frac{d\mathbf{R}_p}{dt} &= \mathbf{V}_p, \\
\frac{d\mathbf{U}_p}{dt} &= \left(1 - \frac{\rho_l}{\rho_p}\right)\mathbf{g} + \frac{\rho_l}{m_p}\int_S \left(\frac{1}{2}|\triangledown\phi|^2\mathbf{n} - \frac{\partial\phi}{\partial n}\triangledown\phi\right)dS.
\end{aligned}
\tag{19}
$$

The program code provides a choice of methods for solving this ODE. The most optimal scheme (from the point of view of time and accuracy of calculations) is a combination of two methods: it is possible to use the fourth-order Runge–Kutta scheme (to preserve history) for the initial time steps, and then the Adams–Bashforth scheme of the sixth order.

10. Is the mesh quality OK? When solving dynamic problems by the BEM, there are problems with the mesh destabilization associated with errors in calculations of geometrical characteristics, such as surface integrals, normal area, tangential component, due to the surface discretization of the triangular elements. At each time step or after several time steps (depending on the time step), it is therefore necessary to check the quality of the surface triangulation according to quality criterion (18). If $q < 0.6$, then a mesh stabilization method should be used.

11. Fix the mesh. To stabilize the mesh, we developed and implemented a new approach based on the use of a spherical filter [21], which removes the arising

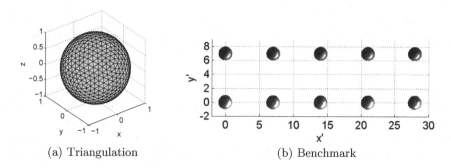

(a) Triangulation   (b) Benchmark

Fig. 2. Structure of a bubble-particle cluster for performance test (Color figure online)

noise from the bubble surface. The method is based on a parametric representation of the surface by spherical harmonics. The use of the spherical filter allows one to calculate several oscillation periods of a strongly collapsing bubble. Moreover, for the mesh stabilization, a correction of the tangential component of the velocity is also applied [21].

4   GPU Acceleration

Since the problem is a nonstationary one, it is necessary to build the BEM matrices at each time step and solve the SLAE (17). As the number of objects increases, the complexity of the problem grows, and more memory is needed to store the BEM matrices. Since the system matrices are of a special type, it is not necessary to store the elements of a matrix, it is sufficient to calculate a matrix-vector product (MVP). In the present study, we calculated all MVPs from Eq. (17) on the GPU. The paralleled codes for the left-hand side (LHS) and the right-hand side (RHS) are implemented using CUDA technology. To solve the SLAE, we use the unpreconditioned general minimal residual method (GMRES) [18], where we redefine the MVP and execute the code for the LHS on the GPU.

Calculations are performed on a workstation equipped with an Intel Xeon 5660 2.8 GHz CPU (12 physical + 12 virtual cores), 12 GB RAM, and one GPU NVIDIA Tesla C2050 (3 GB of global memory). To test the performance of the program code, we consider a benchmark consisting of a cluster containing bubbles (blue color) and particles (red color) under an acoustic field of amplitude $P'_a = 0.7$, and a distance $d' = 7$ between the centers of nearest objects (Fig. 2). Figure 3 demonstrates the dependence between the run time of one MVP for the LHS and RHS of Eq. (17) and the matrix size, using different architectures, precisions and methods. The run time for the MVP calculation and the SLAE solution for different computational points is summarized in Table 1. From the figures and the table, it is evident the speedup of the code implemented on the GPU using the parallel computing platform CUDA (taking into account the

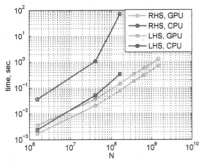

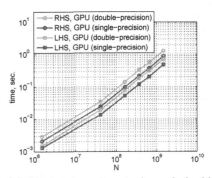

(a) Comparison of codes on CPU and GPU

(b) Comparison for single and double precision

Fig. 3. Wall-clock time of one MVP calculation for the LHS and the RHS

wall-clock time for data transfer) as compared with the code implemented on the CPU using the direct method. Moreover, the MVP for the LHS requires less computational time than for the RHS due to the complexity of calculating the BEM matrix corresponding to H_p (see Eqs. (14) and (17)). Although the calculation of one MVP for the LHS with single precision requires less time than the corresponding calculation with double precision (see Fig. 3(b)), the convergence of the GMRES with single precision is worse, which is seen from the run time of the SLAE. Moreover, in the case using single precision, we could not achieve the same accuracy. As an average, ~12 to 13 (maximum 18) iterations are required for the GMRES with double precision to converge to a relative residual error $\ll 10^{-6}$.

Table 1. Run time of one MVP calculation for the LHS and RHS, and SLAE solution

Bubbles/particles		1/1	5/5	10/10	15/15	20/20	30/30
N		1248	6420	12840	19260	25680	38520
LHS	CPU	0.0024	0.0520	0.3401	–	–	–
	GPU, double prec.	0.0022	0.0208	0.0785	0.1837	0.3123	0.7111
	GPU, single prec.	0.0020	0.0150	0.0550	0.1215	0.2066	0.4870
RHS	CPU	0.0365	1.0574	71.9801	–	–	–
	GPU, double prec.	0.0035	0.0368	0.1439	0.3340	0.5693	1.2967
	GPU, single prec.	0.0031	0.0272	0.0999	0.2209	0.3794	0.8884
SLAE	direct method	0.0728	3.2013	63.7664	–	–	–
	GMRES, double prec.	0.0185	0.0444	0.1420	0.3206	0.5400	1.2069
	GMRES, single prec.	0.1064	0.3178	0.9451	2.4379	4.0785	17.7282

5   Some Physical Results

On the basis of the developed software, it is possible to simulate the interaction between bubbles and rigid spherical particles in the presence of an acoustic field. In this section, we present both some illustrative calculations of the bubble-particle dynamics and a physical analysis of the numerical results obtained.

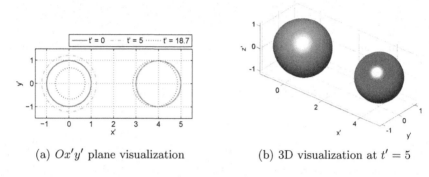

(a) $Ox'y'$ plane visualization   (b) 3D visualization at $t' = 5$

Fig. 4. Bubble-particle interaction

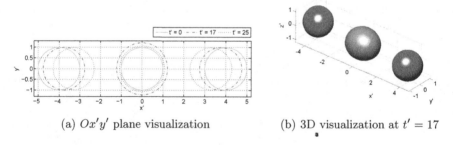

(a) $Ox'y'$ plane visualization   (b) 3D visualization at $t' = 17$

Fig. 5. Particle-bubble-particle interaction

The program code was tested and examined for convergence and calculation accuracy. First, for the interaction of a bubble and a rigid particle at a distance $d' = 9$, we obtained a good agreement between the simulation results and the solution of the Rayleigh–Plesset equation for the radius dynamics of a single spherical bubble in an acoustic field of amplitude $P'_a = 0.5$, neglecting viscosity [23]. The relative error in the L_∞ space for $N_b = 642$ is 2.31%. Then we decreased the distance between bubble and particle. Figure 4 shows the bubble-particle dynamics at $d' = 4$, on the $Ox'y'$ plane (a), and in the three-dimensional case (b). It should be noted that when the bubble expands, the particle is repelled

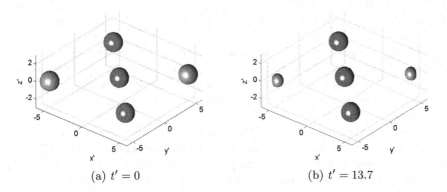

(a) $t' = 0$ (b) $t' = 13.7$

Fig. 6. Dynamics of two bubbles and three particles

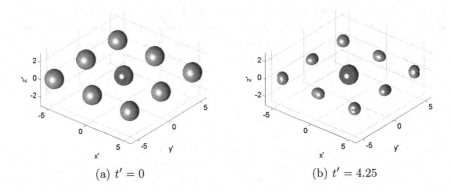

(a) $t' = 0$ (b) $t' = 4.25$

Fig. 7. Dynamics of eight bubbles and one particle

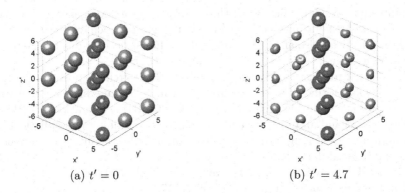

(a) $t' = 0$ (b) $t' = 4.7$

Fig. 8. Dynamics of eighteen bubbles and nine particles

from it, and when the bubble shrinks, the particle is attracted to it. This process is associated with the formation of hydrodynamic flows generated by the oscillations of the bubble.

Second, we considered the dynamics of a system "particle-bubble-particle" in which the centers of all three objects lie on one line. The initial distance between the centers of nearest objects is $d' = 4$. The particle-bubble-particle interaction is shown in Fig. 5, where one can see that the particles move symmetrically towards the bubble.

After small-scale tests, we considered the dynamics of bubble-particle clusters with ordered structure in the presence of an acoustic field of amplitude $P_a' = 0.7$. First, we simulated clusters containing two bubbles and three particles (Fig. 6), and eight bubbles and one particle (Fig. 7), with centers located on the $Ox'y'$ plane. As we can see from the figures, a bubble jet appears pointing towards the central particle. It is worth noting that, in the presence of two or more bubbles, the bubble volume change is significantly different from that in the case of a single-bubble system. This is associated with the secondary Bjerknes forces. The second test involved a cubic cluster containing eighteen bubbles and nine particles. In this case, we obtained the total number of computational nodes $N = 17\,334$. The run time for this test was about 30 minutes for 500 time steps. Figure 8 illustrates the cluster dynamics; here we also observe bubble-jet formation.

6   Conclusions

In this article, we developed an algorithm to simulate the interaction of bubbles and solid particles in the presence of an acoustic field. The algorithm is based on the three-dimensional boundary element method for potential flows. We used a GPU to speedup computations and increase the problem size. Performance tests have shown the effectiveness of the implemented GPU codes.

We considered various cases, including small bubble-particle clusters and more complex systems containing a few bubbles and particles. The analysis of the bubble-particle dynamics shows that hydrodynamic flows generated by oscillating bubbles have a notable influence on the particle behavior. It should be noted that particles also affect the bubble dynamics, especially during the bubble compression stage.

We consider to considerably increase in further research the problem scale and reduce calculation time using algorithmic (fast multipole method) and hardware (heterogeneous computing architecture (CPU + GPU)) acceleration [19–21]. We also plan to expand and test the code for clusters containing non-spherical particles and bubbles, taking into account the torque. Also, qualitative comparison with experimental data shall be conducted.

This study is partially supported by the Skoltech Partnership Program and grant of the President of Russia (MK-3503.2017.8).

References

1. Tomita, Y., Robinson, P.B., Tong, R.P., Blake, J.R.: Growth and collapse of cavitation bubbles near a curved rigid boundary. J. Fluid Mech. **466**, 259–283 (2002). https://doi.org/10.1017/S0022112002001209
2. Brujan, E.A., Keen, G.S., Vogel, A., Blake, J.R.: The final stage of the collapse of a cavitation bubble close to a rigid boundary. Phys. Fluids **14**, 85–92 (2002). https://doi.org/10.1063/1.1421102
3. Miao, H., Gracewski, S.M.: Response of an ultrasonically excited bubble near a fixed rigid object. Acoust. Res. Lett. Online **6**, 144–150 (2005). https://doi.org/10.1121/1.1898344
4. Gracewski, S.M., Miao, H., Dalecki, D.: Ultrasonic excitation of a bubble near a rigid or deformable sphere: implications for ultrasonically induced hemolysis. J. Acoust. Soc. Am. **117**, 1440–1447 (2005). https://doi.org/10.1121/1.1858211
5. Hay, T.A., Hamilton, M.F., Ilinskii, Yu.A., Zabolotskaya, E.A.: Coupled pulsation and translation of a gas bubble and rigid particle. In: AIP Conference Proceedings, vol. 1022, pp. 209–212 (2008). https://doi.org/10.1063/1.2956188
6. Derjaguin, B.V., Dukhin, S.S.: Theory of flotation of small and medium-size particles. Trans. Inst. Min. Metall. **70**, 221–246 (1961)
7. Phan, C.M., Nguyen, A.V., Miller, J.D., Evans, G.M., Jameson, G.J.: Investigations of bubble-particle interactions. Int. J. Miner. Process. **72**, 239–254 (2003). https://doi.org/10.1016/S0301-7516(03)00102-9
8. Dai, Z., Fornasiero, D., Ralston, J.: Particle-bubble collision models - a review. Adv. Colloid Interface Sci. **96**, 54 (2000). https://doi.org/10.1016/S0001-8686(99)00030-5
9. Nguyen, A.V., Evans, G.M.: Attachment interaction between air bubbles and particles in froth flotation. Exp. Therm. Fluid Sci. **28**, 381–385 (2004). https://doi.org/10.1016/j.expthermflusci.2002.12.001
10. Verrelli, D.I., Koh, P.T.L., Nguyen, A.V.: Particle-bubble interaction and attachment in flotation. Chem. Eng. Sci. **66**, 5910–5921 (2011)
11. Basarova, P., Machon, V., Hubicka, M., Horn, D.: Collision processes involving a single rising bubble and a larger stationary spherical particle. Int. J. Miner. Process. **94**, 58–66 (2010). https://doi.org/10.1016/j.minpro.2009.11.004
12. Koh, P.T.L., Schwarz, M.P.: CFD modelling of bubble-particle collision rates and efficiencies in a flotation cell. Miner. Eng. **16**, 1055–1059 (2003). https://doi.org/10.1016/j.mineng.2003.05.005
13. Liu, T.Y., Schwarz, M.P.: CFD-based modelling of bubble-particle collision efficiency with mobile bubble surface in a turbulent environment. Int. J. Miner. Process. **90**, 45–55 (2009)
14. Canot, E., Achard, J.-L.: An overview of boundary integral formulations for potential flows in fluid-fluid systems. Arch. Mech. **43**, 453–98 (1991)
15. Bui, T.T., Ong, E.T., Khoo, B.C., Klaseboer, E., Hung, K.C.: A fast algorithm for modeling multiple bubbles dynamics. J. Comput. Phys. **216**, 430–453 (2006). https://doi.org/10.1016/j.jcp.2005.12.009
16. CUDA C Programming Guide. http://docs.nvidia.com/cuda/cuda-c-programming-guide/index.html
17. NVIDIA Programming Guide. http://docs.nvidia.com/cuda/cuda-c-programming-guide/index.html
18. Saad, Y., Schultz, M.H.: GMRES: a generalized minimal residual algorithm for solving nonsymmetric linear systems. SIAM J. Sci. Stat. Comput. **7**, 856–869 (1986). https://doi.org/10.1137/0907058

19. Itkulova(Pityuk), Yu.A., Solnyshkina, O.A., Gumerov, N.A.: Toward large scale simulations of emulsion flows in microchannels using fast multipole and graphics processor accelerated boundary element method. In: ASME 2012 International Mechanical Engineering Congress and Exposition, pp. 873–881 (2012). https://doi.org/10.1115/IMECE2012-86238
20. Abramova, O.A., Pityuk, Yu.A., Gumerov, N.A., Akhatov, I.Sh.: High-performance BEM simulation of 3D emulsion flow. In: Sokolinsky, L., Zymbler, M. (eds.) PCT 2017. CCIS, vol. 753, pp. 317–330. Springer, Cham (2017). https://doi.org/10.1007/978-3-319-67035-5_23
21. Itkulova(Pityuk), Yu.A., Abramova, O.A., Gumerov, N.A., Akhatov I.S.: Simulation of bubble dynamics in three-dimensional potential flows on heterogeneous computing systems using the fast multipole and boundary element methods. Numer. Methods Program. **15**, 239–257 (2014). (in Russian)
22. Zinchenko, A.Z., Rother, M.A., Davis, R.H.: A novel boundary-integral algorithm for viscous interaction of deformable drops. Phys. Fluid. **9**(6), 1493–1511 (1997). https://doi.org/10.1063/1.869275
23. Plesset, M.S., Prosperetti, A.: Bubble dynamics and cavitation. J. Fluid Mech. **9**, 145–185 (1977). https://doi.org/10.1146/annurev.fl.09.010177.001045

VM2D: Open Source Code for 2D Incompressible Flow Simulation by Using Vortex Methods

Kseniia Kuzmina[1,2], Ilia Marchevsky[1,2(✉)], and Evgeniya Ryatina[1,2]

[1] Bauman Moscow State Technical University, Moscow, Russia
{kuz-ksen-serg,evgeniya.ryatina}@yandex.ru, iliamarchevsky@mail.ru
[2] Ivannikov Institute for System Programming, Russian Academy of Sciences, Moscow, Russia

Abstract. The article describes an open source C++ code we have developed for 2D incompressible flow simulation using vortex methods. The code has a modular structure, and allows users to simulate flows around airfoils (also, around a system of airfoils) and compute unsteady hydrodynamic loads acting the airfoils. It is also possible to simulate hydroelastic regimes of airfoil motion in the flow by using weakly and strongly coupling strategies. The software implements well known algorithms and also original numerical schemes developed by the authors, thereby significantly increasing the accuracy of simulations compared to traditional algorithms. The software makes it possible to run simulations in parallel mode; OpenMP and MPI technologies are supported.

The VM2D source code is available on GitHub under GNU GPL license.

Keywords: Vortex method · Incompressible flow
Hydrodynamic loads · Fluid-structure interaction · Hydroelasticity
MPI · OpenMP · Open source code · Boundary integral equation

1 Introduction

Vortex methods of simulation of flows around airfoils are based on the consideration of vorticity as a primary computed variable and on a fundamental principle discovered by N. E. Zhukovsky in 1906: an immovable airfoil affects an inviscid incompressible flow just as an attached vortex sheet placed on the surface line of the airfoil [1]. Later, this principle was developed and generalized by S. A. Chaplygin and M. W. Kutta. According to these results, it is possible to replace the airfoil with a vortex sheet and then determine somehow the intensity of this vortex sheet. For some airfoils of simple shapes (elliptical airfoils and Zhukovsky wing airfoils), it can be found by using a technique based on conformal mappings. Such solutions can be used as benchmarks for verification of numerical algorithms.

The research is funded by Russian Science Foundation (proj. 17-79-20445).

L. Sokolinsky and M. Zymbler (Eds.): PCT 2018, CCIS 910, pp. 251–265, 2018.
https://doi.org/10.1007/978-3-319-99673-8_18

When solving the Navier–Stokes equations, the classical Zhukovsky approach remains correct in principle both for movable and immovable airfoils. It is necessary, however, to simulate the vorticity flux [2] from the body surface to the flow. This means that the vortex sheet that simulates the airfoil surface line influence should be considered as a free vortex sheet instead of being an attached one. According to Lighthill's approach, this phenomenon can be modelled as a result of vorticity generation on the surface line K. Then the vorticity, which is concentrated in this vortex sheet with intensity $\gamma(r)$, $r \in K$, becomes part of the vortex wake and moves in the flow according to the governing equations.

In order to simulate a movable airfoil in a viscous flow, both attached and free vortex sheets as well as attached source sheets might be considered.

2   Reconstruction of Flow Variables

In meshless Lagrangian vortex methods, which are used for incompressible flow simulation, vorticity is a primary computed variable, while velocity and pressure fields can be reconstructed by using the known vorticity distribution in the flow.

Vorticity distribution in the flow is simulated by a set of vortex elements, namely elementary vorticity fields corresponding to circular vortices with some given shape of vorticity distribution, for example, Rankine's vortex or Lamb's vortex (Fig. 1).

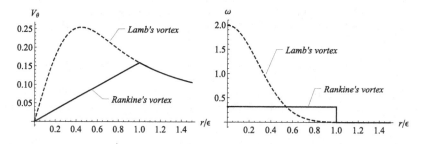

Fig. 1. Circumferential velocity induced by Rankine's and Lamb's vortices, with unit circulation and vorticity distributions; ϵ is Rankine's vortex radius. Lamb's vortex parameters are such that more than 0,998 of the total vorticity is inside an ϵ-circle; the difference of velocities at $r/\epsilon > 1$ is less than 0.2%

2.1   Velocity Reconstruction

The simplest way to compute the flow velocity at an arbitrary point r in the flow domain S is by means of the Biot–Savart law:

$$\boldsymbol{V}(\boldsymbol{r}) = \int\limits_{S} \frac{\Omega(\boldsymbol{\xi})}{2\pi} \frac{\boldsymbol{k} \times (\boldsymbol{r} - \boldsymbol{\xi})}{|\boldsymbol{r} - \boldsymbol{\xi}|^2} dS_{\xi} + \boldsymbol{V}_{\infty}, \tag{1}$$

where $\Omega(\xi)$ is a vorticity field, k is a unit vector orthogonal to the flow plane, V_∞ is the incident-flow velocity. For simplicity, we suppose the airfoils to be immovable.

The computational cost of this method, however, is extremely high; it is proportional to N^2, where N is the number of vortex elements, as it is necessary to take into account the mutual influence of all pairs of vortices. In practice, this fact strongly restricts the capabilities of vortex methods since computation time becomes unacceptably high at $N \sim 10\,000\ldots100\,000$, even if parallel algorithms are used.

Fast approximate methods are being used to solve this problem. The most suitable are the following:

- Barnes–Hut-type fast method, firstly developed for the gravitational N-body problem [3] and then adapted by prof. G. Ya. Dynnikova for vortex methods [4];
- fast method based on the solution of a Poisson auxiliary equation for a stream function on a coarse mesh using the fast Fourier transform (FFT) [5].

The computational complexity of both methods is proportional to $N \log N$, their accuracies are comparable if the parameters choice is optimal. Note, however, that the derivation of quantitative estimations for the accuracy of fast methods is a nontrivial task.

2.2 Pressure Reconstruction and Loads Computation

In order to determine the pressure distribution in the flow domain, an analogue of the Cauchy–Lagrange integral can be used [6]. However, in practice, it is necessary, as a rule, to determine hydrodynamic loads (forces and torque) acting on the airfoil in the flow. It is possible to use for this purpose integral formulae derived by prof. G. Ya. Dynnikova and adapted to several types of problems being solved by means of vortex methods [7]:

- flow around an immovable airfoil;
- flow around a rigid airfoil in translational motion;
- flow around a rigid airfoil in rotational motion;
- flow around a rigid airfoil in arbitrary motion.

These integral formulae have been obtained by integrating the pressure distribution over the airfoil surface line. Approximate formulae have also been derived by prof. G. Ya. Dynnikova to take into account the influence of viscous stresses. In practice, viscous stress influence becomes essential only for flows at Re < 1 000.

3 Vorticity Generation on Airfoil Surface Line

Vorticity in a viscous incompressible flow is generated only at the airfoil surface line. The intensity of the vortex sheet, which is formed during a small time interval, is described by boundary integral equation. There are two possible ways to write down such equation:

– Singular integral equation of the first kind with Hilbert-type kernel Q_n:

$$-\frac{1}{2\pi}\oint_K \underbrace{\frac{(\boldsymbol{r} - \boldsymbol{\xi}) \cdot \boldsymbol{\tau}(\boldsymbol{\xi})}{|\boldsymbol{r} - \boldsymbol{\xi}|^2}}_{Q_n(\boldsymbol{r},\boldsymbol{\xi})} \gamma(\boldsymbol{\xi})\,dl_\xi = f_n(\boldsymbol{r}), \quad \boldsymbol{r} \in K, \tag{2}$$

where $\boldsymbol{\tau}$ is the unit tangent vector at the corresponding point of the airfoil surface line. Specially derived quadrature formulae should be used to compute the Cauchy principal value of the integral. Discrete Vortices Method-type quadrature formulae are used in most cases [8], yet their accuracy is not very high and they require a special surface line discretization which cannot be easily done in cases of airfoils of complex shape, or when they are deformable.

– Fredholm-type integral equation of the second kind:

$$\frac{1}{2\pi}\oint_K \underbrace{\frac{(\boldsymbol{r} - \boldsymbol{\xi}) \cdot \boldsymbol{n}(\boldsymbol{\xi})}{|\boldsymbol{r} - \boldsymbol{\xi}|^2}}_{Q_\tau(\boldsymbol{r},\boldsymbol{\xi})} \gamma(\boldsymbol{\xi})dl_\xi - \frac{\gamma(\boldsymbol{r})}{2} = f_\tau(\boldsymbol{r}), \quad \boldsymbol{r} \in K, \tag{3}$$

where $\boldsymbol{n}$ is the unit normal vector, and the kernel Q_τ is bounded by one half of the maximal curvature of the airfoil (this means that, in case of a smooth airfoil, the kernel is a uniformly bounded function).

For this approach, we develop a hierarchy of numerical schemes of the first and second order of accuracy adapted for flow simulation around smooth and non-smooth airfoils [9–11].

The right-hand sides of both equations are, respectively, the normal and the tangent components of the flow velocity at the corresponding point of the airfoil surface line:

$$f_n(\boldsymbol{r}) = -\boldsymbol{V}_\omega(\boldsymbol{r}) \cdot \boldsymbol{n}(\boldsymbol{r}), \quad f_\tau(\boldsymbol{r}) = -\boldsymbol{V}_\omega(\boldsymbol{r}) \cdot \boldsymbol{\tau}(\boldsymbol{r}), \tag{4}$$

where

$$\boldsymbol{V}_\omega(\boldsymbol{r}) = \boldsymbol{V}_\infty + \sum_{i=1}^N \boldsymbol{Q}^{(i)}(\boldsymbol{r}) \tag{5}$$

is the influence of the incident flow and vortex elements which simulate the vortex wake. It is important to correctly take into account the influence of vortices that are located in close proximity to the airfoil, first of all, in a boundary layer. Numerical experiments show that the second approach and the numerical schemes based on it are much more accurate compared to the first approach [12].

4   Vorticity Evolution in a Flow

The simulation of the vortex wake evolution in an inviscid incompressible flow is rather easy using the vortex method: it is just necessary to find a numerical solution of a system of ordinary differential equations,

$$\frac{d\boldsymbol{r}_i}{dt} = \boldsymbol{V}(\boldsymbol{r}_i), \quad i = 1, \dots, N, \tag{6}$$

where r_i is the position of the i-th vortex element. It means that vorticity is just being translated in the flow domain at flow velocity.

There are several known approaches that take into account viscosity influence: random walk method [13], particle strength exchange (PSE) method [14], diffusive velocity method [15], but it seems that the most suitable method for 2D flows simulation is the viscous vortex domains (VVD) method, developed by G. Ya. Dynnikova and co-authors [16]. Within the framework of the VVD method, the motion of the vortex elements simulating the vortex wake is described by an ODE system,

$$\frac{dr_i}{dt} = V(r_i) + W(r_i), \quad i = 1, \ldots, N, \tag{7}$$

whereas their circulations remain the same.

An efficient approach, which is also considered as an important part of the VVD method [16], is known for the computation of the diffusive velocity,

$$W(r) = -\nu \nabla \ln |\Omega(r)| = -\nu \frac{\nabla \Omega(r)}{\Omega(r)}, \tag{8}$$

which is proportional to the flow viscosity ν and depends both on the vorticity distribution in the flow domain in a neighborhood to the point r and on the shape of the flow region boundary (if there is such in a neighborhood to r).

5 VM2D Open Source Code

The vortex method is a powerful tool for numerical simulations in a wide range of engineering applications: aerodynamical loads estimation for aircrafts and aircraft trail simulation, simulation of hydroelastic oscillations of structural elements interacting with a flow, and solution of problems of industrial aerodynamics. The application of vortex methods is limited to flow regimes with low subsonic Mach numbers, when the influence compressibility can be neglected.

For such problems, especially hydroelastic ones, vortex methods can be very efficient, at least compared to "traditional" mesh methods, where it is necessary to deform or reconstruct the mesh at each time step. Moreover, for vortex methods, the computational cost of solving a hydroelastic problem remains nearly the same as for a flow simulation around an immovable airfoil, and, in fact, it is much easier to implement a numerical scheme with a strong coupling (again, compared to mesh methods).

At the same time, vortex methods have not been implemented in widespread software. Scientific groups developing new modifications of vortex methods have, of course, their own codes, but those are "private", they are not available to other scientists and engineers. As far as we know, there are no open source codes implementing vortex methods (neither 2D nor 3D).

We have developed such a code, named VM2D, for 2D incompressible-flow simulation around movable and immovable airfoils, and also for solving coupled fluid-structure interaction (FSI) problems. This is an open-source software, available on GitHub: https://github.com/vortexmethods/VM2D.

5.1   The Structure of the VM2D Code

The source code is written in C++ and has a modular structure. It is a cross-platform software and can be compiled under Windows and Linux by using MSVC, GCC or Intel C++ Compiler (as well as other compilers supporting the C++11 standard). The Eigen external library is used in VM2D for the numerical solution of linear equations systems and fast Fourier transform.

OpenMP and MPI technologies are used for computation acceleration on multi-core and multiprocessor cluster systems.

5.2   Problems Description in VM2D

It is possible to use VM2D for solving a particular problem or a set of similar (or not similar) problems. For every problem to be solved, a separate directory should be created with the same name as the problem. The list of problems should be saved in the problems text file, which normally has the following structure:

```
problems = {
    wing00deg(np = 1, alpha =  0, tau = 0.015),
    wing05deg(np = 2, alpha =  5, tau = 0.015),
    wing10deg(np = 2, alpha = 10, tau = 0.010)
};
```

In the simplest case, it is enough to leave the parentheses empty, but there are two parameters (pspfile and np) whose default values are the following: 'passport' for pspfile, that is, the problem passport is stored in the file passport inside the directory of the same name as the problem; and 1 for np, which is the number of processors used for the solution in parallel mode through MPI technology. Note that OpenMP technology for parallelization of the algorithm in shared-memory mode is used for every multi-core processor.

All other parameter in curly brackets are the definitions of arbitrary variables which a user can later use inside 'passport' files for their unification and for notational convenience.

The following notations and rules are used:

- "//"—one-line comment;
- "/* ... */"—multi-line comment;
- line break—the same as a blank space;
- ";"—separator between different parameters;
- ","—separator between different parameters inside the list of parameters (in parentheses), or separator between the components of arrays and vectors;
- blank spaces and tabs are ignored;
- parameter names are not case sensitive.

In the above-considered example of problems text file, where a user plans to solve three problems of flow simulation around a wing airfoil (as it follows from the names of the problems) for different angles of incidence, the passport file can be the same for all the problems and it may have the following structure:

```
/*-------------------------------*- VM2D -*-----------------*-----------*\
| ##   ##  ##    ##   ####  #####  |                        | Version 1.0 |
| ##   ##  ### ### ##  ##  ##  ##  | VM2D: Vortex Method    | 2017/12/01  |
| ##   ##  ## # ##      ##  ##  ## | for 2D Flow Simulation *-------------*
| ####  ##   ##   ##    ##  ##  ## | Open Source Code                     |
|  ##   ##   ## ###### #####       | www.github.com/vortexmethods/VM2D    |
|                                                                         |
| Copyright (C) 2017 Ilia Marchevsky, Kseniia Kuzmina, Evgeniya Ryatina  |
*-------------------------------------------------------------------------*
| File name: ./wing00deg/passport                                         |
| Info: Parameters of the problem to be solved                            |
\*-------------------------------------------------------------------------*/

//Physical Properties
rho = 1.0;
vInf = {1.0, 0.0};
timeAccel = 0.5;
nu = 0.001;

//Time Discretization Properties
timeStart = 0;
timeStop = 10.0;
dt = $tau;
deltacntText = 10;
deltacntBinary = 0;

//Wake Discretization Properties
vortexType = vortexRankine;
eps = 0.015;
epscol = 0.010;
distKill = 10.0;
delta = 1e-5;

//Numerical Schemes
linearSystemSolver = eigenLU;
velocityComputation = velocityBiotSavart;
wakeMotionIntegrator = explicitEuler;

//Airfoils and wake
airfoil = {
 wing(basePoint = {0.0, 0.0}, scale = 1.0, angle = $angle,
      panelsType = panelsRectilinear,
      boundaryCondition = boundaryConstLayerAver,
      mechanicalSystem = mechanicsRigidImmovable)  };
wake = {};
```

The meanings of the parameters contained in the passport are described in Table 1. Note that the list of parameters most probably will be modified in future versions of VM2D.

The choice of numerical schemes determines the following numerical methods:

Table 1. Main parameters in `VM2D` passport file

Parameter name	Brief description
Physical properties	
rho	Density of the flow
vInf	Incident flow velocity
timeAccel	Time required for uniform acceleration of the incident flow from zero velocity to `vInf`
nu	Flow kinematic viscosity
Time discretization properties	
timeStart	Physical time at which the simulation starts (normally 0.0)
timeStop	Physical time at which the simulation stops
dt	Time step
deltacntText	Period (in steps) for vortex wake storage to text file
deltacntBinary	The same for binary file (VTK format)
Wake discretization properties	
vortexType	Vortex element type
eps	Typical radius of the vortex element
epscol	Maximum distance at which two vortices can merge into one vortex (if some additional conditions are satisfied)
distKill	Distance at which the vortex wake is removed from the simulation
delta	A small distance at which vortex elements are placed over the airfoil surface line after being generated

- `linearSystemSolver`— method for solution of linear algebraic systems;
- `velocityComputation`—method for computation of velocities in the flow;
- `wakeMotionIntegrator`—method for numerical integration of ODE systems.

Files containing airfoils geometry are stored in the **airfoils** directory; these are text files of a very simple format which becomes clear from examples. The following parameters should be specified in parentheses after the file name:

- `basePoint`—point where the airfoil center should be placed;
- `scale`—scale factor for the airfoil;
- `angle`—angle of incidence;
- `panelsType`—surface-line approximation method;
- `boundaryConditionSatisfaction`—numerical method for boundary condition satisfaction (integral equation solution), which determines the numerical scheme used for computation of the vortex sheet intensity;
- `mechanicalSystem`—numerical scheme for coupling strategy implementation in coupled FSI problems.

All the parameters used in the simulation must be defined in the passport file. For some of them there are predefined default values. Default values with the lowest priority are specified just inside the code; defaults with higher priority can be specified by the user in the **defaults** file. The description (in terms of integer values) for verbal expressions of some options (such as `vortexRankine`, `velocityBiotSavart` and others) is given in the `switchers` file.

Note that two parameters are not defined explicitly in this passport, namely the time step `dt`, and the angle `alpha` of incidence of the airfoil. These parameters are marked with "$", which means that their values are equal to user-defined variables included in the `problems` file, in parentheses after the names of the corresponding problems.

If the flow around a system of airfoils is to be simulated, it is possible to specify more than one airfoil. In this case, the corresponding section of the passport file has the following structure:

```
airfoil =
{ square_160points(basePoint = {0.0,  0.0}, angle = 45.0, scale = 1.0),
  circle_200points(basePoint = {1.2, -0.2}, angle =  0.0, scale = 0.5) };
```

In this example, the interference phenomenon for two airfoils is simulated: a circular airfoil (scaled by a factor of 0.5) placed behind a square airfoil (installed at an angle of incidence of 45°). All other parameters will be set to default values.

If the user wants to use a previously simulated vorticity distribution, it can be uploaded by specifying the corresponding file name in the `fileWake` section of the passport. The files containing the descriptions of vortex wakes should be stored in the directory `wakes`.

5.3 Documentation

The programmer's guide to `VM2D` is generated automatically by using the `doxygen` tool. It includes full information about all the classes implemented in `VM2D`: description of all the class members and methods. Relationships between the classes are shown in graphical mode, as well as execution diagrams of the functions. An html-version is available at http://vortexmethods.github.io/VM2D/, and is updated automatically via `Travis-CI` service after every modification of the source code and its push on GitHub. The pdf-version of the programmer's guide, also generated automatically by `doxygen` and then compiled with LaTeX , is available at https://github.com/vortexmethods/VM2D/pdf/VM2-code.pdf.

5.4 Main Classes in VM2D

The file where the `main` function (the entry point of the C++ program) is determined is `VM2D.cpp`. Its structure is very simple; a `Queue` class instance is created, a list of problems to be solved is loaded and the 'numerical conveyer' is started. It runs until all the problems from the list described in `problems` file are solved. All available processors, which work using MPI, are split into subgroups according to the number of processors required for the simulation of each particular problem. If the number of processors is less than the total number of processors required for the simultaneous solution of all the problems from the list, then the

`Queue` class works as a real queue with 'fifo' discipline. All subgroups of processors work in asynchronous mode; global synchronization is performed every some ΔT seconds (the so-called 'time quantum'), when the queue state is updated, finished problems are replaced with new ones, *etc.* The necessary "universal" constants and functions are defined in the `defs.h` and `defs.cpp` files.

The basic and auxiliary classes defined in `VM2D` are listed in Table 2.

Table 2. Classes defined in `VM2D`

Class name	Brief description
Basic classes	
Queue	List of problems to be solved, organizes its solution in parallel mode according to the number of required and available processors
Task	The state of the problem in `Queue` and its description (passport)
Passport	Full definition of the problem (its passport)
World2D	All the properties and current state of a particular problem from the queue; the instance of this class is the main object in the numerical simulation of the flow around airfoils
Airfoil	*abstract class*—the geometry of the airfoil
Wake	Describes vortex wake
Sheet	Determines attached and free vortex sheets and attached source sheets placed on the surface lines of airfoils
Boundary	*abstract class*—the numerical scheme used for the integral equation solution with respect to the unknown free vortex sheet intensity
Velocity	*abstract class*—the numerical method for computation of velocities in the flow
Mechanics	*abstract class*—the coupling strategy for hydroelastic problems
Auxiliary classes	
numvector	*template class* which determines an array of specified length consisting of variables of a specified type; for 'numerical' vectors, all basic arithmetic operations are defined, including the vector product (`operator^`) for 3-dimensional vectors consisting of `float` and `double` variables
Point2D	Inherits `numvector<double, 2>` and has the necessary MPI-descriptor
Vortex2D	Properties of the vortex element (its position and circulation), including the corresponding MPI-descriptor
Parallel	Properties of the MPI-communicator created for a particular problem
Preprocessor	Tools for input files preprocessing: comments exclusion, replacement of line-breaks with blank spaces, multiple spaces and tabs removal; the result is used as input data for `StreamParser`
StreamParser	Tools for input files parsing after preprocessing; it is used for reading all text files, including `problems`, `defaults`, `passport`, *etc.*
Times	Class with structures for time statistics assembling Ant tools for its saving to the `timestat` file

5.5 Abstract Classes Implementations

There are four main abstract classes in VM2D:

Airfoil,   Boundary,   Velocity,   Mechanics,

whose implementations correspond to different modifications of vortex methods; some of them are briefly described in the beginning of this paper.

The inheriting classes have names consisting of the name of the parent class and an additional word which specifies the particular method implemented there. The most important implementations of the abstract classes are given in Table 3.

Table 3. Abstract classes implementations

Class name	Brief description
For the Airfoil class	
Rectilinear	Piecewise rectilinear (polygonal) approximation of the airfoil surface line
Elliptic	Piecewise elliptical approximation of the airfoil surface line
For the Boundary class	
MDV	'Classical' discrete vortex method, singular integral equation solution
VortColl	Similar to the 'classical' discrete vortex method, but with Fredholm-type integral equation solution
ConstLayerAver	Piecewise constant vortex sheet approximation, Fredholm-type integral equation solution using Galerkin approach
LinearLayerAver	Discontinuous piecewise linear vortex sheet approximation, Fredholm-type integral equation solution
FEMLayerAver	Continuous (except for specified angle points) piecewise linear vortex sheet approximation, Fredholm-type integral equation solution
For the Velocity class	
BiotSavart	Direct velocity computation using the Biot–Savart law; $O(N^2)$ computational complexity
BarnesHut	Fast multipole method based on a Barnes–Hut-type hierarchial tree construction; $O(N \log N)$ computational complexity
Fourier	Fast method based on the solution of a Poisson auxiliary problem for stream function on a coarse mesh using fast Fourier transform; $O(N \log N)$ computational complexity
For the Mechanics class	
RigidImmovable	Flow simulation around a rigid immovable airfoil
RigidGivenLaw	Flow simulation around a rigid airfoil in arbitrary motion prescribed according to a given law
RigidWCoupled	Flow simulation around a rigid airfoil in a weakly coupled FSI problem
RigidSCoupled	Flow simulation around rigid airfoil in a strongly coupled FSI problem

Note that some of the described classes are not implemented in the current version of the VM2D code; they will be available in future versions, maybe in a slightly different state than it is described here.

5.6 Results of Simulation

The results of simulation are saved in files in the directory with the same name as the problem being solved. Files containing the description of the vortex wake at particular time steps (every `deltacnt` steps) are saved in the `snapshots` subdirectory in text or/and VTK formats.

Hydrodynamic loads acting the airfoils are saved in `forces-airfoil-n` for all time steps; their positions and velocities are saved in `position-airfoil-n` files; time statistics is stored in the `timestat` file. The program log is shown on screen. It can be redirected to a file by using the standard command prompt/shell operator `>`.

5.7 VM2D Compilation and Execution

In order to compile the `VM2D` code, it firstly should be downloaded or cloned from the `GitHub` repository:

https://github.com/vortexmethods/VM2D

Then a subdirectory `build` should be created in the directory where the repository has been cloned, and from there, the command "`cmake ..`" should be executed (or with the necessary `CMake` parameters). The MPI implementation and the `Eigen` library should be preliminarily installed and configured in your system, of course, additionally to some C++ compiler (C++11 standard compatible).

After such preparation, the code can be compiled. In order to run a simulation, it is necessary to execute the program from the folder where the `problems` file containing a list of problems description is placed. If it is necessary, it can be executed in parallel mode by using `mpirun`/`mpiexec`.

6 Some Results of Flow Simulation

Some results of flow simulation using VM2D are shown below.

6.1 Development of an Unsteady Flow Behind a Circular Cylinder

The initial (symmetrical) phase of the wake development behind a circular cylinder, at Reynolds number Re $= 1\,000$, is shown in Fig. 2. The next phase (when instability becomes apparent) is shown in Fig. 3.

The computed time dependencies of the unsteady drag force and lift force coefficients for a circular cylinder are shown in Fig. 4.

Drag force and lift force dimensionless coefficients are in good agreement with experimental data, as well as the Strouhal number which corresponds to the dimensionless vortex shedding frequency.

Fig. 2. Vortex wake behind a cylinder at the initial phase of its development

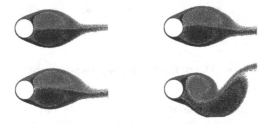

Fig. 3. Instability development of the vortex wake behind a circular cylinder

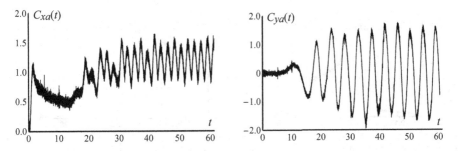

Fig. 4. Unsteady drag and lift force coefficients for a circular cylinder

6.2   The Blasius Solution

When the simulation physical time is rather high, viscous flow around a thin plate becomes steady (Fig. 5). The simulation results (horizontal and vertical components of the flow velocity in a boundary layer) are in excellent correspondence with the analytical solution obtained by Blasius.

Fig. 5. Steady regime of the flow around a thin plate

6.3 Circular Airfoil Wind Resonance

This occurs when the natural frequency of a mechanical subsystem (Fig. 6) is close to the vortex shedding frequency. The problem has been solved in the framework of the weakly coupled ("step-by-step") approach.

Fig. 6. The coupled FSI problem statement

The dependency between the oscillation amplitude and the natural frequency Sh_ω is shown in Fig. 7a. The results for maximum amplitude, resonance frequency and hysteresis properties (Fig. 7b) are in good agreement with experiments.

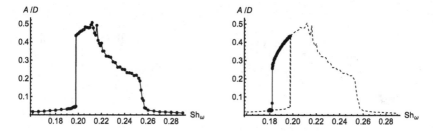

Fig. 7. Amplitude of a circular airfoil oscillations: initial state in equilibrium position (left), initial state is close to resonance oscillations (right): hysteresis loop is shown

6.4 Autorotation Simulation of the Savonius Rotor

Initially, the rotor has zero angular velocity and it starts spinning under hydrodynamic torque. The simulation has been performed as a strongly coupled FSI problem. The initial phases and time dependency for the angular velocity is shown in Fig. 8. The steady regime parameters are the same as those computed by other researchers.

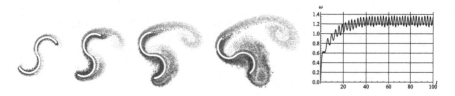

Fig. 8. The Savonius rotor spinning process in a flow, and its angular velocity

References

1. Tokaty, G.: A History and Philosophy of Fluid Mechanics. Courier Corporation, Dover (1994)
2. Lighthill, M.J.: Introduction. Boundary layer theory. In: Rosenhead, J. (ed.) Laminar Boundary Layers, pp. 54–61. Oxford University Press, New-York (1963)
3. Barnes, J., Hut, P.: A hierachical $O(N \log N)$ force-calculation algorithm. Nature **324**, 446–449 (1986). https://doi.org/10.1038/324446a0
4. Dynnikova, G.Ya.: Fast technique for solving the N-body problem in flow simulation by vortex methods. Comput. Math. Math. Phys. **49**, 1389–1396 (2009). https://doi.org/10.1134/S0965542509080090
5. Morgenthal, G., Walther, J.H.: An immersed interface method for the Vortex-In-Cell algorithm. Comput. Struct. **85**, 712–726 (2007). https://doi.org/10.1016/j.compstruc.2007.01.020
6. Dynnikova, G.Y.: An analog of the Bernoulli and Cauchy-Lagrange integrals for a time-dependent vortex flow of an ideal incompressible fluid. Fluid Dyn. **35**(1), 24–32 (2000). https://doi.org/10.1007/BF02698782
7. Dynnikova, G.Ya.: The integral formula for pressure field in the nonstationary barotropic flows of viscous fluid. J. Math. Fluid Mech. **16**(1), 145–162 (2014). https://doi.org/10.1007/s00021-013-0148-z
8. Lifanov, I.K.: Singular Integral Equations and Discrete Vortices. VSP, Utrecht (1996)
9. Kuzmina, K.S., Marchevsky I.K., Milani, D., Ryatina, E.P.: Accuracy comparison of different approaches for vortex sheet discretization on the airfoil in vortex particles method. In: Proceedings of V International Conference on Particle-Based Methods – Fundamentals and Applications, Hannover, Germany, pp. 691–702 (2017)
10. Kuzmina, K.S., Marchevskii, I.K., Moreva, V.S., Ryatina, E.P.: Numerical scheme of the second order of accuracy for vortex methods for incompressible flow simulation around airfoils. Russ. Aeronaut. **60**(3), 398–405 (2017). https://doi.org/10.3103/S1068799816030114
11. Kuzmina, K.S., Marchevsky, I.K., Ryatina, E.P.: Exact analytical formulae for linearly distributed vortex and source sheets influence computation in 2D vortex methods. J. Phys.: Conf. Ser. **918**, 012013 (2017). https://doi.org/10.1088/1742-6596/918/1/012013
12. Kuzmina, K.S., Marchevsky, I.K., Moreva, V.S.: Vortex sheet intensity computation in incompressible flow simulation around airfoil by using vortex methods. Math. Models Comput. Simul. **10**(3), 276–287 (2018). https://doi.org/10.1134/S2070048218030092
13. Chorin, A.J.: Numerical study of slightly viscous flow. J. Fluid Mech. **57**, 785–796 (1973). https://doi.org/10.1017/S0022112073002016
14. Degond, P., Mas-Gallic, S.: The weighted particle method for convection-diffusion equations. Part I: the case of an isotropic viscosity. Math. Comput. **53**, 485–507 (1989). https://doi.org/10.2307/2008716
15. Ogami, Y., Akamatsu, T.: Viscous flow simulation using the discrete vortex model-the diffusion velocity method. Comput. Fluids. **19**, 433–441 (1991). https://doi.org/10.1016/0045-7930(91)90068-S
16. Dynnikova, G.Y.: The Lagrangian approach to solving the time-dependent Navier-Stokes equations. Doklady Phys. **49**(11), 648–652 (2004). https://doi.org/10.1134/1.1831530

Modeling of Nonstationary Two-Phase Flows in Channels Using Parallel Technologies

Yury Perepechko⬤, Sergey Kireev$^{(\boxtimes)}$⬤, Konstantin Sorokin⬤, and Sherzad Imomnazarov⬤

N.A. Chinakal Institute of Mining, SB RAS,
54, Krasniy Prospekt, Novosibirsk 630091, Russia
`perep@igm.nsc.ru`, `kireev@ssd.sscc.ru`

Abstract. The work is devoted to the modeling of a nonstationary flow of compressible two-phase mixtures. We construct a continuum approximation to a thermodynamically consistent model of two-velocity hydrodynamics under the assumption of phase equilibrium with respect to temperature. The parallel implementation of the control volume method is based on the methods of domain decomposition. The numerical algorithm stability is studied using problems of two-velocity flows in inclined channels that inhomogeneous with respect to the volume fraction of phases. The results of numerical simulation prove the effectiveness of the approach used.

Keywords: Mathematical modeling · Viscous compressible fluid
Two-phase mixture · Numerical simulation
Control volume method · Domain decomposition methods

1 Introduction

Interest in mathematical modeling of nonstationary nonisothermal two-phase mixtures is determined by the need to describe the hydrodynamics of such media as suspensions, emulsions, aerosols, and granular media, regarding technological processes in various heat-exchange systems, transportation of oil and gas, and micro-tunneling in soil, as well as by the necessity to study natural processes, for example, in the mantle-crust fluid, and fluid-magmatic systems. In the description of the mentioned technological processes and natural systems, researchers use various mathematical models obtained in the framework of such methodological approaches as the averaging method, the variational method, and the method of conservation laws. The equations of the models of two-speed hydrodynamics obtained by averaging are nonhyperbolic [1], which leads to an

---

This work was supported by the Russian Scientific Foundation under Project No. 17-77-20049 (Russia).

L. Sokolinsky and M. Zymbler (Eds.): PCT 2018, CCIS 910, pp. 266–279, 2018.
https://doi.org/10.1007/978-3-319-99673-8_19

ill-posed Cauchy problem. By the variational methods [2–4] and the method of conservation laws [5–8], one obtains confined models of heterogeneous media, where equations are thermodynamically consistent. This property ensures the physical correctness of the obtained solutions, a point that is of special importance in the analysis of compressible two-phase media. This property also allows using efficient numerical methods. The control volume method is currently one of the most widespread in numerical simulation of two-phase media. Difference algorithms built on the control volume method are formally divided into two groups: those based on the calculation of the density [9] with explicit approximation in time, and those based on the calculation of the pressure [10,11] partly or fully implicit in time. The schemes of the second group are more suitable for the analysis of a hydrodynamic system taking into account dissipative processes of various types.

In this paper, we consider the hydrodynamics of two-phase compressible media using on the example of a nonstationary nonisothermal flow of two-phase mixtures of viscous liquids in the presence of an impurity and taking into account thermal and diffusion effects. The equations of the thermodynamically consistent model of a compressible two-phase medium were obtained by the method of conservation laws [6,7]. Their differential implementation is based on the control volume method [10,11] which guarantees the integral preservation of the basic parameters of the model. The parallel implementation is based on the methods of domain decomposition. The stability of the numerical algorithm and the efficiency of the parallel algorithm are considered for the problem of a two-phase homogeneous flow in a vertical channel. The numerical simulation results are given for the problem of a two-phase flow in arbitrary inclined channels with nonuniform distribution of the volume content of phases in the flow. In this approach, we considered beforehand such models of two-velocity media as filtration in saturated porous and granular media [6,7,12], and the convection of a compressible two-phase medium [13].

2 The Mathematical Model

This paper considers a model of the dynamics of mixtures of compressible viscous fluids to a two-speed hydrodynamic approximation in the presence of impurities and taking into account the surface tension of the dispersed phase. A two-phase medium has the following structure: the particles of each phase move relative to each other and together constitute a two-speed continuum whose unit volume is characterized by two densities, two velocities, entropy, specific surface (a number of drops of the dispersed phase), and impurity concentration. As mentioned above, the governing equations for such a medium are derived from the conservation laws [6,7]. The method is based on fundamental physical principles: the harmonization of the first and the second laws of thermodynamics, the laws of conservation of mass, energy, and momentum, and the group invariance of the equations. Setting the form of the internal energy of the two-phase medium, namely $E_0 = E_0\left(\rho, \rho_\alpha, J, \mathbf{j}, \mathbf{j_0}, S\right)$ determines its thermodynamic properties. The first law of thermodynamics can thus be formulated as

$$dE_0 = TdS + \mu d\rho + \mu_\alpha d\rho_\alpha + \varsigma\sigma dJ + (\mathbf{u}_1 - \mathbf{u}_2, d\mathbf{j}_0), \tag{1}$$

where $\rho = \rho_1 + \rho_2$; $\mathbf{j} = \rho_1\mathbf{u} + \rho_2\mathbf{v}$ are, respectively, the density and the momentum of the two-velocity medium; μ, μ_α are the chemical potentials of the two-phase medium and the impurities; σ is the surface tension tensor; ς is the surface area of the dispersed phase; and $\mathbf{j}_0 = \mathbf{j} - \rho\mathbf{u}_2$ is the relative density of the pulse phases. The impurity concentration is determined by the ratio $c = \rho_\alpha/\rho$.

The pressure is determined by the thermodynamic formula [14]

$$p = -E_0 + TS + \mu_1\rho_1 + \mu_\alpha\rho_\alpha + \varsigma\sigma J + (\mathbf{u}_1 - \mathbf{u}_2)\mathbf{j}_0, \tag{2}$$

$$dp = SdT + \rho d\mu_1 + \rho_\alpha d(\mu_\alpha - \mu_1) + \varsigma J d\sigma + \mathbf{j}_0 d(\mathbf{u}_1 - \mathbf{u}_2). \tag{3}$$

Since we are considering here a two-phase medium flowing in a gravity field, the expressions of the internal energy and the chemical potential must contain a term conditioned by the additional energy of the system in the gravity field: $\rho\varphi = (\rho_1 + \rho_2 + \rho_\alpha)\varphi$. Here φ is the gravity field potential. The state of thermodynamic equilibrium of this medium is determined by the following conditions:

$$\nabla(\mu_1 + \varphi) = 0, \ \nabla\left(\mu_2 + \frac{\varsigma J}{\rho_2}\sigma + \varphi\right) = 0,$$
$$\nabla(\mu_\alpha + \varphi) = 0, \ \nabla T = 0, \ \mathbf{u}_1 = \mathbf{u}_2 = 0. \tag{4}$$

The full system of constitutive equations of the hydrodynamics of a two-phase mixture with impurities, taking into account dissipative effects may be written as

$$\frac{\partial\rho}{\partial t} + \operatorname{div}\mathbf{j} = 0, \ \frac{\partial J}{\partial t} + \operatorname{div}(J\mathbf{u}_1) = 0, \tag{5}$$

$$\frac{\partial\rho_\alpha}{\partial t} + \operatorname{div}(c_1\rho_1\mathbf{u}_1 + c_2\rho_2\mathbf{u}_2 + \mathbf{L}_\alpha) = 0 \tag{6}$$

$$\frac{\partial j_i}{\partial t} + \partial_k\left(\rho_1 u_{1i}u_{1k} + \rho_2 u_{2i}u_{2k} + P\delta_{ik} - (\zeta_1 + \zeta_{12})\delta_{ik}\operatorname{div}\mathbf{u}_1\right.$$
$$- (\zeta_2 + \zeta_{12})\delta_{ik}\operatorname{div}\mathbf{u}_2 - (\eta_1 + \eta_{12})\left(\partial_k u_{1i} + \partial_i u_{1k} - \frac{2}{3}\delta_{ik}\operatorname{div}\mathbf{u}_1\right) \tag{7}$$
$$\left.- (\eta_2 + \eta_{12})\left(\partial_k u_{2i} + \partial_i u_{2k} - \frac{2}{3}\delta_{ik}\operatorname{div}\mathbf{u}_2\right)\right) = \rho\mathbf{g},$$

$$\frac{\partial u_{2i}}{\partial t} + (\mathbf{u}_2, \nabla)u_{2i} = -\frac{1}{\rho}\partial_i P + \varsigma\frac{J}{\rho}\partial_i\sigma + \frac{\rho_1}{2\rho}\partial_i\mathbf{w}^2 - \frac{1}{\rho_2}b(j_i - \rho u_{1i})$$
$$+ \frac{1}{\rho_2}\partial_k\left(\eta_2\left(\partial_k u_{2i} + \partial_i u_{2k} - \frac{2}{3}\delta_{ik}\operatorname{div}\mathbf{u}_2\right)\right. \tag{8}$$
$$\left.+ \eta_{12}\left(\partial_k u_{1i} + \partial_i u_{1k} - \frac{2}{3}\delta_{ik}\operatorname{div}\mathbf{u}_1\right)\right) + \frac{1}{\rho_2}\partial(\zeta_2\operatorname{div}\mathbf{u}_2 + \zeta_{12}\operatorname{div}\mathbf{u}_1) + g_i,$$

$$\frac{\partial S}{\partial t} + \operatorname{div}\left(S_1\mathbf{u}_1 + S_2\mathbf{u}_2 + \frac{\mathbf{L}_q}{T} - \frac{\mu_\alpha\mathbf{L}_\alpha}{T}\right) = \frac{R}{T}. \tag{9}$$

The dissipative function R is given by the formula

$$
R = \frac{1}{\rho_2} b \left(j_i - \rho u_{1i} \right)^2 + \kappa \left(\frac{\nabla T}{T} \right)^2 + 2\nu \nabla \left(\frac{\mu_\alpha}{T} \right) \nabla T
$$
$$
+ DT^2 \left(\nabla \left(\frac{\mu_\alpha}{T} \right) \right)^2 + \frac{1}{2}\eta_1 \partial_k u_{1i} + \frac{1}{2}\eta_2 \partial_k u_{2i} + \eta_{12} \partial_k u_{1i} \partial_k u_{2i}
$$
$$
+ \zeta_1 \left(\mathrm{div}\, \mathbf{u}_1 \right)^2 + \zeta_2 \left(\mathrm{div}\, \mathbf{u}_2 \right)^2 + \zeta_{12} \,\mathrm{div}\, \mathbf{u}_1 \,\mathrm{div}\, \mathbf{u}_2.
\tag{10}
$$

Dissipative flows associated with thermal and diffusion phenomena have the form

$$
\mathbf{L}_q = -\kappa \frac{1}{T}\nabla T - \nu T \nabla \left(\frac{\mu_\alpha}{T} \right), \quad \mathbf{L}_\alpha = -\nu \frac{1}{T}\nabla T - DT\nabla \left(\frac{\mu_\alpha}{T} \right).
\tag{11}
$$

Here and above, p is the pressure; S is the entropy density, and $\mathbf{g}$ is the acceleration of free fall. The impurity concentration in the phase is given by $c_1\rho_1 = \frac{\rho_\alpha}{\rho}\rho_1 + 2\lambda_1\rho_2$, $c_2\rho_2 = \frac{\rho_\alpha}{\rho}\rho_2 - 2\lambda_1\rho_2$ [15]. The entropy density of the phases is expressed as $S_2 = \frac{\rho_2}{\rho}S - 2\left(\lambda_2 - \lambda_1\frac{\mu_\alpha}{T}\right)\frac{\rho_2}{\rho}$, $S_1 = \frac{\rho_1}{\rho}S + 2\left(\lambda_2 - \lambda_1\frac{\mu_\alpha}{T}\right)\frac{\rho_2}{\rho}$. The kinetic coefficients of interfacial friction b, shear viscosity of the phases η_i, thermal conductivity of the two-phase medium κ, and the coefficient ν are all functions of the thermodynamic parameters. Effects of bulk viscosity are not taken into account.

The equation of state of the two-phase medium to a linear approximation is given by

$$
\delta\rho = \rho\alpha\delta p - \rho\beta\delta T, \quad \delta s = \frac{c_p}{T}\delta T - \frac{1}{\rho}\beta\delta p.
\tag{12}
$$

The coefficients of volumetric compression α, thermal expansion β, and specific heat capacity c_p are supposed additive over the phases. The presence of impurities is considered to an approximation of the ideal solution. The surface tension and the chemical potential of the impurities are determined by the ratios

$$
\sigma = \sigma_0 \frac{(T_c - T)}{(T_c - T_{ref})} - a_2 \ln\left(1 + a_3 c\right), \quad \mu_\alpha = d_1 P + d_2 T + \bar{R}T\ln\left(c\right),
\tag{13}
$$

where $\bar{R}$ is the universal gas constant.

3   The Numerical Algorithm

The computational algorithm is based on the application of the control volume method for the integration of the original governing equations, along with an iterative procedure IPSA [16] for the calculation of a pressure field consistent with the flow field. For the calculation of the flows through the faces of the control volumes (approximation of the convective terms), we use an implementation of the HLPA scheme of second order of accuracy [17]. A scheme fully implicit in time is used. For the diffusion terms, we use a finite-difference approximation. All calculations are performed on uniform rectangular grids with shifted distribution of the nodes. The scheme of the numerical algorithm is shown in Fig. 1.

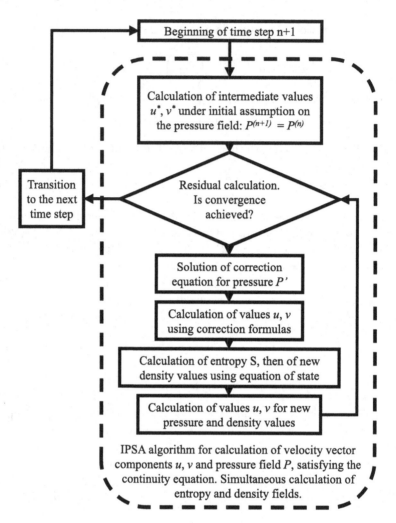

Fig. 1. Scheme of the numerical algorithm

The main computational complexity of the algorithm is in the solution of the systems of linear algebraic equations (SLAE) that result from discretization of the equations of the system. For solving most of these equations, it is possible to apply an alternating direction method (ADM), which is an alternate tridiagonal sweep along the coordinate directions until the conditions of convergence are met. In addition, the study has shown that for the considered class of problems in two-velocity hydrodynamics, the sweep in one of the directions has virtually no effect on the process of convergence, so it was excluded. This algorithm was implemented for solving most of the system equations. In the case of the pressure-correction equation, however, the ADM may diverge. There is therefore a need to use other methods of solution.

To solve the pressure-correction equation, it was decided to use the PETSc library [18]. This choice is based on the fact that the PETSc library contains a large number of subprograms for solving systems of linear algebraic equations by various methods, including interfaces to external solvers, and supports the use of parallel computing systems with distributed memory (MPI library). After comparing the solvers implemented in the PETSc library, we chose the biconjugate gradients stabilized (BiCGStab) method to be used in a parallel program. In addition, the usability of the solver for solving the remaining equations of the system was built into the parallel program.

4   Parallel Implementation

A parallel program based on the above model and methods was developed. The program makes it possible to simulate nonstationary two-phase flows in channels on distributed memory supercomputers (clusters). The parallel program was written in Fortran and parallelized using the MPI library. The parallel implementation is based on a domain decomposition method. The simulation domain is divided into subdomains along both directions. In this case, all arrays of grid values are appropriately cut and distributed between parallel processes (Fig. 2). The number of subdomains in each direction is an implementation parameter. The shadow edges of subdomains have a width of two cells, which corresponds to the size of the difference schemes used.

The parallel program developed uses two solvers: our own ADM implementation, and an implementation of the BiCGStab method in the PETSc library. Each of these implementations has certain advantages and disadvantages, which are discussed below.

The advantage of the parallel implementation of the ADM is that it does not require that the sparse matrix be explicitly built, i.e., no data redistribution between parallel processes is necessary. The decomposition of data structures (2D arrays) applied by this method is done in a natural way (Fig. 2). Possible imbalances of computational load due to the sequential nature of the sweep algorithm are leveled through the use of a pipelined algorithm [19,20]. The essence of the algorithm lies in the fact that when you run the sweep for a large number of rows, each process starts its part of the calculation of the next row before the end of the calculation of the previous one by other processes. At that, the adjacent processes transmit the sweep coefficients of the respective rows in the sweep direction.

The disadvantages of the existing implementation of the ADM are the following. Firstly, the ADM requires a sweep to be carried on in the direction of the Y-axis, which leads to inconsistent memory traversal and, consequently, to loss of productivity. Secondly, the current implementation of the ADM does not use vectorization, and the auto-vectorization is impossible owing to the sequential nature of the algorithm.

In contrast to the ADM, the BiCGStab method is implemented in the optimized PETSc library, in turn, using the optimized Intel MKL library. This to some extent ensures that at a low level all the necessary optimizations, such as efficient use of cache and vectorization, are fulfilled. Another feature of using PETSc is the requirement to build the SLAE matrix explicitly, i.e. some sort of redistribution of data between processes is still required. To minimize data exchanges, the PETSc library supports various methods of distribution of the elements of the global matrix SLAE between processes, including those consistent with the used 2D decomposition of the solution vector. This consistent distribution was used in the parallel program.

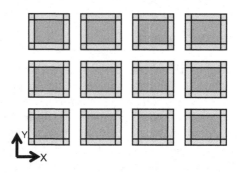

Fig. 2. 2D decomposition of the simulation domain

5 Performance Evaluation

To evaluate the performance of the parallel program the following computing resources were used:

- MVS-10P cluster [21]. Each cluster node contains 2 × 8-core Intel Xeon E5-2690 2.9 GHz (16 cores per node, 2 threads per core), 64 GB RAM, and 2 × MIC accelerators (not used in tests). From now on, these nodes will be referred to as SandyBridge.
- NKS-1P cluster [22], consisting of two different partitions:
  - KNL partition with nodes containing 1 × 72-core Intel Xeon Phi 7290 1.5 GHz (72 cores per node, 4 threads per core), 96 GB RAM, 16 GB MCDRAM. These nodes will henceforward be referred to as KNL.
  - Broadwell partition with nodes containing 2 × 16-core Intel Xeon E5-2697Av4 2.6 GHz (32 cores per node, 2 threads per core), 128 GB RAM. As of now, these nodes will be referred to as Broadwell.

The following cluster software was used:

- On MVS-10P cluster: Intel Fortran Compiler v.14.0.1, PETSc v.3.8.2 (built with Intel MKL v.11.1 update 1), Intel MPI library v.4.1.
- On NKS-1P cluster: Intel Fortran Compiler v.17.0.4, PETSc v.3.8.2 (built with Intel MKL v.2017.0 update 3), Intel MPI library v.2017 update 3.

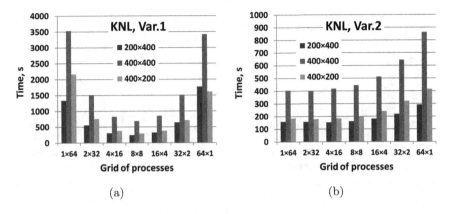

Fig. 3. Comparison of two variants of solvers using different mesh sizes

PETSc, similar to Intel MKL, may be run using several OpenMP threads. However, only MPI parallelization was used in this work, and the number of OpenMP threads was explicitly set to one.

The following two variants of solvers were used in the parallel program and compared:

- Variant 1: ADM solver is used for solving the equation of motion and the transfer equation, BiCGStab solver from PETSc library is used for solving the pressure-correction equation.
- Variant 2: BiCGStab solver from PETSc library is used for solving all equations.

As a test problem we consider the problem of a pressure flow of compressible fluids in a vertical channel in a gravity field. The dimensions of the computational domain were $LX = 1.0\,\mathrm{m}$, $LY = 0.15\,\mathrm{m}$. In all cases, the calculation time for 10 time steps is given.

The first step in the study was to determine the optimal parameters of decomposition of the region for two different versions of the program. Figure 3 presents the calculation results on the KNL nodes for two of the solvers used.

A more detailed view of the execution time is shown in Fig. 4. One can see that the largest components of the execution time correspond to the solution of the equations of motion ("V" on graphs) and the solution of the pressure-correction equation ("P" on graphs). The other computation steps are almost always negligible. The times for the solution of the pressure-correction equation are equal in both variants, as expected, because the same BiCGStab solver is used. But the solution of the equations of motion by the ADM solver is slower than by the BiCGStab solver (more than twice for the fastest cases). One more thing is that the ADM solver is much more dependent on the domain decomposition parameters than the BiCGStab.

The next step of the parallel program analysis is the evaluation of the scalability within cluster node. In Fig. 5 execution times for different cluster nodes

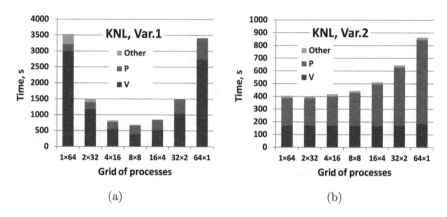

Fig. 4. Components of execution time for 400×400 mesh size on KNL, for two different solvers. Here V – solution of the equations of motion, P – solution of the pressure-correction equation, Other – all the other computations

depending on the number of MPI processes are shown. The mesh size 400×400 is used. The KNL node supports 288 hardware threads (72 cores, 4 threads per core) and is capable of running 288 MPI processes. The Fig. 5a shows that the least execution time is obtained for 64 MPI processes. A more thorough analysis showed that the optimal number of MPI processes for the KNL node is 72 (one thread per core). The Broadwell node supports 64 hardware threads (32 cores, 2 threads per core) and is capable of running 64 MPI processes. The Fig. 5b shows that the optimal number of MPI processes for the Broadwell node depends on the solver used. For variant 1 the optimal number is 32, whereas for variant 2 it is 32 or 64. The SandyBridge node supports 32 hardware threads (16 cores, 2 threads per core) and is capable of running 32 MPI processes. The Fig. 5c shows that the optimal number of MPI processes for the SandyBridge node, equally to the Broadwell node, depends on the solver used. For variant 1 the optimal number is 16, and for variant 2, it is 16 or 32. These results will be taken into account in further tests. A comparison of best execution times for different cluster nodes is presented in Table 1.

Table 1. Best execution times for different types of cluster nodes and different variants of solvers (mesh size 400×400, 10 time steps)

	KNL node	Broadwell node	SandyBridge node
Variant 1 (ADM + BiCGStab)	687 s	151 s	835 s
Variant 2 (BiCGStab)	406 s	86 s	273 s

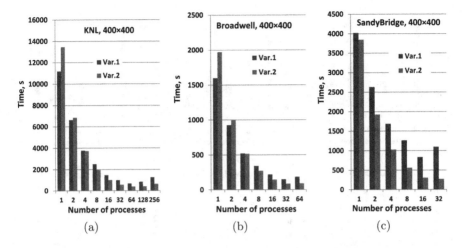

Fig. 5. Execution time depending on the number of MPI processes for one KNL node (a), one Broadwell node (b), and one SandyBridge node (c)

Parallelization speedup for all three types of nodes is shown in Fig. 6. As can be seen, the BiCGStab solver (variant 2) scales better than the ADM solver (variant 1). The maximum parallelization efficiency achieved within the nodes is 20 to 30% for variant 1, and 50 to 80% for variant 2.

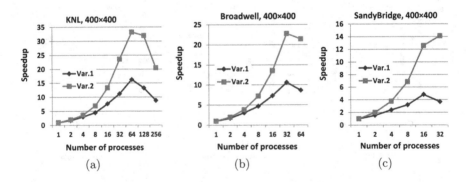

Fig. 6. Parallelization speedup depending on number of MPI processes for one KNL node (a), one Broadwell node (b), and one SandyBridge node (c)

The final step is the performance evaluation using several cluster nodes. Figures 7 and 8 show the results for cluster nodes of different types for two mesh sizes: 400×400 (Fig. 7) and 800×800 (Fig. 8). The number of cores shown at the X axis corresponds to the number of MPI processes used.

For the KNL nodes the limit of scalability for the 400×400 grid was reached on 8 nodes, whereas for the 800×800 grid, on 12 nodes. The use of a large number of MPI processes is adduced to explain these results. To push forward

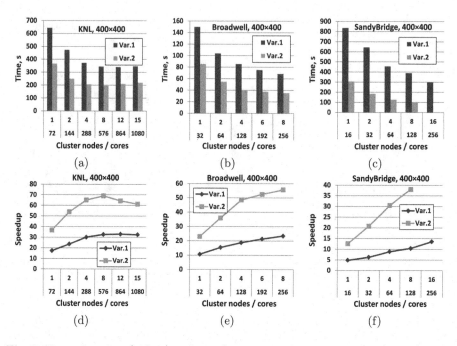

Fig. 7. Execution time (a, b, c) and speedup compared to a single core (d, e, f) using several cluster nodes of different types, mesh size: 400 × 400

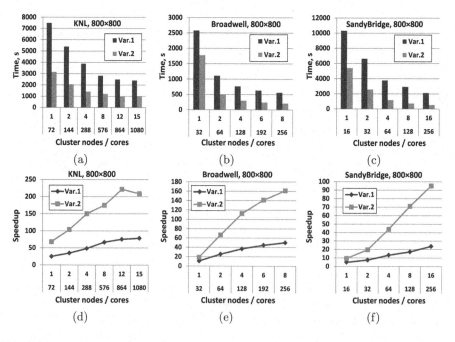

Fig. 8. Execution time (a, b, c) and speedup compared to a single core (d, e, f) using several cluster nodes of different types, mesh size: 800 × 800

the border of scalability, one can use parallelization in OpenMP threads inside a node. On the nodes Broadwell and SandyBridge, we did not note the approach of the scalability limit for the number of nodes considered.

In Fig. 9, a weak scaling efficiency for cluster nodes of different types is presented. A strong decrease in efficiency is observed for all types of nodes. The reason is that, besides usual parallel execution overhead, the number of iterations of the underlying numerical methods increases when the mesh size grows. For example, the number of IPSA iterations in the first time step equals 18 at a mesh size of 200×200, 28 at 400×400 mesh size and 69 at 800×800.

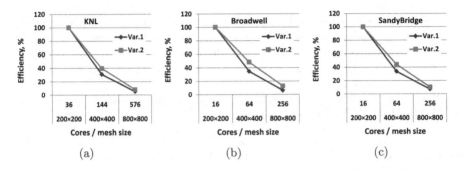

Fig. 9. Weak scaling efficiency for KNL (a), Broadwell (b), and SandyBridge (c) cluster nodes

Numerical simulation results are given for problems related to a nonstationary flow of compressible fluid mixture in an arbitrarily inclined channel with nonuniform distribution of phases across the section of the channel. The studied problem relates to a pressurized flow in an inclined channel of dimensions $LX = 1.0\,\text{m}$, $LY = 0.15\,\text{m}$. Figure 10 shows the distribution of the dispersed phase density for various angles of channel inclination.

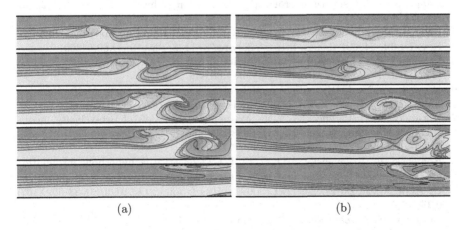

Fig. 10. Distribution of droplets of the dispersed phase for 20k, 30k, 40k, 50k, 100k steps: (a) channel inclination $0°$, (b) channel inclination $25°$

6   Conclusion

On the basis of the study into the problem of a nonstationary nonisothermal flow of a mixture of viscous compressible fluids, we can draw the following conclusions:

- The highest acceleration in the problem considered was specific for the nodes KNL, namely 221 times on 12 nodes (864 cores) on an 800×800 grid. Primarily, this is explained by the fact that only on the KNL nodes it became possible to run the task on such a large number of cores. Another record of acceleration is 161 times on eight Broadwell nodes (256 cores) on an 800×800 grid. The Broadwell nodes have shown a higher parallelization efficiency.
- Broadwell nodes showed the minimum work time on all grid sizes and for both versions of solvers used. Of the remaining node types, the KNL node in most cases shows a better performance than the SandyBridge node. However, when using multiple nodes, the SandyBridge nodes achieve less computation time due to better scalability.
- Of the two versions of solvers used, the fastest in almost all cases is the BiCGStab from the PETSc library.

Future plans: to optimize the ADM solver, to use parallelization in both solvers with OpenMP threads and to study the possibility of using other solvers, e.g. PARDISO in Intel MKL.

References

1. Kraiko, A.N., Sternin, L.E.: Theory of flows of a two-velocity continuous medium containing solid or liquid particles. Appl. Math. Mech. **29**, 418–429 (1965). https://doi.org/10.1016/0021-8928(65)90059-6
2. Bedford, A., Drumheller, D.S.: A variational theory of immiscible mixture. Arch. Ration. Mech. Anal. **68**, 37–51 (1978). https://doi.org/10.1007/BF00276178
3. Gavrilyuk, S.L., Saurel, R.: Mathematical and numerical modeling of two-phase compressible flows with microinertia. J. Comput. Phys. **175**, 326–360 (2002). https://doi.org/10.1006/jcph.2001.6951
4. Prix, R.: Variational derivation of Newtonian multi-fluid hydrodynamics. Phys. Rev. D. **69**, 043001 (2004). https://doi.org/10.1103/PhysRevD.69.043001
5. Landau, L.D.: The theory of superfluidity of helium II. J. Phys. (USSR) **5**, 71–79 (1941). https://doi.org/10.1103/PhysRev.60.356
6. Dorovsky, V.N.: Mathematical models of two-velocity media. I. Math. Comput. Model. **21**(7), 17–28 (1995). https://doi.org/10.1016/0895-7177(95)00028-Z
7. Dorovsky, V.N., Perepechko, Y.V.: Mathematical models of two-velocity media. II. Math. Comput. Model. **24**(10), 69–80 (1996). https://doi.org/10.1016/S0895-7177(96)00165-3
8. Romenski, E., Resnyansky, A.D., Toro, E.F.: Conservative hyperbolic formulation for compressible two-phase flow with different phase pressures and temperatures. Q. Appl. Math. **65**(2), 259–279 (2007). https://doi.org/10.1090/S0033-569X-07-01051-2
9. Godunov, S.K., Zabrodin, A.V., Ivanov, M.I., Kraiko, A.N., Prokopov, G.P.: Numerical solution of multidimensional problems of gas dynamics. Moscow Izdatel Nauka (1976). (in Russian)

10. Patankar, S.V., Spalding, D.B.: A calculation procedure for heat, mass and momentum transfer in three-dimensional parabolic flows. Int. J. Heat Mass Transf. **15**, 1787–1806 (1972). https://doi.org/10.1016/0017-9310(72)90054-3
11. Date, A.W.: Introduction to Computational Fluid Dynamic. Cambridge University Press, New York (2005)
12. Dorovsky, V.N., Perepechko, Y.V.: A hydrodynamic model for a solution in fracture-porous media. Rus. Geol. Geophys. **37**(9), 117–128 (1996)
13. Perepechko, Y.V., Sorokin, K.E.: Two-velocity dynamics of compressible heterophase media. J. Eng. Thermophys. **22**(3), 241–246 (2013). https://doi.org/10.1134/S1810232813030089
14. Dorovsky, V.N., Perepechko, Y.V., Sorokin, K.E.: Two-velocity flow containing surfactant. J. Eng. Thermophys. **26**(2), 160–182 (2017). https://doi.org/10.1134/S1810232817020047
15. Dorovsky, V.N.: The hydrodynamics of particles suspended in a melt with the self-consistent concentration field of an admixture. I. Comput. Math. Appl. **35**(11), 27–37 (1998). https://doi.org/10.1016/S0898-1221(98)00081-9
16. Yeoh, G.H., Tu, J.: Computational Techniques for Multi-phase Flows. Butterworth-Heinemann, Oxford (2010). https://doi.org/10.1016/B978-0-08-046733-7.00003-5
17. Wang, J.P., Zhang, J.F., Qu, Z.G., He, Y.L., Tao, W.Q.: Comparison of robustness and efficiency for SIMPLE and CLEAR algorithms with 13 high-resolution convection schemes in compressible flows. Numer. Heat Transf. Part B **66**, 133–161 (2014). https://doi.org/10.1080/10407790.2014.894451
18. PETSc - Portable, Extensible Toolkit for Scientific Computation. https://www.mcs.anl.gov/petsc
19. Povitsky, A.: Parallelization of the pipelined thomas algorithm. ICASE Report No. 98-48, NASA Langley Research Center, Hampton (1998)
20. Sapronov, I.S., Bykov, A.N.: A parallel-pipelined algorithm. Atom **4**, 24–25 (2009)
21. MVS-10P cluster, JSCC RAS. http://www.jscc.ru
22. NKS-1P cluster, SSCC SB RAS. http://www.sscc.icmmg.nsc.ru

Supercomputer Simulation of Cathodoluminescence Transients in the Vicinity of Threading Dislocations

Karl K. Sabelfeld and Anastasiya Kireeva$^{(\boxtimes)}$

Institute of Computational Mathematics and Mathematical Geophysics,
6, Prospekt Lavrentjeva, Novosibirsk 630090, Russia
karl@osmf.sscc.ru, kireeva@ssd.sscc.ru

Abstract. The article deals with the implementation of the Monte Carlo method for simulation of cathodoluminescence transients in the vicinity of threading dislocations in semiconductors. The Monte Carlo algorithm is based on the random-walk-on-spheres method proposed by K. K. Sabelfeld for solving drift-diffusion-reaction parabolic equations. The cathodoluminescence intensity depends on the dislocation density, the recombination of excitons on dislocations and semiconductor surface, and the exciton diffusion length. To investigate the behavior of the cathodoluminescence characteristics for long times, we use a parallel implementation of the code involving the distribution of simulated exciton trajectories between MPI processes and OpenMP threads. The simulation results are contrasted with the exact solution of the equation.

Keywords: Cathodoluminescence · Threading dislocations
Random-walk-on-spheres method · Monte Carlo method
Parallel implementation · MPI with OpenMP programming

1 Introduction

Cathodoluminescence microscopy is a powerful technique to examine the internal structure of semiconductors. In particular, it is widely used to study the optical properties of gallium nitride (GaN) and indium gallium nitride (InGaN) [1], on the basis of which blue light-emitting diodes (LEDs), which are efficient lighting sources for energy saving, are constructed [2]. The cathodoluminescence (CL) and electron-beam induced current (EBIC) techniques are applied to analyze dislocations in crystals. Understanding a carrier dynamics around dislocations can help to achieve a high efficiency of LEDs. Threading dislocations act as non-radiative centers for excitons (bound pairs of an electron and hole) and are seen as dark spots in CL images [3,4]. Transient behavior of CL intensity depends mainly on the dislocation density, diffusion length, and exciton life time.

The support of the Russian Science Foundation under grant No. 14-11-00083 is kindly acknowledged.

L. Sokolinsky and M. Zymbler (Eds.): PCT 2018, CCIS 910, pp. 280–293, 2018.
https://doi.org/10.1007/978-3-319-99673-8_20

Thus, the CL intensity transients can be used for determination of the parameters of these materials. However, the investigation of CL around dislocations requires picosecond temporal and nanoscale spatial resolutions, which is difficult to achieve [1]. Computer simulation of CL allows to calculate CL intensity transients for various parameters values with the required resolution.

In [1], a model of exciton diffusion and recombination on a single dislocation is proposed based on the two-dimensional transient diffusion equation. The authors determined the effective radius of the dislocation, the diffusion length and the effective exciton lifetime by adjusting the model parameters for the experimental data. In [5], a three-dimensional Monte Carlo-based model of the steady-state CL emission is presented. The electron-hole pair generation under an incident electron beam is simulated by means of the Monte Carlo method. The resulting steady-state carrier distribution in the semiconductor volume is calculated using Berz and Kuiken's formulation [6] of the diffusion and surface-recombination processes.

In [7], we suggested a random-walk-on-spheres (RWS) method for the calculation of CL and EBIC maps. This stochastic method is based on a reciprocity theorem formulated and proven in [7] for a general case of the Robin boundary condition. This approach was used in [4], where we derived exact solutions in the case of one threading dislocation with general boundary conditions on the dislocation and plane surfaces. In [8], this algorithm was generalized to the drift-diffusion-reaction problem. In [9], we extended the RWS method to transient drift-diffusion-reaction equations. In [9], we derived the probability density of the first passage time and the probability density of the exit-position distribution on a sphere, and also formulated and proved the reciprocity theorem for the general case of the transient drift-diffusion-reaction equation with general Robin boundary conditions. This allows to directly calculate the flux to the boundary, which appears in many practical problems, for instance, in the simulation of fluxes on threading dislocations and surface (EBIC) in crystals.

In the present paper, we are interested in the simulation of CL and EBIC intensity transients for various threading dislocation densities, i.e. the number of dislocations per unit area. The diffusion and surface recombination of excitons are described by the transient diffusion-reaction equation, which is solved by the Monte Carlo RWS method with the probability densities of the first passage time and the exit point on the sphere that were derived in [9]. According to the Monte Carlo approach [10], many independent exciton trajectories must be simulated to obtain the CL and EBIC intensity with sufficient accuracy. For the reduction of computation time, a parallel implementation of the Monte Carlo method simulating CL and EBIC transients is developed. The general approach to the parallel implementation of Monte Carlo algorithms is to distribute independent tasks among the processors of a supercomputer [11,12]. We provide a parallel implementation of the Monte Carlo RWS method using hybrid programming with MPI and OpenMP. The parallel code efficiency has been studied with respect to the number of MPI processes and OpenMP threads in the "NKS-1P" cluster of the Siberian Supercomputer Center of the SB RAS.

The paper is organized as follows. In the second section, we give the formulation of the problem, the transient diffusion-reaction equation and the RWS method for solving this equation. Section 3 presents the parallel implementation of the Monte Carlo RWS algorithm and its efficiency. The CL transients obtained by simulation for various threading dislocation densities are shown in Sect. 4. This section also presents a comparison of the simulation results and the exact solution of the transient diffusion-reaction equation in a special case.

2   Algorithm for Simulation of CL and EBIC Intensity Transients

2.1   The Transient Diffusion-Reaction Equation

The problem of the simulation of CL and EBIC transients can be formulated as follows [4,9]. The half space $G = R_+^3 = \{(x, y, z) : z \geq 0\}$ is considered as crystal domain containing a family of dislocations. The dislocations are assumed to be circular semicylinders with axes perpendicular to the plane $z = 0$. The boundary Γ of the domain G consists thus of the plane $\Gamma_z = \{(x, y, z) : z = 0\}$ and the surfaces of the semicylinders Γ_d. An instantaneous impulse of excitons is injected at the initial time $t = 0$ into the domain according to a generation function $f(\mathbf{r})$. The change of the exciton concentration u in the domain G with respect to time t is governed by the equation

$$\frac{\partial u}{\partial t} = D\Delta u(\mathbf{r}, t) - \lambda^2 u + f(\mathbf{r})\delta(t), \ \mathbf{r} \in G, \ t \in [0, T]. \tag{1}$$

Here $\mathbf{r}$ is a space coordinate, D is a constant diffusion coefficient, and $\lambda^2 = 1/t_{\text{live}}$, where t_{live} is the mean life time of the exciton. The following initial and Robin (third type) boundary conditions are imposed:

$$u(\mathbf{r}, 0) = 0, \ \mathbf{r} \in G,$$
$$\left(D\nabla u \cdot \boldsymbol{\nu} + Su\right)\big|_\Gamma = 0, \tag{2}$$

where $\boldsymbol{\nu}$ is the outward normal unit vector to the surface, and S denotes the recombination velocity on the boundary Γ. The Robin boundary conditions simulate a partial reflection of particles from the boundary. The fraction of particles adsorbed onto the boundary is determined by the recombination velocity S.

When we use Eq. (1), we assume that the conditions of existence and uniqueness of the regular solution of Robin boundary value problem (2), given in [13], are satisfied. In the case considered here, they are satisfied because all the equation coefficients are bounded.

The EBIC intensity is defined as the flux of excitons to the boundary $\Gamma_z = \{(x, y, z) : z = 0\}$:

$$I_e(\mathbf{r}, t) = \int_0^t d\tau \left\{ -D \int_{\Gamma_z} \frac{\partial u}{\partial \boldsymbol{\nu}} \, dx \, dy \right\}. \tag{3}$$

Analogously, the total flux $I_d(\mathbf{r}, t)$ to the surface of the dislocations is defined. The CL intensity is defined as the total concentration of excitons annihilated in the domain G: $I_{cl}(\mathbf{r}, t) = 1 - I_e(\mathbf{r}, t) - I_d(\mathbf{r}, t)$.

According to the reciprocity theorem formulated and proved in [9, pp. 201–202], to calculate the flux of the solution u to a boundary part Γ_i over time interval $[0, T]$, from an instantaneously released at time $t = 0$ point source positioned at $\mathbf{r} = \mathbf{r_0}$, one has to find at this point a solution of the adjoint homogeneous equation

$$\frac{\partial w}{\partial \tau} = D\Delta w(\mathbf{r}, \tau) - \lambda^2 w, \quad \mathbf{r} \in G, \ \tau = T - t, \ \tau \in [T, 0], \quad (4)$$

vanishing at time $t = T$: $w(\mathbf{r}, T) = 0$, and on the boundary parts Γ_k for all $k \neq i$, and having value S_i on the boundary Γ_i:

$$(D\nabla w \cdot \boldsymbol{\nu_k} + S_k w)|_{\Gamma_k} = S_k \delta_{ik}, \quad \delta_{ik} = \begin{cases} 0, & \text{if } i \neq k, \\ 1, & \text{otherwise}, \end{cases} \quad (5)$$

where δ_{ik} is the Kronecker delta, $\boldsymbol{\nu_k}$ is the outward normal unit vector to the surface, and S_k is the recombination velocity on the boundary part Γ_k. The adjoint equation (4) is solved backward in time, starting with zero "initial" condition at time $\tau = T$. Based on the reciprocity theorem, the transient fluxes $I_e(\mathbf{r}, t)$ and $I_d(\mathbf{r}, t)$ can be calculated by the RWS algorithm described below in the next section.

A more efficient technique to compute the CL intensity when there are many boundaries is to calculate directly the concentration of the absorbed particles, without calculating the total flux to all boundaries. This can be done by using the reciprocity theorem for the absorbed concentration, formulated in ([9], pp. 202–203). This theorem claims that the concentration of particles absorbed inside the domain satisfies the inhomogeneous adjoint equation

$$\frac{\partial w}{\partial \tau} = D\Delta w(\mathbf{r}, \tau) - \lambda^2 w + \lambda^2, \quad \mathbf{r} \in G, \ \tau = T - t, \ \tau \in [T, 0], \quad (6)$$

with zero initial and boundary conditions. The solution of Eq. (6) can be found by using the RWS algorithm.

2.2 Random-Walk-on-Spheres Algorithm for the Transient Diffusion-Reaction Equation

According to [7,9], a random diffusion-reaction walk-on-spheres process in a domain G, starting in a point $\mathbf{x_0} \in G$ at time $t_0 = 0$, is constructed as a Markov chain of points $\{(\mathbf{x_k}, \tau_k)\}$, $k = 0, 1, \ldots$:

$$\mathbf{x_{k+1}} = \mathbf{x_k} + \omega_\mathbf{k} R_k, \quad t_{k+1} = t_k + \tau_k, \quad k = 0, 1, \ldots, \quad (7)$$

where $\omega_\mathbf{k} R_k$ is a random point uniformly distributed on a sphere $S(\mathbf{x_k}, R_k)$, with the center placed at $\mathbf{x_k}$ and radius R_k equal to the minimal distance to the boundary Γ. Time τ_k is the first time exit, that is, the random time at which a diffusing particle first leaves the sphere $S(\mathbf{x_k}, R_k)$. The time τ_k has the distribution density $p_t(\tau)$ derived in [9]:

$$p_t(\tau) = \frac{2}{Q} \sum_{n=1}^{\infty} (-1)^{n+1} \frac{\pi^2 n^2 D}{R_k^2} \exp\left\{ -\left(\frac{\pi^2 n^2 D}{R_k^2} + \lambda^2 \right) \tau \right\}, \tag{8}$$

where $Q = \dfrac{\frac{\lambda R_k}{\sqrt{D}}}{\sinh\left(\frac{\lambda R_k}{\sqrt{D}}\right)}$.

In each sphere, the survival probability Q_{R_k} is calculated by the formula (9). This is the probability for a particle to survive inside the sphere and reach the surface of this sphere independently from time:

$$Q_{R_k} = \frac{R_k/L}{\sinh(R_k/L)}, \tag{9}$$

where $L = \sqrt{D \cdot t_{\text{live}}}$ is the diffusion length.

The process is stopped with a probability $F_{R_k} = 1 - Q_{R_k}$. The conditional probability density of the life time of a particle before it is absorbed inside the sphere can be derived using a Green-function approach suggested in [9]. We omit the long calculations and give the final result:

$$p_{\text{ads}}(\tau) = \frac{2\lambda^2}{F_{R_k}} \sum_{n=1}^{\infty} (-1)^{n+1} \exp\left\{ -\left(\frac{\pi^2 n^2 D}{R_k^2} + \lambda^2 \right) \tau \right\}. \tag{10}$$

The walking process is stopped if the diffusing particle enters a neighborhood Γ_ϵ of the boundary Γ. The set Γ_ϵ is defined as

$$\Gamma_\epsilon = \{\mathbf{x} \in G : \rho(\mathbf{x}, \mathbf{y}) < \epsilon, \ \mathbf{y} \in \Gamma\}, \tag{11}$$

where $\rho(\mathbf{x}, \mathbf{y})$ is the minimum distance between a point $\mathbf{x} \in G$ and a point $\mathbf{y} \in \Gamma$.

To calculate the CL and EBIC transients, we start the RWS process in the source position $\mathbf{r_0}$ at time $t_0 = T$, and the time runs backwards according to the reciprocity theorem. The scorers for the CL and EBIC intensity throughout time $t \in [t_0, 0]$ are written during simulation.

Based on [9], the algorithm for CL and EBIC simulation conforming to the diffusion-reaction equation (1) with conditions (2) can be formulated as follows:

Algorithm **A1**:

1. Initiation:
   Set all scores to zero: $I_{\text{cl}}^t := 0$ for the CL intensity, $I_{\text{e}}^t := 0$ for the EBIC, $I_{\text{d}}^t := 0$ for the flux on dislocations.
2. Set the index of the walking step of the trajectory to 1: $i := 1$;
   generate the coordinates of the starting position of exciton: $\mathbf{x_i} = \mathbf{r_0}$, according to the generation function $f(\mathbf{r})$;
   set $\tau := 0$ for the exciton current time.

3. Construct a sphere $S(\mathbf{x_i}, R_i)$ centered at $\mathbf{x_i}$ with radius R_i equal to the minimal distance from $\mathbf{x_i}$ to the boundaries Γ_z and Γ_d.
4. Calculate the survival probability Q_{R_k} using formula (9).
5. If the exciton does not survive:
   simulate the time of absorption τ_{ads} inside the sphere, with density $p_{ads}(\tau)$ (10) and using the Devroye algorithm [14];
   increase the exciton current time: $\tau := \tau + \tau_{ads}$;
   increment by 1 the CL intensity for time τ: $I_{cl}^\tau := I_{cl}^\tau + 1$;
   terminate the trajectory;
   go to step 2 to start a new trajectory.
6. If the exciton survives:
   simulate the first passage time τ_s, with density $p_t(\tau)$ (8) and using the Devroye algorithm [14] and the exit position $\mathbf{x_{i+1}}$ on the sphere $S(\mathbf{x_i}, R_i)$, which is uniformly distributed on the surface of $S(\mathbf{x_i}, R_i)$;
   increase the exciton current time: $\tau := \tau + \tau_s$;
   set $\mathbf{x_{i+1}}$ as new coordinate for the exciton.
7. If $\mathbf{x_{i+1}}$ hits an ϵ-neighborhood Γ_ϵ of one of the boundary parts Γ_k, then calculate the recombination probability using the formula

$$P_{\Gamma_k} = \frac{hS_k}{D + hS_k},\qquad(12)$$

where h is the step for the reflection from the boundary Γ_k.
   (a) If the exciton recombines on the boundary part Γ_k:
      add 1 to the score of the flux to the boundary part Γ_k for time τ;
      terminate the trajectory;
      go to step 2 to start a new trajectory.
   (b) If the exciton does not recombine on the boundary:
      a random walk trajectory is reflected in the direction opposite to the normal direction ν_k to the point $\mathbf{x_{i+1}} - h\nu_k$, i.e. the new coordinate of the exciton is $\mathbf{x_{i+1}} := \mathbf{x_{i+1}} - h\nu_k$.
8. If $\tau \geq T$, i.e. the exciton current time τ exceeds the preset time T, then terminate the trajectory;
   go to step 2 to start a new trajectory.

9. If $\tau < T$, then increase the walking step: $i := i + 1$, and go to step 3 to continue.

The CL, EBIC and flux on the dislocations are obtained by averaging the scores over N independent trajectories.

3   Parallel Implementation of the Algorithm for Simulation of CL and EBIC Intensity Transients

The number of trajectories N for the Monte Carlo simulation of the CL and EBIC must be quite large since the number of steps of the RWS algorithm that are required to achieve a statistical accuracy ϵ is $O(|\ln\epsilon|/\epsilon^2)$ and $N \sim 1/\epsilon^2$ [9]. We developed a parallel implementation of the Monte Carlo RWS method using the OpenMP and MPI standards. We applied the general approach to the parallel implementation of Monte Carlo algorithms [11,12]. At first, N independent trajectories are distributed among n_{mpi} MPI processes. Then, for each MPI process, (N/n_{mpi}) trajectories are distributed among n_{omp} OpenMP threads. The OpenMP threads simultaneously calculate the values of the characteristics I_{cl}^t, I_{e}^t, I_{d}^t using their own arrays. Each MPI process sums up the values obtained by its child OpenMP threads. After computing all the trajectories, the root MPI process gathers the values obtained for the characteristics and averages them. The code uses the parallel 128-bit linear congruential generator of pseudo-random numbers implemented in [15].

The efficiency of the parallel implementation of the CL and EBIC simulation code was evaluated in a computer experiment with the following values of the parameters: diffusion length: $L = 100\,\mathrm{nm}$; exciton lifetime: $t_{\mathrm{live}} = 1\,\mathrm{ns}$; dislocation radius: $R_{\mathrm{dis}} = 3\,\mathrm{nm}$; dislocation density: $\rho_{\mathrm{dis}} = 10^{-5}\,\mathrm{nm}^{-1}$; recombination velocity on the boundary Γ: $S = 10^5\,\mathrm{nm/ns}$; size of the neighborhood for the plain Γ_z: $\epsilon_z = 0.01\,\mathrm{nm}$, and for the dislocations Γ_{d}: $\epsilon_{\mathrm{d}} = 0.001\,\mathrm{nm}$; number of trajectories: $N = 10^8$. Simulations were performed on the "NKS-1P" cluster, put into operation in 2017 in the Siberian Supercomputer Center of the Siberian Branch of the Russian Academy of Sciences (SSCC SB RAS)[1]. The "NKS-1P" cluster consists of two partitions: 20 "Broadwell" nodes and 16 "KNL" nodes. The "Broadwell" computation node is equipped with two Intel Xeon E5-2697v4 processors having 16 cores and 32 threads each, which amounts to a total of 64 threads on the node. The "KNL" computation node consists of one Intel Xeon Phi 7290 processor having 72 cores with four threads each, in all, 288 threads on the node.

At first, the efficiency of the parallel implementation of the code was assessed using a single computational node. We tested three distinct distributions of trajectories among MPI processes and OpenMP threads:

1. using only MPI processes, each having a single OpenMP thread;
2. using only OpenMP threads created by a single MPI process;
3. using a combination of MPI processes and OpenMP threads in such a way that all node resources are busy.

In Fig. 1, the characteristics of the parallel implementation for the first distribution are shown: computational time $T(n_{\mathrm{mpi}})$, speedup $S(n_{\mathrm{mpi}}) = T(1)/T(n_{\mathrm{mpi}})$, and efficiency $Q(n_{\mathrm{mpi}}) = S(n_{\mathrm{mpi}})/n_{\mathrm{mpi}}$.

The "Broadwell" partition computational time for any n_{mpi} is less than that of the "KNL" partition (Fig. 1a). The "Broadwell" speedup and efficiency fall when $n_{\mathrm{mpi}} > 32$ (Fig. 1b, c); this behavior can be explained by the fact that the "Broadwell" node has two processors with 16 cores each. So, when all cores of the node are involved, a further increase of n_{mpi} leads to a drop in efficiency. A similar result is obtained for the "KNL" partition. A sharp drop in S and

---

[1] SSCC SB RAS website: http://www.sscc.icmmg.nsc.ru.

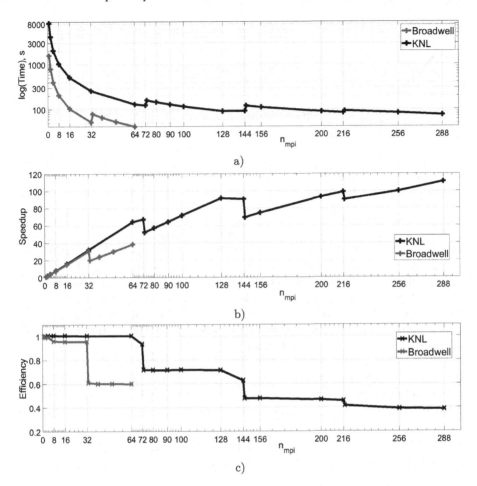

Fig. 1. Characteristics of the code parallel implementation using only MPI processes, each having a single OpenMP thread: (a) computational time, (b) speedup, (c) efficiency.

Q occurs when moving to the next level of core loading. For example, when $n_{\text{mpi}} \leq 72$, only one MPI process is executed per core. When $n_{\text{mpi}} = 73$, one of the cores executes two MPI processes, while the other cores execute only one MPI process and then are idle. The root MPI process waits for completion of all processes to calculate the average values of the characteristics, so the load imbalance leads to an increase in waiting time. A further increase in the number of MPI processes up to $n_{\text{mpi}} = 144$ leads to a more uniform load of the cores, so that the efficiency increases. The drop in S and Q for $n_{\text{mpi}} = 145$ and $n_{\text{mpi}} = 217$ can be explained in the same way.

In the case of the second distribution, when the trajectories are distributed only among OpenMP threads created by a single MPI process in each node, the efficiency and speedup smoothly decrease when the number of threads n_{omp} increases without sharp jumps (Fig. 2).

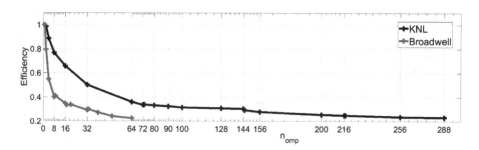

Fig. 2. Efficiency of the code parallel implementation using only OpenMP threads created by a single MPI process.

Tables 1 and 2 summarize the characteristics of the code parallel implementation for the third case of distribution, for the "Broadwell" and "KNL" partitions. The trajectories are distributed among MPI processes and OpenMP threads, so that all node resources are busy, i.e. $n_{mpi} + n_{omp} = 64$ for "Broadwell", and $n_{mpi} + n_{omp} = 288$ for "KNL". The tables show the characteristic values averaged over 10 code executions. For the "Broadwell" partition, the minimal computational time was achieved for the parallel code implementation using 32 MPI processes and 2 OpenMP threads. For the "KNL" partition, the minimal computational time was achieved for $n_{mpi} = 74$ with $n_{omp} = 4$. These results are adequate to the architecture of computational nodes: the "Broadwell" node includes 32 cores with two threads each, while the "KNL" node has 72 cores with 4 threads each.

Table 1. Characteristics of the code parallel implementation executed on a single computational node of the "Broadwell" partition.

n_{mpi}; n_{omp}	1;64	2;32	4;16	8;8	16;4	**32;2**	64;1
T, s	105.05	73.60	47.88	44.20	43.98	**41.11**	41.23
S	1	1.43	2.19	2.38	2.39	**2.56**	2.55

Table 2. Characteristics of the code parallel implementation executed on a single computational node of the "KNL" partition.

n_{mpi}; n_{omp}	1;288	2;144	4;72	8;36	9;32	16;18	18;16	32;9	36;8	**72;4**	144;2	288;1
T, s	125.48	97.98	84.22	78.75	77.66	72.61	75.56	76.56	75.29	**71.93**	74.64	77.17
S	1	1.28	1.49	1.59	1.62	1.73	1.66	1.64	1.67	**1.74**	1.68	1.63

Furthermore, the efficiency of the parallel code was investigated with regards to the number of computational nodes when the optimal ratio between n_{mpi} and n_{omp} is used on each node, namely $n_{mpi} = 32$ with $n_{omp} = 2$ for "Broadwell", and $n_{mpi} = 74$ with $n_{omp} = 4$ for "KNL". The speedup of the code parallel implementation is close to linear for the both partitions (Fig. 3). As shown in Fig. 3, the plots of the speedup for "Broadwell" and "KNL" overlap.

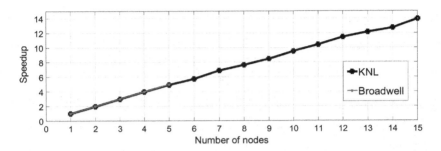

Fig. 3. Speedup of the code parallel implementation depending on the number of computational nodes. (Note that the lines for "Broadwell" and "KNL" overlap).

4   Results of the Simulation of CL, EBIC and Flux on Dislocations

The CL intensity, the EBIC signal and the flux on the dislocations are simulated in a parallel implementation of the code using for each node the optimal number of MPI processes and OpenMP threads that were found above.

First, to verify the code that implements the Monte Carlo RWS algorithm (**A1**), we compare the simulation results against the exact solution of the diffusion-reaction equation (1) in the case of recombination only on the top plane Γ_z, for which we obtained the result explicitly:

$$u(t)^* = \lambda^2 \cdot \exp(-\lambda^2 t) \cdot \exp\left(\frac{Dt}{z_s^2}\right) \operatorname{erfc}\left(\frac{\sqrt{Dt}}{z_s}\right), \tag{13}$$

where z_s is a size of the exciton source along the axis Z.

The following model parameters are taken for the calculations: diffusion coefficient: $D = 10000\,\text{nm/ns}^2$; exciton lifetime: $t_{\text{live}} = 1\,\text{ns}$; diffusion length: $L = \sqrt{Dt_{\text{live}}} = 100\,\text{nm}$; recombination velocity on the top plane: $S_z = 10^5\,\text{nm/ns}$, and on the dislocation surface: $S_d = 0\,\text{nm/ns}$; dislocation radius: $R_{\text{dis}} = 3\,\text{nm}$; value of the neighborhood of the boundary Γ_z: $\epsilon = 0.01\,\text{nm}$, and that of Γ_d: $\epsilon = 0.001\,\text{nm}$; number of trajectories: $N = 10^9$. The exciton position is generated at random. The coordinates x and y are uniformly chosen on the XY plane; the z-coordinate has exponential distribution $z \sim \text{Exp}(0.01)$, so the mean size of the exciton source along the axis Z is $z_s = 100$. The size of the domain is set equal to $X_s \times Y_s = 1000 \times 1000\,\text{nm}^2$ along the axes X and Y, and infinity along the axis Z.

As shown in Fig. 4, the CL intensity obtained by simulation for a domain without dislocations is in a good agreement with the exact solution $u(t)^*$.

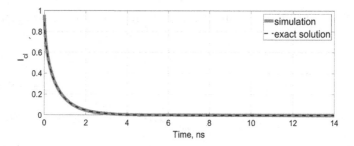

Fig. 4. Comparison of the simulation result with the exact solution for a domain without dislocations.

The next task is to investigate the CL intensity transients depending on the dislocation density. In this test, the recombination velocity on the dislocation surface is $S_d = 10^5$ nm/ns, the number of dislocations N_{dis} is increased from 1 to 1000, while the other parameters remain the same as in the previous test. The dislocation density is calculated as the ratio $N_{dis}/(X_s \cdot Y_s)$ nm^{-2}. Figure 5 shows the transient CL intensity and the flux on the dislocations in a logarithmic scale. When the dislocation density increases, the excitons recombine faster on the dislocation surface, rather than self-annihilate, so the CL intensity decreases. The larger N_{dis} is, the larger the recombination rate is. Thus, at the beginning,

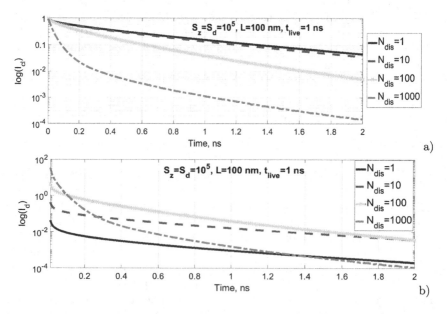

Fig. 5. Influence of the number of dislocations on the CL intensity (a), and the flux on dislocations (b).

the flux on dislocations for larger N_{dis} is greater; later, both the number of excitons and the flux I_d decrease with time.

In addition to the dislocation density, the recombination velocity on the dislocation surface also influences the CL intensity. To analyze the recombination velocity effect, the number of dislocations is set at $N_{\mathrm{dis}} = 10$, and S_d is increased from 10 to 10^5 nm/ns. The other parameters do not change. The character of the recombination velocity influence on the CL intensity is similar to that of the dislocation density effect (Fig. 6a), although it is less pronounced. So I_{cl} slightly decreases when S_d increases, whereas the differences between I_{cl} curves for various N_{dis} are quite clear (Fig. 5a). The flux on the dislocation surface visibly increases (Fig. 6b), but there is no such a change in recombination rate with time, as it is the case when N_{dis} increases.

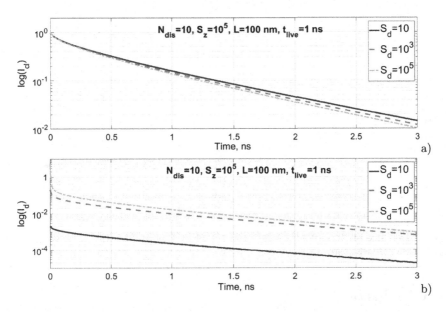

Fig. 6. Influence of the surface recombination velocity on the CL intensity (a), and the flux on dislocations (b).

Another parameter influencing the CL intensity and boundary fluxes is the diffusion length. Figure 7 shows the CL intensity, the EBIC signal and the flux on the dislocations calculated for $N_{\mathrm{dis}} = 10$, $S_d = 10^5$ nm/ns and the same values as above for the rest of the parameters. When the diffusion length L increases, the excitons reach the boundaries faster, so the recombination rate on both boundaries, Γ_z and Γ_d, increases. Thus, for larger L, a larger fraction of excitons is recombined on the boundaries and the CL intensity decreases.

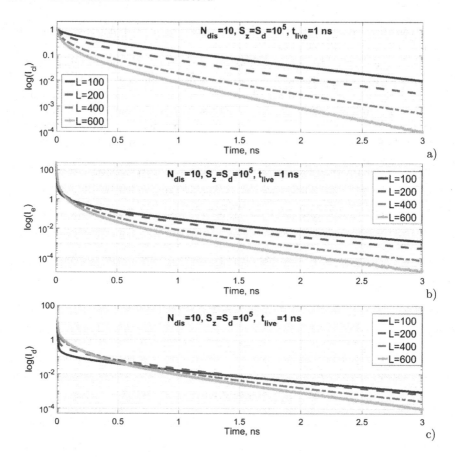

Fig. 7. Influence of the diffusion length on the CL intensity (a), the EBIC signal (b), and the flux on dislocations (c).

5   Conclusions

We have presented in this paper parallel implementations of the Monte Carlo method for simulation of cathodoluminescence transients in the vicinity of dislocations. The simulation algorithm is based on the random-walk-on-spheres method for solving transient diffusion-reaction equations [9]. The parallel implementation uses MPI and OpenMP programming. The parallel code efficiency was analyzed for different numbers of MPI processes and OpenMP threads within a single node for both partitions, "Broadwell" and "KNL", of the "NKS-1P" cluster. Moreover, a linear speedup was obtained for the code executed on multiple computational nodes, each using the optimal ratio of MPI processes to OpenMP threads.

The CL intensity, the EBIC signal and the flux on the dislocations were simulated using the parallel code. The simulation results were compared against the exact solution of the transient diffusion-reaction equation without dislocations, and it was demonstrated that the calculated CL intensity is in good agreement with the theoretical value. We also investigated the dependence of the CL intensity for various dislocation densities, recombination velocities on the dislocation surface, and diffusion lengths.

References

1. Liu, W., Carlin, J.F., Grandjean, N., Deveaud, B., Jacopin, G.: Exciton dynamics at a single dislocation in GaN probed by picosecond time-resolved cathodoluminescence. Appl. Phys. Lett. **109**(4), 042101-1–042101-5 (2016). https://doi.org/10.1063/1.4959832. Article ID 042101
2. Weisbuch, C., Piccardo, M., Martinelli, L., Iveland, J., Peretti, J., Speck, J.S.: The efficiency challenge of nitride light-emitting diodes for lighting. Phys. Status Solidi A **212**(5), 899 (2015). https://doi.org/10.1002/pssa.201570427
3. Rosner, S.J., Carr, E.C., Ludowise, M.J., Girolami, G., Erikson, H.I.: Correlation of cathodoluminescence inhomogeneity with microstructural defects in epitaxial GaN grown by metalorganic chemical-vapor deposition. Appl. Phys. Lett. **70**(4), 420–422 (1997). https://doi.org/10.1063/1.118322
4. Sabelfeld, K.K., Kaganer, V.M., Pfüller, C., Brandt, O.: Dislocation contrast in cathodoluminescence and electron-beam induced current maps on GaN(0001). J. Phys. D **50**, 405101 (2017). https://doi.org/10.1088/1361-6463/aa85c8
5. Phang, J.C.H., Pey, K.L., Chang, D.S.H.: A simulation model for cathodoluminescence in the scanning electron microscope. IEEE Trans. Electron Dev. **39**(4), 782–791 (1992). https://doi.org/10.1109/16.127466
6. Berz, F., Kuiken, H.K.: Theory of life time measurements with the scanning electron microscope: steady state. Solid-State Electron. **19**, 437–445 (1976). https://doi.org/10.1016/0038-1101(76)90003-4
7. Sabelfeld, K.K.: Splitting and survival probabilities in stochastic random walk methods and applications. Monte Carlo Methods Appl. **22**(1), 55–72 (2016). https://doi.org/10.1515/mcma-2016-0103
8. Sabelfeld, K.K.: Random walk on spheres method for solving drift-diffusion problems. Monte Carlo Methods Appl. **22**(4), 265–275 (2016). https://doi.org/10.1515/mcma-2016-0118
9. Sabelfeld, K.K.: Random walk on spheres algorithm for solving transient driftdiffusion-reaction problems. Monte Carlo Methods Appl. **23**(3), 189–212 (2017). https://doi.org/10.1515/mcma-2017-0113
10. Sabelfeld, K.K.: Monte Carlo Methods in Boundary Value Problems. Springer, Berlin (1991)
11. Rosenthal, J.S.: Parallel computing and Monte Carlo algorithms. Far East J. Theor. Stat. **4**, 207–236 (2000)
12. Esselink, K., Loyens, L.D.J.C., Smit, B.: Parallel Monte Carlo simulations. Phys. Rev. E **51**(2), 1560–1568 (1995). https://doi.org/10.1103/physreve.51.1560
13. Friedman, A.: Partial Differential Equations of Parabolic Type. Courier Dover Publications, Mineola (2008)
14. Devroye, L.: The series method for random variate generation and its application to the Kolmogorov-Smirnov distribution. Am. J. Math. Manag. Sci. **1**(4), 359–379 (1981). https://doi.org/10.1080/01966324.1981.10737080
15. Marchenko, M.: Parallel pseudorandom number generator for large-scale Monte Carlo simulations. In: Malyshkin, V. (ed.) PaCT 2007. LNCS, vol. 4671, pp. 276–282. Springer, Heidelberg (2007). https://doi.org/10.1007/978-3-540-73940-1_28

Supercomputer Simulation of Promising Nanocomposite Anode Materials for Lithium-Ion Batteries: New Results

Vadim M. Volokhov, Dmitry A. Varlamov$^{(\boxtimes)}$, Tatyana S. Zyubina,
Alexander S. Zyubin, Alexander V. Volokhov, Elena S. Amosova,
and Gennady A. Pokatovich

Institute of Problems of Chemical Physics of the RAS, Chernogolovka, Russia
{vvm,dima,zyubin,vav,aes,pga}@icp.ac.ru

Abstract. Following a considerable number of computational experiments on various supercomputer resources, we constructed a quantum-chemical and molecular-dynamics model of various nanocomposite components of Si–C-based anode materials for Li-ion batteries. Also, we simulated various aspects of processes occurring inside Li-ion batteries during multiple charge-discharge cycles. Simulation results of stable Si–C nanorods and flexible 3D Si–C fibrous-paper electrodes are presented as the most promising electrodes for Li-ion batteries.

Keywords: Computer simulation · Silicon–carbon nanocomposites
Li-ion batteries · VASP applied package · Quantum chemistry
Molecular dynamics · 3D Si–C fibrous-paper electrodes · Si–C nanorods

1 Introduction

In this work, we present new results obtained within the framework of the project "Computer simulation of absorption and transport properties of solid electrolytes and nanostructured electrodes based on carbon and silicon in Li-ion batteries", which aims for the supercomputer simulation of quantum-chemistry and molecular-dynamics processes in new nanocomposite materials (based on silicon and carbon) and solid electrolytes with high ionic conductivity and transport. Structural and energetic processes occurring in the modeled nanostructures and on the "interface" between them have been described in detail in a number of our previous publications [1–9]. Computer simulation of nanostructures and processes occurring in them has been carried out to improve previous models. Here is a brief description of new results obtained during this simulation.

Li-ion batteries (LIB) are currently the most promising and common types of power sources and batteries. LIB are based on the transport of Li-ions through a liquid or solid electrolyte from cathode to anode (and back when charging). The design of new types of LIB is required to improve their efficiency parameters, such as energy capacity, number of charge-discharge cycles, resistance to external

© Springer Nature Switzerland AG 2018
L. Sokolinsky and M. Zymbler (Eds.): PCT 2018, CCIS 910, pp. 294–305, 2018.
https://doi.org/10.1007/978-3-319-99673-8_21

conditions (temperature), production and disposal safety from the environmental point of view, and cost (prime cost of materials in main components).

LIB are to date the most promising power sources for use in electric vehicles, portable electronics, portable power tools, etc. [10,11]. However, some characteristics of LIB need to be improved. One of them is the discharge capacity of the negative electrode (NE). The most popular material for NE is graphite, with a theoretical capacity of 372 mAh $\cdot$ g^{-1} (or 837 mAh $\cdot$ cm^{-3}) for LiC_6, which is not very high. Silicon is regarded as a prospective candidate for negative-electrode active component. Compared with carbon, solid Si is able to accommodate a tenfold amount of lithium, with a specific capacity of $\sim$4200 mAh$\cdot$g^{-1} (or 9785 mAh $\cdot$ cm^{-3}) for $Li_{4.2}Si$. However, this advantage cannot be exploited, owing to deep structural transformations leading to rapid changes in volume and resulting in electrode destruction. One of the possible solutions to the problem is searching for noncrystalline forms of silicon (such as nanoparticles, thin films/fibers, nanowires, etc.) capable of maintaining its shape after multiple lithiation-delithiation processes. To avoid stress-induced fracturing, it is necessary to increase the electrode stability during Li insertion and extraction, so the silicon particle size must be below $\sim$1 μm. There are a number of different types of silicon nanostructures that can resist the stress caused by changes in silicon volume during lithium insertion without impairing the electrochemical properties of the electrode: silicon nanopowders, nanowires, nanofibers, nanotubes, nanorods, thin films, and so on. Quasi-one-dimensional materials based on nanostructured silicon, such as nanorods and nanowires, seem to be preferable for various reasons. First of all, they can provide a significant volumetric expansion of silicon and reduce the lithium migration path in the volume owing to small axial dimensions. Furthermore, in some cases, they yield improved results in electron transport.

Simulated materials might be the basis for the design and creation of new types of electrochemical and ecologically safe LIB. These power sources would be able to operate at low and medium temperatures, provide significantly higher energy densities, and improve operational and cost characteristics. The simulation of lithiation processes (saturation of the anode by lithium = discharge process) and delithiation (lithium ions return to the electrolyte and cathode = charge process) is a basic operation for the comprehension of processes related to LIB functioning in general, estimation of limiting factors and prediction of the most promising nanocomposite materials.

The synthesis of new nanocomposite materials and the study of their properties and predictable applications are only possible within the framework of detailed computer simulations of crystalline composite structures, elementary processes and mechanisms of chemical reactions and transport processes at the molecular level.

Experimental studies of factors having a major influence on the solution of the issues listed above are complex, expensive, not always possible and, in most cases, do not determine clearly the mechanisms of ongoing physical and chemical processes, the reasons for their differences depending on the composition of the system and external conditions, possible directions of reactions, etc. At the same time, the experimental (analogue) simulation of the influence of a number of factors on the properties of the constituent components of LIB and processes occurring in them are labor-intensive and costly tasks.

Since experiments give only initial and final information about processes, it is quite difficult to build a genuine analytical model. Such tasks can be solved partially in laboratory conditions, where analytic experiments give incomplete or indirect information about mechanisms and structures of experimental components. Modern numerical methods of quantum-chemistry and molecular-dynamics simulation can provide a substantial assistance in determining the characteristics of processes and assessing the impact of individual factors with a high degree of accuracy. These methods allow obtaining new theoretical data on the structure and properties of both nanostructured cathode-anode systems and ion-conducting solid electrolytes, making it possible to subsequently develop new highly effective materials for electrochemical devices. A detailed simulation of elementary processes as well as mechanisms of lithiation/delithiation and ion-transport processes in LIB at the micro level may lead to a better control over chemical reactions occurring in them, allowing to design the most appropriate anode materials in terms of electricity generation efficiency, lithiation processes, stability during multiple charge-discharge cycles, cost of LIB and environmental recycling.

Also, the models created can be reviewed for adequacy by comparing them (and the properties of materials modeled on their basis) with observable analytical, experimental and theoretical data published in specialized literature.

For this task, we carried out a detailed quantum-chemical and molecular-dynamics simulation of various nanosystems based on carbon and silicon, both in cluster approximation and for periodic boundary conditions with projector-augmented wave (PAW), using VASP and Gaussian on a number of high-performance computing resources [1–3].

The objects of the computer simulation are composites based on carbon and silicon that are able to repeatedly absorb Li without damage and are promising materials for Li-ion power sources (Si–C nanorods, 3D Si–C fibrous-paper nanopapers with active crystal surfaces).

Silicon nanorods can be used for producing stable high-capacity negative electrodes. The use of this quasi-one-dimensional material can also help to save the initially high silicon capacity by distributing distortion and stress among degrees along structural units. The discharge capacity of such electrodes is $1038\,\text{mAh·g}^{-1}$ after 170 charge-discharge cycles, which is higher than that of porous or single-crystal silicon. The discharge capacity of negative electrodes based on silicon nanorods is $2000\,\text{mAh·g}^{-1}$ after 25 lithium insertion/extraction cycles. The high capacity and good cyclability of silicon nanorods can be explained by structural features and a large specific surface, which ensures contact of the electrode with the electrolyte, thereby facilitating lithium transport and its accumulation on the boundary. However, capacity decreases during cycling. Degradation of a silicon-based negative electrode occurs when it is charged to 100% of its capacity. Degradation of silicon nanowires is caused by silicon loss and reduction of the diameter of nanowires during cycling. Thus, the optimization of the cycling mode based on variation of the insertion depth of lithium into silicon particles can stabilize the cycling performance of Si-based negative electrodes. Another way to avoid rapid degradation of Si-based negative electrodes consists in using composites with carbon, aluminum or copper. This helps to increase conductivity and flexibility of the negative-electrode material. Thus, understanding the processes that occur both on the surface and in the volume of silicon nanorods during multiple lithium insertion/extraction cycles facilitates the development of stable high-capacity materials for negative electrodes of LIB.

In previous works [4–9], we studied lithium-absorbing composites based on silicon deposited on the surface of nitrogen-doped silicon carbide, as well as silicon nanoclusters placed on a grid of amorphous carbon. Another promising structure was considered in [4,6], namely silicon nanorods, which are superior in their stability to nanoclusters: as the diameter of nanorods increases, their specific energy approaches that of a crystal. This can contribute to cycling performance, owing to the stability of the initial Si nanorod structure and the presence of free space around the rods. A part of the present work is devoted to modeling lithium insertion and extraction processes by such silicon nanorods.

The most promising are flexible 3D Si–C fibrous-paper electrodes [13], synthesized by simultaneous electrospraying of Si and PAN (polyacrilinitrile) nanoclusters followed by their electrospinning and carbonization. Batteries of flexible 3D Si/C fibrous paper are extremely attractive for use in electrical vehicles, flexible electronic devices, space and military applications, owing to easily scalable and simple synthetic methods, good mechanical properties and their excellent electrochemical characteristics at high loading with Si. Combined technologies allow for a uniform incorporation of Si nanoparticles into textile carbon matrices to form fibrous paper of nano-Si-C-composite. Flexible 3D Si–C electrodes of fibrous paper demonstrate a very high total capacity of $\approx 1600\,\mathrm{mAh} \cdot \mathrm{g}^{-1}$, with a loss in power lower than 0.079% per cycle for 600 cycles at stable power. Their extraordinary efficiency is explained by the unique architecture of flexible 3D Si/C fibrous paper, which is a stable network of ion-conduction channels consisting of Si/C clusters uniformly distributed in a carbon fibrous matrix (with strong adhesion between carbon fibers and Si nanoparticles). It can be assumed that during lithiation, the silicon clusters break up into nanoclusters of a definite size and retain this size in the subsequent (de)lithiation to form a reproducible $^2\mathrm{Li}/^1\mathrm{Si}$ structure. In the present work, we model this process for translated nanofibers such as $[\mathrm{Si}_n \mathrm{C}_m]k$ ($k = \infty$) with n ranging from 12 to 16 and m ranging from 8 to 19.

2 Simulation Methods

For modeling the systems under study, we employ the same approach as in earlier works [1–9], combining the use of density functional theory (DFT) with periodic boundary conditions using VASP (Vienna *Ab initio* Simulation Package, https://www.vasp.at). The basis set consists of PAW projector plane-waves with an appropriate pseudo-potential and PBE (Perdew–Burke–Ernzerhof) functional. The energy limit (E_c), which determines the completeness of the basis set, is 400 eV.

The use of this approach for modeling crystalline silicon gives a unit-cell parameter of 5.47 Å and a dissociation energy D_e/n of 4.57 eV, which is quite consistent with experimental data: 5.43 Å and 4.64 eV, respectively. For modeling one- and two-dimensional structures (surfaces, tubes, rods, etc.), the sizes of multiplied cells are chosen in order that the distances between the nearest surfaces of a periodically repeated system are no less than 10 Å.

For simulation of atom redistribution after Li insertion or extraction, we used nonempirical molecular dynamic modeling, namely the MD-VASP approximation, which implements the same algorithms as in ordinary structure optimization but with rougher criteria of calculation accuracy ($E_c = 200$ eV, prec = *low*). Initially, the system is heated to the preset temperature (as a rule, 600 K), then it is equilibrated with a Nosé thermostat at this temperature (when equilibrium is attained, the potential energy fluctuations cease, which occurs usually within 10 000 femtosecond) and then cooled to 0 K. The final structure is refined by means of optimization with an ordinary accuracy ($E_c = 400$ eV, prec = *normal*).

The energetic stability of the combination system was evaluated as $D_e/n(\text{Li})$, which is calculated as the difference between its energy and the energy of the substrate (a silicon rod) and isolated lithium atoms, divided by the number of lithium atoms n:

$$\frac{D_e}{n(\text{Li})} = -\frac{1}{n}[E(\text{Si}_m\text{Li}_n) - E(\text{Si}_m) - nE(\text{Li})].$$

For metallic lithium, this value in terms of the approximation used is 1.61 eV, whereas the experimental value is 1.64 eV.

2.1   Estimation of Adequacy of the Models

To assess the adequacy, reliability and accuracy of the constructed computer models of nanocomposite materials and processes with their participation, we test these models by various methods. Fist of all, there should be neither obvious contradictions between a model and the physical and chemical effects observed during the evolution of real simulated systems (for example, during heating-cooling cycles), nor inconsistencies in the physical and chemical states of simulated substances (for example, formation of metallic lithium or decomposition of electrolytes). The test program of a model includes the following methods:

1. Comparison of data obtained in the simulation with independent external data (analytical, experimental, theoretical, reference) by contrasting the parameters obtained (for example, average bond energy of atoms in a crystal, parameters of a crystal cell, photoelectronic absorption spectra, etc.) with those already known from literature or reference sources.
2. Comparison of simulation data acquired during operation of the model as a whole with data obtained earlier during operation of separate components or processes of the model.

3. Verification of the correctness and stability of the model using various combinations of simulated substances, e.g., nanocomposite electrodes and solid/polymeric electrolytes with various external parameters and multiple lithiation/delithiation cycles.
4. Assessment of the correctness and independence of the model when carrying out computational experiments on various high-performance resources with various configurations of computer equipment (random access and disk memory, number of compute nodes, versions of the application packages used).

The main test method consists of mass computational experiments on various computing resources using the created computer model with a wide range of input parameters, and subsequent analysis of the results to select data for comparison. Later, we carry out a consistency analysis of the results obtained from the point of view of physical and chemical criteria, verify their correctness and compare them with independent external data or with an array of previous results of model operation obtained for separate components. A fairly accurate numerical estimation of the level of compliance with directly observable data is possible for calculated quantitative parameters.

This test program allows to estimate both the general adequacy of the model to the processes simulated and the correctness of using the results obtained for analytical conclusions regarding nanocomposite electrodes on the basis of carbon-silicon and solid/polymeric electrolytes for new types of LIB.

We used the Gaussian package (http://gaussian.com) for comparing and estimating the accuracy of the simulation of some nano-objects at the DFT/B3LYP level. By comparing different levels of calculation, we noted that the computed values used in VASP and Gaussian for average bond energies and distances of identical objects give consistent results with an accuracy of ranging from 0.02 to 0.04 eV and, respectively, from 0.005 to 0.01 Å.

It should be noted that the difference between the computation results at B3LYP/6-31G(d,p), PBE/6-31G(d,p) and PBE/PAW levels are in the range from 0 to 2% for distances, and from 1 to 13% for energies. The calculation level chosen yields the following calculation accuracy in computer models: the Si crystal lattice calculated parameters $a = b = c$ are 5.48 Å (experimental: 5.43 Å), the Si–Si distance is 2.37 Å (experimental: 2.34 Å), and the energy of the crystal is 4.44 eV (experimental: 4.52 eV).

The adequacy of the computer models was also evaluated by comparing the values calculated on the basis of their physico-chemical characteristics (optical and X-ray spectra, thermodynamic measurements, energy parameters) with those observed in physical experiments. For example, the calculated structural parameters for crystal electrolytes ($a = b = 8.79$ Å and $c = 12.80$ Å) are in good agreement with those in X-ray experiments ($a = b = 8.72$ Å and $c = 12.63$ Å).

3   Computational Complexity and Efficiency of Calculations

In the past, such computer simulations were hindered by a catastrophic lack of computing resources, since calculating the behavior of small/medium atomic clusters of Si_{7-126} type, even in a simplified form, would require months, and modeling systems as a whole (containing thousands of atoms) would take approximately $n \cdot 10^6$ CPU-hours per year.

These simulations have become feasible only in recent times, using high-performance supercomputing centers and grid polygons. Currently, the use of computing resources with speeds of the order of teraflops and petaflops allows to make sufficiently detailed simulations of geometrical and energy characteristics of modeled nanostructures. It is also possible to study the effects of various factors and processes occurring in these nanostructures for a variety of conditions determining the efficiency of LIB created.

Let us summarize the computational complexity and use efficiency of computational resources in the process of quantum-chemical simulation of the structures we have studied. We used the IPCP cluster (176 dual-node HP Proliant, making a total of 1472 cores based on 4- and 6-core Intel Xeon processors 5450 and 5670 3 GHz, 8 and 12 GB of RAM per node; InfiniBand DDR communication network; Gigabit Ethernet transport and network management; hard drives: no less than 36 GB per node), and the "Lomonosov-1,2" supercomputing installations at the SCC of MSU, having various pools of processors (8 to 128 CPU) with obligatory presence of local drives and no less than 2 GB of RAM per core [12].

A sufficiently effective speedup of VASP for this type of tasks was observed for 40 to 48 CPU cores. The subsequent growth of the efficiency of parallelization of the task is limited (or even reduced) by the rate of data exchange due to a significant increase in the amount of data being transferred between nodes. Thus, increasing the number of CPU over 48 (at least for this task variant) is meaningless for the moment. If the number of processors is more than 64, the dependence between the speedup and the number of processors is practically absent or even falls [3].

The average effective time for calculation of Si_n clusters (from $n = 2$ to 350) and $C_n Si_m$ nanofibers increases as the dimension of the silicon-carbon fragment increases, taking up to 4 days (78 h on a pool based on 4-core Intel Xeon 5450 3 GHz processors) and even more (owing to complications of the structure). The calculation of lithiated large mesostructures of silicon and aggregates reinforced with nanorods or 3D Si–C nanofibrous paper takes tens of days to complete.

The most critical calculation parameter is the amount of memory per core, which has an effect of acceleration of calculations when the number of allocated cores is decreased since the amount of RAM per core increases. For molecular-dynamics calculations, we used 14 000 steps per calculation (for example, heating up to 400 K during 2000 steps, holding at 400 K during 10 000 steps, cooling down to 10 K during 2000 steps, and optimizing the structure in standard mode; the model time step was 1 femtosecond). The calculation of complex structures, such as Li-saturated nanofibers with solid electrolytes, requires up to 30 000 CPU-hours.

The total number of computing experiments run at all stages of the work exceeds 2000. The use of computing resources is estimated as follows: "Lomonosov-1,2" SC: about 30 to 40% of the total number of experiments; IPCP cluster: 50 to 60%; IPCP workstations: 2 to 3%.

4   Simulation Results

4.1   Computer Simulation of Various Types of Porous Nanocomposite Materials Based on Carbon and Silicon

We have constructed computer models of the following types of Si–C nanocomposites (in addition to the previously simulated):

- silicon clusters with silicon carbide core (rod-shaped), 1.2 to 2.8 nm in diameter, and nanofibers of Si_nC_m type, $n/m = 1$ to 3;
- infinite carbon nanofibers coated with silicon nanoclusters and translated to mesostructures (for example, "nanopapers").

The structure of the inner part of the silicon rods is close to that of a crystal, and the faces correspond to its reduced surface, as it was considered in [6].

Lithiation/Delithiation of Silicon Rods. Already with the presence of only two layers of silicon around the axis (L2, diameter 12.5 Å; Fig. 1), the D_e/n(Si) value for an endless rod turns out to be quite high (4.14 eV). When the number of layers increases to $L = 3$ and $L = 4$ with a diameter of 19 and 26 Å, respectively, the D_e/n(Si) value increases up to 4.28 and 4.35 eV (and with a further increase in the number of layers, it tends to the corresponding value for a crystal), which is higher than that for isolated polyhedral Si clusters of similar diameter. Thus, for a Si_{350} cluster with a diameter of ~23 Å, this value reaches 4.04 eV.

When lithium is inserted into quasi-crystalline silicon rods, their regular structure is disturbed and does not recover even when the metal is completely extracted. Thus, the probability of distortion of the rod shape, as well as rod agglomeration and destruction in the process of cycling is high. These effects can be avoided either by reinforcing the rods with a rigid material that does not interact with lithium (for example, silicon carbide) or by encircling them with insulating layers that prevent agglomeration (for example, amorphous carbon). With lithium saturation, the transverse dimensions of the rods increase significantly, at least by a factor of two.

If Li/Si $> 1/2$, lithium easily redistributes in the volume, i.e. it is freely inserted into the rod or extracted from it. At a lower concentration, Li distribution within a silicon rod becomes substantially heterogeneous, and its transition between the outer layer and the inner part of the rod is difficult. Thus, when using such systems as a negative electrode material, the Li/Si ratio should not be reduced to less than 0.5.

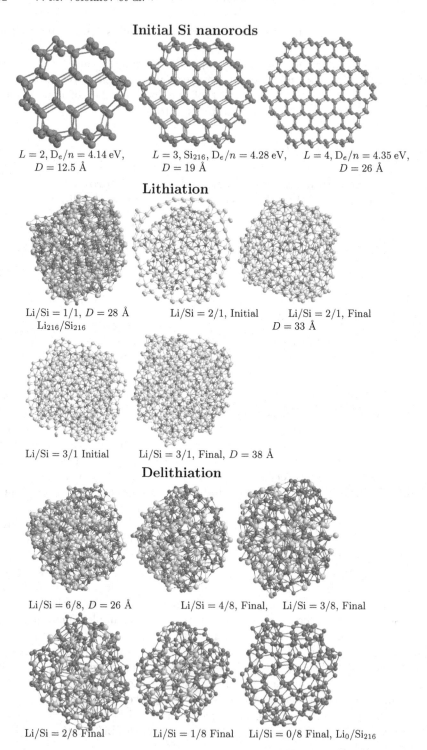

Fig. 1. Cross sections of Si rods with various diameters (D) and lithiation/delithiation Li/Si ratios (Li–yellow balls, Si–green balls) (Color figure online)

4.2 Quantum-Chemical Simulation of Transport Processes of Lithium Ions in Nanocomposite Materials Based on Carbon and Silicon

On the basis of the constructed models of nanocomposites (see above), we made [1–9] a quantum-chemical simulation of various processes occurring during charge-discharge cycles of LIB (i.e. processes of lithiation and delithiation on electrodes based on the above described nanostructures).

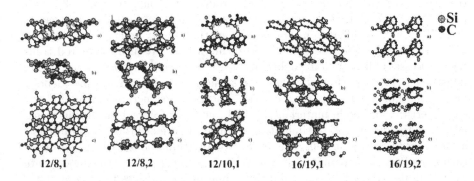

Fig. 2. The structure of space-replicated fragments of $[Si_{12}C_8]$, $[Si_{12}C_{10}]$, and $[Si_{16}C_{19}]$ nanofibers from three different angles (a, b, c). In the notation of the structures, the first number corresponds to the number of silicon atoms (n), the second number is the number of carbon atoms (m), and the third number (after the comma) is the isomer number on the energy scale starting from the isomer with the lowest energy

A majority of characteristics of these processes have been established, including:

1. Li-ion transport processes and processes of lithium consistent implementation in Si–C nanostructures of various types and dimensions.
2. Structural and energetic changes of nano-objects in processes of absorption of lithium atoms.
3. Possible paths and migration barriers for lithium atoms in the process of nanoparticle saturation.
4. Construction of models of sequential removal of lithium atoms from lithiated nanoparticles and determination of structural and energetic changes identified in this process.
5. Determination of the limits of resistance to fracturing for nanoparticles during delithiation processes.

Thus, the results of quantum-chemical modeling (Fig. 2) allow us to conclude that infinite space-replicated nanofibers, such as $[Si_nC_m]k$ $(k = \infty)$ with n ranging from 12 to 16 and m from 8 to 19, can be used as anodes in lithium-ion batteries, since they change their volume during cycling by 6 to 8% starting from the third cycle. For practical use, we recommend anodes subjected to no less than 3 to 4 lithiation-delithiation cycles, with already stabilized relative changes in their structure and volume on lithiation not exceeding 10% of the initial value.

5   Conclusions

On the basis of a large number (more 200) of numerical experiments involving computer quantum-chemistry and molecular-dynamics simulation, we modeled the structures and surfaces of solid electrodes (Si–C nanorods and 3D Si–C nanofibers) for LIB, their interaction with various nano-objects based on carbon and silicon having different morphologies, spatial rigidity values, power characteristics, and saturation potentials with lithium ions. We also calculated transport processes of lithium ions (delithiation-lithiation) in nanocomposites, including structural energy characteristics and structures evolving over time (depending on the number of lithiation cycles). The model structures calculated, the characteristics of anode materials for LIB and their interaction during charge and discharge were used to simulate the whole picture of lithiation and delithiation processes in Li-ion cells, the interaction of lithium ions with the surface of carbon and silicon nanomaterials, to determine the capacity of the anode materials, and also to model both components and new LIB types in general. These results will be used to determine optimal conditions for the synthesis and production of the most energetically favorable and industrially suitable electrolytes and anode materials for new types of Li-ion power sources.

References

1. Volokhov, V.M., Varlamov, D.A., Zyubina, T.S., Zyubin, A.S., Volokhov, A.V., Amosova, E.S.: The supercomputer simulation of nanocomposite components and transport processes in the Li-ion power sources of new types. In: Voevodin, V., Sobolev, S. (eds.) RuSCDays 2017. CCIS, vol. 793, pp. 299–312. Springer, Cham (2017). https://doi.org/10.1007/978-3-319-71255-0_24
2. Volokhov, V., Varlamov, D., Zyubina, T., Zyubin, A., Volokhov, A., Amosova, E.: Supercomputer simulation of components and processes in the new type Li-ion power sources. In: Sokolinsky, L., Zymbler, M. (eds.) PCT 2017. CCIS, vol. 753, pp. 275–287. Springer, Cham (2017). https://doi.org/10.1007/978-3-319-67035-5_20
3. Volokhov, V.M., Varlamov, D.A., Zyubina, T.S., Zyubin, A.S., Volokhov, A.V.: The supercomputer simulation of processes of interaction silicon-carbonic nanostructured electrodes and solid electrolytes in new types of Li-ion power sources. In: Russian Supercomputing Days 2016, Proceedings of the International Scientific Conference, 26–27 September 2016, Moscow, pp. 690–699 (2016). (in Russian)
4. Zyubin, A.S., Zyubina, T.S., Dobrovol'skii, Y.A., Volokhov, V.M.: Silicon- and carbon-based anode materials: a quantum-chemical modeling. Russ. J. Inorg. Chem. **61**(1), 48–54 (2016). https://doi.org/10.1134/s0036023616010241
5. Zyubina, T.S., Zyubin, A.S., Dobrovolsky, Yu.A., Volokhov, V.M.: Quantum chemical modeling of nanostructured silicon Si_n (n = 2-308). The snowball-type structures. Russ. Chem. Bull. **65**(3), 621–630 (2016). https://doi.org/10.1007/s11172-016-1346-7

6. Zyubin, A.S., Zyubina, T.S., Dobrovol'skii, Y.A., Volokhov, V.M.: Quantum-chemical modeling of lithiation of a silicon-silicon carbide composite. Russ. J. Inorg. Chem. **61**(11), 1423–1429 (2016). https://doi.org/10.1134/s0036023616110231

7. Zyubina, T.S., Zyubin, A.S., Dobrovolsky, Yu.A., Volokhov, V.M.: Lithiation-delithiation of infinite nanofibers of the $Si_n C_m$ type - the possible promising anodic materials for lithium-ion batteries. Quantum-chemical modeling. Russ. J. Electrochem. **52**(10), 988–991 (2016). https://doi.org/10.1134/S1023193516100141

8. Zyubina, T.S., Zyubin, A.S., Dobrovolsky, Yu.A., Volokhov, V.M.: Quantum-chemical modeling of lithiation-delithiation of infinite fibers $[Si_n C_m]k$ $(k = \infty)$ for $n = 12$–16 and $m = 8$–19 and small silicon clusters. Russ. J. Inorg. Chem. **61**(13), 1677–1687 (2016). https://doi.org/10.1134/s0036023616130040

9. Zyubina, T.S., Zyubin, A.S., Dobrovolsky, Yu.A., Volokhov, V.M.: Migration of lithium ions in a nonaqueous nafion-based polymeric electrolyte: quantum-chemical modeling. Russ. J. Inorg. Chem. **61**(12), 1545–1553 (2016). https://doi.org/10.1134/s0036023616120238

10. Armand, M., Tarascon, J.M.: Building better batteries. Nature **451**, 652–657 (2008). https://doi.org/10.1038/451652a

11. Kang, B., Ceder, G.: Battery materials for ultrafast charging and discharging. Nature **458**, 190–193 (2009). https://doi.org/10.1038/nature07853

12. Voevodin, Vl.V., et al.: Practice of a supercomputer "Lomonosov". Open Syst. **7**, 36–39 (2012). (inRussian)

13. Xu, Y., Zhu, Y., Han, F., Luo, C., Wang, C.: 3D Si-C fiber paper electrodes fabricated using a combined electrospray/electrospinning technique for Li-ion batteries. Adv. Energy Mater. **5**(1), 1400753 (2014). https://doi.org/10.1002/aenm.201400753

Parallel Solution of Sediment and Suspension Transportation Problems on the Basis of Explicit Schemes

Alexander I. Sukhinov[1], Alexander E. Chistyakov[1],
and Valentina V. Sidoryakina[2(✉)]

[1] Don State Technical University, Rostov-on-Don, Russia
sukhinov@gmail.com, cheese_05@mail.ru
[2] Taganrog University, Named After A. P. Chekov – Branch of Rostov State
University of Economics, Taganrog, Russia
cvv9@mail.ru

Abstract. The article has been devoted to construction and investigation of parallel algorithms for the numerical realization of 3D models of suspended matter transportation and deposition and 2D models of bottom sediment transportation in sea coastal systems on the basis of explicit schemes with regularization terms that provide improved stability quality. The developed models take into account coastal currents and stress near the bottom caused by wind waves, turbulent spatial-three-dimensional motion of the water medium, particle size distribution and porosity of bottom sediments and hydraulic size of suspended particles, complicated shoreline shape and bottom relief and other factors. The numerical realization of the suspension transportation problem is carried out on the basis of explicit regularized difference schemes. The discrete model is constructed by means of including additional term according to idea of B. Chetverushkin – a discrete analogue of a second-order difference derivative with a small factor has been inserted in right side diffusion-advection equation. The value of the small factor determined on the basis of physical considerations and stability conditions. Compared with traditional parallel algorithms oriented to the use of implicit schemes, the use of explicit regularized algorithms allows to reduce the time of numerical solution of problems on a multi-core computing system with distributed memory containing 2048 cores and a peak performance of 18 Tflops in 12–80 times. The program package constructed by the authors for parallel realization given models has practical significance: it will allow to improve the accuracy of the real-time forecast and the validity of the engineering solutions taken for coastal infrastructure projects.

Keywords: Coastal zone · Mathematical model · Nonlinear task
Linearized task · Difference scheme

This paper was partially supported by the grant No. 17-11-01286 of the Russian Science Foundation.

L. Sokolinsky and M. Zymbler (Eds.): PCT 2018, CCIS 910, pp. 306–321, 2018.
https://doi.org/10.1007/978-3-319-99673-8_22

1   Introduction

Transportation of sediment, siltation and deformation of the bottom in coastal and shallow water systems in the South of Russia significantly effect the safety of navigation, the conditions for the reproduction of marine bioresources in shallow water reservoirs, changes in recreational zones, etc. The creation and application of precision models of these processes with predictive accuracy is an actual problem of mathematical modeling of water systems, since it allows predicting both the results of anthropogenic impact associated with the construction and reconstruction of coastal infrastructure in the South of Russia, and the consequences of the evolution of weather and climate phenomena – an increase frequency and intensity of storms, precipitation of extreme precipitation, etc. [1–3]. In connection with the foregoing, it is topical to create a set of predictive interrelated models of sediment transport, suspension and deformation of the bottom, methods for their numerical implementation, allowing in the operational mode to perform predictive modeling of these phenomena [4,5]. Existing models of sediment transport in the coastal zone do not have the necessary predictive accuracy and do not fully take into account the spatial-three-dimensional nature of the water movement and the complex form of the shoreline [6,7].

A complex of proposed transport models for multicomponent bottom sediment and suspensions (weighed particles) is taking into account:

- turbulent 3D motion of water medium;
- complicated form of coast line;
- real shape of water basin bottom, in particular, presence of relatively deep and narrow ship canals;
- presence of several types- components of particles- in suspension flow as well as in bottom sediment with different sizes and hydrophysical characteristics.

The method of full parallel algorithm constructing is based on idea of B.N. Chetverushkin of explicit schemes regularization [8]. The second order time derivatives with small multipliers have been included in left side of transport equations. It allows to improve (to raise) the acceptable time step - $O(h^2)$ to the value - $O(h^{3/2})$, where h - is norm of step spacing.

The proposed schemes allow efficient implementation on high-performance computing systems, which in turn will allow improving the accuracy of the operational forecast and the validity of the engineering decisions taken when creating coastal infrastructure objects.

2   Complex of Mathematical Models

The set of interrelated mathematical models is used to describe the hydrodynamic processes of the coastal zone, including models for the transport of suspensions, sediments, water movement and turbulence.

2.1   Continuous 3D Model of Diffusion-Convection-Aggregation of Suspensions

Let us consider a continuous mathematical model for the propagation of suspensions of various types in an aqueous medium, taking into account the diffusion and convection of suspensions, the action on the suspension of gravity, the mutual transformation of particles of various types, the presence of a bottom and a free surface.

We will use a rectangular Cartesian coordinate system Oxy, where the axis Ox passes over the surface of the unperturbed water surface and is directed towards the sea. Let $h = H + \eta$ – total depth of the water area, [m]; H – depth at the unperturbed surface of the reservoir, [m]; η – elevation of the free surface relative to the geoid (sea level), [m] (Fig. 1).

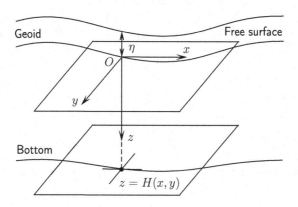

Fig. 1. Introduction of the coordinate system.

In the model, we move from the z-coordinate system to the θ-coordinate system for which we use the Descartes coordinate system in the horizontal plane, and as a vertical coordinate a dimensionless variable θ, $\theta \in [0; 1]$.

In the θ-coordinate system, the water column is divided into the same number of layers at each point, regardless of depth, so when we use the 'new' coordinate system, some problems associated with adding and subtracting layers are solved [9, 10].

When we go to the θ-coordinate system, we use formula:

$$\theta = a - \frac{(a - b)(z - \eta)}{h}, \quad x_\theta = x, \quad y_\theta = y, \tag{1}$$

where $\theta = a = 0$ on the free surface of the reservoir (upper boundary), $\theta = b = 1$ at the bottom (Fig. 2).

Next, instead of expressions (1), we use

$$\theta = \frac{z - \eta}{h}, \quad x_\theta = x, \quad y_\theta = y. \tag{2}$$

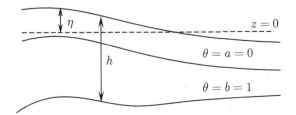

Fig. 2. The θ-coordinate system.

Suppose that there are R types of particles in the water volume $V = \{0 \leq x \leq L_x,\ 0 \leq y \leq L_y, 0 \leq \theta \leq 1\}$, which at the point (x, y, θ) and at the time t have a concentration $c_r = c_r(x, y, \theta, t)$, [mg/l]; t – time variable, [sec]; $r = 1, 2, \ldots, R$ (Fig. 3).

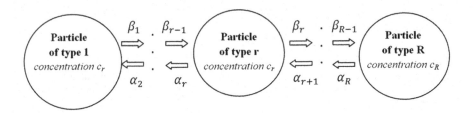

Fig. 3. Scheme of transformation of particles of different types.

The system of equations describing the behavior of particles will look like this:

$$
\begin{cases}
\dfrac{\partial c_r}{\partial t} + \dfrac{\partial (u c_r)}{\partial x} + \dfrac{\partial (v c_r)}{\partial y} + \dfrac{a - b}{h}\dfrac{\partial ((w + w_{g,r}) c_r)}{\partial \theta} = \\[2mm]
\qquad = \mu\left(\dfrac{\partial^2 c_r}{\partial x^2} + \dfrac{\partial^2 c_r}{\partial y^2}\right) + \left(\dfrac{a - b}{h}\right)^2 \dfrac{\partial}{\partial \theta}\left(\nu \dfrac{\partial c_r}{\partial \theta}\right) + F_r, \\[2mm]
F_1 = (\alpha_2 c_2 - \beta_1 c_1) + \Phi_1(x, y, \theta, t), \\
\ldots \\
F_r = (\beta_{r-1} c_{r-1} - \alpha_r c_r) + (\alpha_{r+1} c_{r+1} - \beta_r c_r) + \Phi_r(x, y, \theta, t), \\
\ldots \\
F_R = (\beta_{R-1} c_{R-1} - \alpha_R c_R) + \Phi_R(x, y, \theta, t), \quad r = 2, \ldots, R - 1,
\end{cases}
\tag{3}
$$

where u, v, w – the components of the velocity vector U of the fluid, [m/sec]; $w_{g,r}$ – the hydraulic size or the rate of deposition of particles of the r-th type, [m/sec]; μ, η – the coefficients of horizontal and vertical diffusion of particles of the r-th type, [m^2/sec]; α_r, β_r – particle conversion rates of the r-th type into $(r - 1)$-th and $(r + 1)$-th type, $\alpha_r \geq 0$, $\beta_r \geq 0$ [m/sec]; Φ_r – power of sources of particles of the r-th type, [mg/l sec].

The terms on the left side (apart from the time derivative) of the first equation of system (3) describe the convection of particles: their transport under the action of fluid flow and gravity. The terms on the right-hand side describe the diffusion of suspensions and their transformation from one type to another. The vertical diffusion coefficient is chosen to be different from the horizontal diffusion coefficient due to the fact that the effect of the difference in these coefficients is often observed in different media and can be caused by various factors.

Suppose that D is the domain, where the process takes place, and S – its boundary, which is a piecewise smooth line. The domain of setting the system (3) is the cylinder $C_T = D \times (0, T)$ heights T with base $D(x, y) = \{0 < x < L_x,\ 0 < y < L_y\}$. Its boundary consists of a lateral surface $S \times [0, T]$ and two bases: the lower $\bar{D} \times \{0\}$ – bottom and top $\bar{D} \times \{T\}$ – the unperturbed surface of water (Fig. 4).

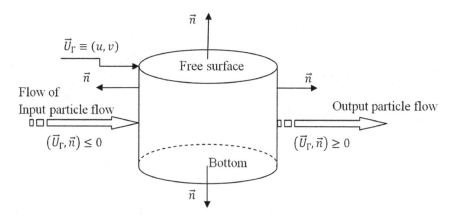

Fig. 4. The area of solution of the problem of transportation of suspended matter.

Add to the system (3) the initial and boundary conditions (assuming that the deposition of particles on the bottom is irreversible):

– initial conditions at time $t = 0$

$$c_1(x, y, \theta, 0) \equiv c_{10}(x, y, \theta),\ \ldots,\ c_r(x, y, \theta, 0) \equiv c_{r0}(x, y, \theta),\ \ldots,$$
$$c_R(x, y, \theta, 0) \equiv c_{R0}(x, y, \theta),\quad r = 2, \ldots, R - 1; \tag{4}$$

– boundary conditions on the cylindrical (lateral) boundary of the region S cylinder C_T

$$\frac{\partial c_1}{\partial n} = \cdots = \frac{\partial c_r}{\partial n} = \cdots = \frac{\partial c_R}{\partial n} = 0, \quad \text{if } (\boldsymbol{U}_\Gamma, \boldsymbol{n}) \leq 0, \tag{5}$$

$$\frac{\partial c_1}{\partial n} = -\frac{u_\Gamma}{\mu} c_1, \ \ldots, \ \frac{\partial c_r}{\partial n} = -\frac{u_\Gamma}{\mu} c_r, \ \ldots, \ \frac{\partial c_R}{\partial n} = -\frac{u_\Gamma}{\mu} c_R, \ \text{if } (\boldsymbol{U}_\Gamma, \boldsymbol{n}) \geq 0, \tag{6}$$

where n – the outer normal to the boundary of the domain S, U_Γ – the velocity vector of the fluid at the boundary S, u_Γ – the velocity vector projection U_Γ on the direction of the normal n on the border of the region S;
– boundary conditions on the water surface

$$\frac{\partial c_1}{\partial \theta} = \cdots = \frac{\partial c_r}{\partial \theta} = \cdots = \frac{\partial c_R}{\partial \theta} = 0; \tag{7}$$

– boundary conditions at the bottom [11]

$$\frac{\partial c_1}{\partial n} = -\frac{w_{g,1}}{\nu}c_1, \ \ldots, \ \frac{\partial c_r}{\partial n} = -\frac{w_{g,r}}{\nu}c_r, \ \ldots, \ \frac{\partial c_R}{\partial n} = -\frac{w_{g,R}}{\nu}c_R. \tag{8}$$

2.2 Nonlinear 2D Model of Sediment Transport

For simplicity, we assume that in the equation of transport of sediment transport, the axes Ox, Oy are consistent with the directions of the coordinate axes of the models of the hydrodynamic block, in which the components of the velocity vector of the water medium and the coefficient of turbulent exchange along the vertical direction are calculated. Further, for simplicity, the case is considered when the normal to the shoreline is directed to the north, coinciding with the axis Ox; the axis Oy is directed to the east.

The reformation of the coastal zone of the water areas due to the movement of water and solid particles will be described for the case when the sediment particles move in one direction (the side of the shore). In the work, the assumption is made that the sediments move only in one direction – the resultant transfer. The motion of the particles in the direction opposite to the direction of the resulting transfer will be neglected.

Let the sediments that participate in sediment transport consist Q of fractions, each of which has a relative fraction V_q in the total volume and density ρ_q, $q = 1, 2, \ldots, Q$.

The equation of sediment transport, which generalizes the known equation (see [12–14]) and takes into account the complex granulometric composition of the bottom material, will be written in the form

$$(1 - \bar{\varepsilon})\frac{\partial H}{\partial t} + \text{div}\left(\sum_{q=1}^{Q} V_q k_q \tau_b\right) = \text{div}\left(\sum_{q=1}^{Q} V_q k_q \frac{\tau_{bc,q}}{\sin\varphi_0}\text{grad}\,H\right) + \\ + \sum_{r=1}^{R} \frac{w_{g,r}}{\rho_r^*}c_r, \tag{9}$$

where $\bar{\varepsilon} = \sum_{q=1}^{Q} V_q \varepsilon_q$ – the averaged over fractions porosity of bottom materials; ε_q – porosity of q-type sediment fraction; τ_b – the vector of tangential stress sat the water bottom; $\tau_{bc,q}$ – the critical value of the tangential stress for the q-th fraction; $\tau_{bc,q} = a_q \sin\varphi_0$, φ_0 – an angle of repose of soil in the water; ρ_r^* – density of particles of suspended matter of the r-th type, which move in accordance with Eq. (3); $k_q = k_q(H, x, y, t)$ – the nonlinear coefficient, determined by the relation:

$$k_q \equiv \frac{A\bar{\omega}d_q}{((\rho_q - \rho_0)gd_q)^\beta} \left| \tau_b - \frac{\tau_{bc,q}}{\sin\varphi_0} \operatorname{grad} H \right|^{\beta-1}, \tag{10}$$

(ρ_q, d_q – density and characteristic particle size of the q-th fraction, respectively; ρ_0 – density of the aquatic environment; g – the gravity acceleration; $\bar{\omega}$ – the averaged wave frequency; A and β – dimensionless constants).

As with the equation of transport of suspensions, the region of specifying Eq. (9) is the cylinder $C_T = D \times (0,T)$, $D(x,y) = \{0 < x < L_x,\ 0 < y < L_y\}$.

We supplement Eq. (9) by the initial condition assuming that the function of the initial conditions belongs to the corresponding class of smoothness:

$$H(x,y,0) = H_0(x,y). \tag{11}$$

Let us formulate the conditions on the boundary of the region, starting from physical considerations:

$$\left| \tau_b \right| \big|_{y=0} = 0, \tag{12}$$

$$H(L_x, y, t) = H_2(y,t), \quad 0 \le y \le L_y, \tag{13}$$

$$H(0, y, t) = H_1(y,t), \quad 0 \le y \le L_y, \tag{14}$$

$$H(x, 0, t) = H_3(x), \quad 0 \le x \le L_x, \tag{15}$$

$$H(x, L_y', t) = H_4(x,t), \quad 0 \le x \le L_x, \quad L_y' < L_y. \tag{16}$$

We assume that there is always a layer of liquid of finite thickness in the considered region and for the indicated time interval there is no dehydration of the region, that is

$$H(x,y,t) \ge C \equiv \mathrm{const} > 0, \quad 0 \le x \le L_x, \quad 0 \le y \le L_y, \quad 0 \le t \le T. \tag{17}$$

The mathematical models of the transport of suspensions and sediments described above are also supplemented by models of the motion of the aquatic environment and turbulence, the equations of which are solved in the hydrodynamic block by the method of correction to pressure.

3   Linearization of 2D Sediment Transport Model

We construct a uniform grid ω_t in increments of τ (i.e. the set of points $\omega_t = \{t_n = n\tau,\ n = 0, 1, \ldots, N,\ N\tau = T\}$) and we realize the linearization of the initial-boundary value problem (9)–(17) [15–17] to create a linearized model on the time interval $0 \le t \le T$.

We linearize the term $\operatorname{div}\left(\sum_{q=1}^{Q} V_q k_q \tau_b\right)$ and the coefficient k_q by choosing their values at the time $t = t_n$, $n = 0, 1, \ldots, N$ and considering Eq. (9) in the time interval $t_{n-1} < t \le t_n$, $n = 1, 2, \ldots, N$. It is assumed that we know the function $H^{(n)}(x, y, t_{n-1}) \equiv H^{(n-1)}(x, y, t_{n-1})$ and its partial derivatives at spatial variables.

In the case, if $n = 1$, it is enough to take the function with the initial conditions $H^{(1)}(x, y, t_0)$, i.e. $H^{(1)}(x, y, t_0) \equiv H_0(x, y)$. If $n = 2, \ldots, N$, the function $H^{(n)}(x, y, t_{n-1}) = H^{(n-1)}(x, y, t_{n-1})$ is assumed to be known, since it is assumed that the problem (9)–(17) for the previous time interval $t_{n-2} < t \leq t_{n-1}$ is solved.

We introduced the notation:

$$(k_q)^{n-1} \equiv \frac{A \bar{\omega} d_q}{((\rho_q - \rho_0)g d_q)^\beta} \left| \tau_b - \frac{\tau_{bc,q}}{\sin \varphi_0} H^{n-1}(x, y, t_{n-1}) \right|^{\beta - 1}, \qquad (18)$$

$n = 1, 2, \ldots, N$, then we write Eq. (3) after linearization in the form:

$$(1 - \bar{\varepsilon}) \frac{\partial H^n}{\partial t} + \mathrm{div} \left(\sum_{q=1}^{Q} (V_q k_q)^{n-1} \tau_b \right) = \mathrm{div} \left(\sum_{q=1}^{Q} (V_q k_q)^{n-1} \frac{\tau_{bc,q}}{\sin \varphi_0} \, \mathrm{grad}\, H^n \right) +$$

$$+ \sum_{r=1}^{R} \frac{w_{g,r}}{\rho_r^*} c_r, \quad t_{n-1} < t \leq t_n, \quad n = 1, \ldots, N \qquad (19)$$

and with the initial conditions:

$$H^1(x, y, t_0) = H_0(x, y), \quad \ldots, \quad H^n(x, y, t_{n-1}) = H^{n-1}(x, y, t_{n-1}), \quad \ldots,$$

$$H^N(x, y, t_{N-1}) = H^{N-1}(x, y, t_{N-1}), \quad n = 2, \ldots, N - 1. \qquad (20)$$

The equation component $\mathrm{div} \left(\sum_{q=1}^{Q} (V_q k_q)^{n-1} \tau_b \right)$ is a well-known function of the right side at such linearization. The boundary conditions (11)–(17) are expected to be completed for all time intervals $t_{n-1} < t \leq t_n$, $n = 1, 2, \ldots, N$.

Note that the coefficients $(V_q k_q)^{n-1}$, $n = 1, 2, \ldots, N$ depend on the spatial coordinates x, y and the time variable t_{n-1}, $n = 1, 2, \ldots, N$, that are defined by the choice of the interval τ of the grid ω_t i.e. $(k_q)^{n-1} = (k_q)^{n-1}(x, y, t_{n-1})$, $n = 1, 2, \ldots, N$.

Using the methods described in [18], an estimate is obtained guaranteeing the continuous convergence of positive solutions $\tilde{H}(x, y, t)$ linearized initial-boundary value problem to the solution of the original nonlinear problem in the norm of a Hilbert space $L_1(D)$ at $\tau \to 0$, $N\tau = T$

$$\|\tilde{H}(T) - H(T)\|_{L_1(D)} \leq \frac{1}{2C(1 - \tau)} \tau L_x L_y C^* T, \qquad (21)$$

where C^* – constant, $C^* > 0$.

4 Discretization of Models

4.1 Discretization of the Suspension Transport Model

The problem was solved on the basis of an explicit scheme. Advantages of the explicit scheme are insignificant computational complexity for determining the solution on the time layer, natural parallelism and relative simplicity of software implementation on multiprocessor systems, greater physical proximity to the simulated processes. To increase the stability margin of the explicit scheme, it is proposed in [8] to use regularized schemes. For the construction of an explicit regularized scheme according to the idea of B. Chetverushkin we supplement the first equation of the system (3) by the term $\frac{\tau^*}{2}\frac{\partial^2 c_r}{\partial t^2}$ with the second time derivative:

$$
\frac{\partial c_r}{\partial t} + \frac{\tau^*}{2}\frac{\partial^2 c_r}{\partial t^2} + \frac{\partial(u c_r)}{\partial x} + \frac{\partial(v c_r)}{\partial y} + \frac{a-b}{h}\frac{\partial((w+w_{g,r})c_r)}{\partial \theta} =
$$

$$
\mu\left(\frac{\partial^2 c_r}{\partial x^2} + \frac{\partial^2 c_r}{\partial y^2}\right) + \left(\frac{a-b}{h}\right)^2 \frac{\partial}{\partial \theta}\left(\nu\frac{\partial c_r}{\partial \theta}\right) + F_r, \tag{22}
$$

where $\tau^* \sim \tau/c$ – the regularization parameter, τ – the grid spacing, c – the speed of sound in an aquatic environment.

Select an option τ^* is determined by the need to ensure an acceptable step in time when calculating the three-layer explicit scheme, as well as the proximity of the solutions of the newly obtained equation and the original equation. Parameter τ^* is determined on the basis of an estimate of the minimum time for the movement of particles in a cell of a spatial grid from one boundary to another. For example, for shallow coastal areas with a spatial mesh cell size $20\,\text{m} \times 20\,\text{m} \times 0.1\,\text{m}$ depending on the granulometric composition of suspensions, this value is in the range 1–50 s.

For the numerical realization of the discrete mathematical model of the hydrodynamic problem posed, a uniform grid is introduced:

$$
\bar{\omega} = \omega_t \times \omega_x \times \omega_y \times \omega_\theta,
$$

$$
\omega_t = \{t^n = n\tau,\ n = 0,\ldots,N,\ N\tau = T\},
$$

$$
\omega_x = \{x_i = ih_x,\ i = 0,\ldots,N_x,\ N_x h_x = L_x\},
$$

$$
\omega_y = \{y_j = jh_y,\ j = 0,\ldots,N_y,\ N_y h_y = L_y\},
$$

$$
\omega_\theta = \{\theta_k = kh_\theta,\ k = 0,\ldots,N_\theta,\ N_\theta h_\theta = 1\},
$$

where τ – the time step, h_x, h_y, h_θ – steps in space, N – the number of time layers, T – the upper time limit, N_x, N_y, N_θ – the number of steps in space, L_x, L_y – borders on space.

The condition for the stability of an explicit regularized scheme is $\tau \leq O(\|h\|^{3/2})$, $\|h\| = \sqrt{h_x^2 + h_y^2 + h_\theta^2}$, where τ is the time step, which is a less stringent requirement in comparison with the condition for explicit scheme, without regularization term – $\tau \leq O(\|h\|^2)$.

On the basis of the balance method, Eq. (22) can be approximated, with the discrete analogue of the regularized equation for calculating the transport of a suspension takes the form [10]:

$$
\frac{(c_r)_{i,j,k}^{n+1} - (c_r)_{i,j,k}^n}{\tau} + \frac{\tau^*}{2}\frac{(c_r)_{i,j,k}^{n+1} - 2(c_r)_{i,j,k}^n + (c_r)_{i,j,k}^{n-1}}{\tau^2}
$$

$$+\frac{(uc_r)_{i+1/2,j,k}^n - (uc_r)_{i,j,k}^n}{h_x} + \frac{(uc_r)_{i,j,k}^n - (uc_r)_{i-1/2,j,k}^n}{h_x}$$

$$+\frac{(vc_r)_{i,j+1/2,k}^n - (vc_r)_{i,j,k}^n}{h_y} + \frac{(vc_r)_{i,j,k}^n - (vc_r)_{i,j-1/2,k}^n}{h_y}$$

$$+\frac{a-b}{h}\frac{((w+w_{g,r})c_r)_{i,j,k+1/2}^n - ((w+w_{g,r})c_r)_{i,j,k}^n}{h_\theta}$$

$$+\frac{a-b}{h}\frac{((w+w_{g,r})c_r)_{i,j,k}^n - ((w+w_{g,r})c_r)_{i,j,k-1/2}^n}{h_\theta}$$

$$= \mu\left(\frac{(c_r)_{i+1,j,k}^n - (c_r)_{i,j,k}^n}{h_x^2} - \frac{(c_r)_{i,j,k}^n - (c_r)_{i-1,j,k}^n}{h_x^2}\right)$$

$$+\mu\left(\frac{(c_r)_{i,j+1,k}^n - (c_r)_{i,j,k}^n}{h_y^2} - \frac{(c_r)_{i,j,k}^n - (c_r)_{i,j-1,k}^n}{h_y^2}\right)$$

$$+\left(\frac{a-b}{h}\right)^2 \nu_{i,j,k+1/2}\frac{(c_r)_{i,j,k+1}^n - (c_r)_{i,j,k}^n}{h_\theta^2}$$

$$-\left(\frac{a-b}{h}\right)^2 \nu_{i,j,k-1/2}\frac{(c_r)_{i,j,k}^n - (c_r)_{i,j,k-1}^n}{h_\theta^2} + (F_r)_{i,j,k}^n. \tag{23}$$

To calculate the components of the velocity vector of the aquatic environment, a three-dimensional model of hydrodynamic flow around the bottom relief is used, taking into account the bottom friction and the level rise.

For an unregularized analogue of the scheme (23), there are restrictions on space steps

$$\frac{uh_x}{\mu} \le 2, \quad \frac{vh_y}{\mu} \le 2, \quad \frac{wh_\theta}{\mu} \le 2$$

and the time step limitation $\tau \le \left(\frac{\mu}{h_x^2} + \frac{\mu}{h_y^2} + \frac{\mu}{h_\theta^2}\right)^{-1}$.

4.2   Discretization of Sediment Transport Model

To construct an explicit scheme, we perform the regularization of Eq. (19):

$$(1-\bar{\varepsilon})\frac{\partial H^n}{\partial t} + \frac{\tau^{**}}{2}\frac{\partial^2 H^n}{\partial t^2} + \text{div}\left(\sum_{q=1}^Q (V_q k_q)^{n-1}\tau_b\right) =$$

$$= \text{div}\left(\sum_{q=1}^Q (V_q k_q)^{n-1}\frac{\tau_{bc,q}}{\sin\varphi_0}\,\text{grad}\,H^n\right) + \sum_{r=1}^R \frac{w_{g,r}}{\rho_r^*}c_r, \tag{24}$$

$$t_{n-1} < t \le t_n, \quad n = 1,\dots,N.$$

where τ^{**} – the small factor introduced in the same way as the formulation of the suspension transport problem. The value of τ^{**} has been determined on the basis of evaluation of the time for the movement of particles within the cell of the spatial grid.

We construct a finite-difference scheme that approximates Eq. (24) using the computational grid $\overline{\overline{\omega}}$, considering that the grid is in time ω_t – previously defined (see point 3), and the region D, as in the case of discretization of the suspension transport model, we cover with a uniform rectangular grid in increments h_x, h_y on spatial directions Ox, Oy respectively:

$$\overline{\overline{\omega}} = \omega_t \times \omega_x \times \omega_y.$$

Using the balance method, we obtain a difference scheme that approximates the regularized continuous problem of sediment transport:

$$(1 - \bar{\varepsilon}) \frac{H_{i,j}^{n+1} - H_{i,j}^n}{\tau} + \frac{\tau^{**}}{2} \frac{H_{i,j}^{n+1} - 2H_{i,j}^n + H_{i,j}^{n-1}}{\tau^2}$$

$$+ \sum_{q=1}^{Q} \left(\frac{(V_q k_q \tau_{b,x})_{i+1/2,j}^n - (V_q k_q \tau_{b,x})_{i-1/2,j}^n}{h_x} \right.$$

$$\left. + \frac{(V_q k_q \tau_{b,y})_{i,j+1/2}^n - (V_q k_q \tau_{b,y})_{i,j-1/2}^n}{h_y} \right)$$

$$= \frac{\tau_{bc,q}}{\sin \varphi_0} \sum_{q=1}^{Q} \left((V_q k_q)_{i+1/2,j}^n \frac{H_{i+1,j}^n - H_{i,j}^n}{h_x^2} - (V_q k_q)_{i-1/2,j}^n \frac{H_{i,j}^n - H_{i-1,j}^n}{h_x^2} \right)$$

$$+ \frac{\tau_{bc,q}}{\sin \varphi_0} \sum_{q=1}^{Q} \left((V_q k_q)_{i,j+1/2}^n \frac{H_{i,j+1}^n - H_{i,j}^n}{h_y^2} - (V_q k_q)_{i,j-1/2}^n \frac{H_{i,j}^n - H_{i,j-1}^n}{h_y^2} \right)$$

$$+ \sum_{r=1}^{R} \frac{w_{g,r}^n}{(\rho_r^*)^n} c_r^n, \tag{25}$$

where

$$(V_q k_q \tau_{b,x})_{i+1/2,j}^n = \frac{(V_q k_q \tau_{b,x})_{i+1,j}^n + (V_q k_q \tau_{b,x})_{i,j}^n}{2},$$

$$(V_q k_q \tau_{b,y})_{i,j+1/2}^n = \frac{(V_q k_q \tau_{b,y})_{i,j+1}^n + (V_q k_q \tau_{b,y})_{i,j}^n}{2},$$

$$(k_q)_{i+1/2,j}^n = \frac{A \bar{\omega} d_q \left| (\tau_b)_{i+1/2,j}^n - \frac{\tau_{bc,q}}{\sin \varphi_0} (\text{grad } H)_{i+1/2,j}^n \right|}{((\rho_q - \rho_0) g d_q)^\beta}.$$

Values $\text{grad } H \big|_{i+1/2,j}^n$, $\text{grad } H \big|_{i,j+1/2}^n$ will be written in the form

$$(\text{grad } H)_{i+1/2,j}^n = \frac{H_{i+1,j} - H_{i,j}}{h_x} \boldsymbol{i} + \frac{H_{i+1/2,j+1} - H_{i+1/2,j-1}}{2h_y} \boldsymbol{j},$$

$$(\operatorname{grad} H)_{i,j+1/2}^{n} = \frac{H_{i+1,j+1/2} - H_{i-1,j+1/2}}{2h_x} \boldsymbol{i} + \frac{H_{i,j+1} - H_{i,j}}{h_y} \boldsymbol{j},$$

The approximation of the boundary conditions for brevity of the exposition is not reducible.

The estimation, which guarantees implicit difference scheme stability in mesh space $c_{h,\tau}$ may be presented in the form:

$$\|H^n\|_{c_{h,\tau}} \le \|H_0\|_{c_h} + \max\left(\|H_1\|_{c_{h,\tau}}, \|H_2\|_{c_{h,\tau}}, \|H_3\|_{c_{h,\tau}}\right) +$$

$$(1 - \bar{\varepsilon})\left(\sum_{m=0}^{n}\sum_{q=1}^{Q}\left(\left\|(V_q k_q(t_m)\tau_{b,x})_{\overset{\circ}{x}}^{m}\right\|_{c_{h,\tau}} + \left\|(V_q k_q(t_m)\tau_{b,y})_{\overset{\circ}{y}}^{m}\right\|_{c_{h,\tau}}\right) + \quad (26)$$

$$\frac{\max\limits_{r}\|w_{g,r}^n\|}{\min\limits_{r}(\rho_r^*)^n}\sum_{r=1}^{R}\|c_r^n\|_{c_{h,\tau}}\Bigg).$$

It is found that the approximation error of the discrete sediment transport model is a quantity of order $O(\tau + h_x^2 + h_y^2)$.

We note that in the numerical realization of the coupled model of suspension-sediment transport the value of permissible time step is less than minimum among regularizing factors τ^*, τ^{**} for explicit schemes (23) and (25).

5   Estimation of Computational Complexity Suspension Transport Model Realization on a Multiprocessor Computer System

Since in the numerical solution of the problems of sediment transport and transport of suspended particles, the latter is the main difficulty, its consideration will be presented in the framework of this paper.

Computational complexity algorithm Q_{neiavn} for the suspension transport problem numerical realization on the basis implicit scheme is estimated:

$$Q_{\text{neiavn}} = n_{\tau n}(\varepsilon)N_x N_y q_{\text{PTM}},$$

where q_{PTM} – the number of arithmetic operations for one iteration MATM (modified alternating-triangular method) ($q_{\text{PTM}} \sim 50$); $n(\varepsilon) = O(N_{\max})$ – the number of iterations; $N_{\max}$ – the number of nodes in space; $N_{\max} = \max\{N_x, N_y\}$, N_x, N_y – the number of steps along the coordinate axes O_x, O_y respectively; $n_{\tau n} = T/\tau_n$ – the number of time layers for the implicit scheme, T – calculated time interval, τ_n – the time step for an implicit scheme.

For an explicit scheme such an estimate has the form:

$$Q_{\text{iavn}} = n_{\tau n}(\varepsilon)N_x N_y q_{\text{iavn}},$$

where q_{iavn} – the number of operations for the transition to the next time layer by an explicit regularized scheme ($q_{iavn} \sim 14$); $n_{\tau n} = T/\tau_n$ – the number of time layers for an explicit scheme, τ_r – the time step for an explicit scheme.

Numerical experiments have been fulfilled for grids $kN_x \times kN_y \times N_z$, $N_x = 122$, $N_y = 102$, $N_z = 13$, where $k = 1, 2, 4$. Table 1 shows the values of the time steps. The accuracy of the calculations is about one percent of the solution for explicit regularized and implicit schemes for the different number of nodes of grids.

Table 1. The values of steps for a temporary variable for explicit and implicit schemes.

Grid size	Explicit scheme		
	$101 \times 101 \times 11$	$201 \times 201 \times 11$	$401 \times 401 \times 11$
Step by time	0.072	0.036	0.025
Grid size	Implicit scheme		
	$101 \times 101 \times 11$	$201 \times 201 \times 11$	$401 \times 401 \times 11$
Step by time	0.2	0.1	0.075

Based on the results of numerical experiments, the following estimate is obtained, showing the time gain for the explicit scheme with respect to the implicit scheme, in the case of grid sizes with $101 \times 101 \times 11$ (the number of iterations 8): $Q_{neiavn}/Q_{iavn} \approx 10.286$, and in the case of grid sizes $201 \times 201 \times 11$ (the number of iterations 10): $Q_{neiavn}/Q_{iavn} \approx 12.857$, in the case of mesh sizes $401 \times 401 \times 11$ (the number of iterations 12): $Q_{neiavn}/Q_{iavn} \approx 14.286$.

Table 2 shows the times for executing the transitions between layers for an explicit scheme and the execution of one iteration by an implicit scheme, as well as the values of the acceleration and efficiency of parallel algorithms. The calculated grid consisted of $101 \times 101 \times 11$ knots. From the numerical estimates of the ratio of the times for solving the model problem to the explicit and implicit schemes, it can be concluded that when the size of the computational grid is increased, the gain in the calculation time of the explicit scheme only increases.

Table 2. Comparison of parallel algorithms based on explicit and implicit schemes.

Number of calculators		1	2	4	8	16	32	64
Explicit scheme	Time, sec	0.00271	0.00074	0.00052	0.00029	0.00025	0.00060	0.00125
	Acceleration	1	3.662	5.212	9.345	**10.84**	4.517	2.168
	Efficiency	1	1.831	1.303	1.168	0.677	0.141	0.034
Implicit schema	Time, sec	0.01183	0.00446	0.00232	0.00179	0.00231	0.00365	0.00642
	Acceleration	1	2.652	5.099	**6.609**	5.121	3.241	1.843
	Efficiency	1	1.326	1.275	0.826	0.32	0.101	0.029
The time gain for an explicit scheme		10.286	14.201	10.513	14.544	**21.772**	14.334	12.102

Table 2 shows the times of execution for one time step for explicit scheme and one iteration for implicit schemes as well as the values of the acceleration and efficiency of parallel algorithms. The calculated grid consisted of $101 \times 101 \times 11$ knots. From the numerical estimates of the ratio of the times for solving the model problem to the explicit and implicit schemes, it can be concluded that when the size of the computational grid is increased, the gain in the calculation time of the explicit scheme only increases.

When solving a problem on a grid containing $101 \times 101 \times 11$ knots, the maximum acceleration for an explicit scheme was achieved on 16 cores and equal to 10.84. For the implicit scheme, the maximum acceleration, equal to 6.609, was achieved on 8 cores. Thus, the gain in time for the explicit scheme in relation to the implicit scheme was 16.871 times. Table 3 shows the times for executing the transitions between layers for the explicit scheme and the execution of one iteration by an implicit scheme on the computational grid $5001 \times 5001 \times 101$ nodes. In this case, the gain of the explicit scheme in time on 512 cores of the supercomputer system was 71.547 times.

Table 3. Time of execution of time steps by an explicit scheme and one iteration by an implicit scheme and values Acceleration and efficiency of parallel algorithms.

Number of calculators		1	2	4	8	16	32	64	128	256	512
Explicit scheme	Time, sec	163	82	53	18	14	7.8	4.4	1.7	0.987	0.715
	Acceleration	1	1.98	3.06	8.68	11.5	20.93	37.1	96.05	165.434	**228.36**
	Efficiency	1	0.99	0.76	1.08	0.72	0.654	0.58	0.75	0.646	0.446
Implicit schema	Time, sec	370	188	127	48.9	47.2	31.8	18.2	7.6	6.318	5.8805
	Acceleration	1	1.967	2.924	7.555	7.837	11.61	20.29	48.33	58.563	62.921
	Efficiency	1	0.984	0.731	0.944	0.49	0.363	0.317	0.378	0.229	0.123
The time gain for an explicit scheme		19.74	19.94	20.76	23.63	29.33	35.47	36.38	38.89	55.691	71.547

6   Conclusion

The complex of coupled multicomponent mathematical models have been presented for sediment bottom material and suspended particles transport in coastal systems. In these model many types of particles with different densities and medium sizes have been included. These models are satisfying the basic conservation laws, including mass conservation law. For the indicated mathematical models, the formulation of the initial and boundary conditions is described. Conservative stable difference schemes are constructed and investigated. Parallel algorithms, based on implicit and regularizes explicit schemes have been presented. A comparative analysis of the efficiency of the use of implicit and explicit regularized difference schemes in the numerical implementation of the problems under consideration is given. It is shown that the use of explicit regularized schemes leads to significant time savings (more than 10 times), compared to previously used algorithms based on implicit schemes.

References

1. Sukhinov, A.A., Sukhinov, A.I.: 3D model of diffusion-advection-aggregation suspensions in water basins and its parallel realization. In: Parallel Computational Fluid Dynamics, Multidisciplinary Applications, Proceedings of Parallel CFD 2004 Conference, Las Palmas de Gran Canaria, Spain, pp. 223–230. Elsevier, Amsterdam (2005). https://doi.org/10.1016/B978-044452024-1/50029-4

2. Sukhinov, A.I., Sukhinov, A.A.: Reconstruction of 2001 ecological disaster in the Azov sea on the basis of precise hydrophysics models. In: Parallel Computational Fluid Dynamics, Multidisciplinary Applications, Proceedings of Parallel CFD 2004 Conference, Las Palmas de Gran Canaria, Spain, pp. 231–238. Elsevier, Amsterdam (2005). https://doi.org/10.1016/B978-044452024-1/50030-0

3. Alekseenko, E., Roux, B., Sukhinov, A., Kotarba, R., Fougere, D.: Coastal hydrodynamics in a windy lagoon. J. Comput. Fluids **77**, 24–35 (2013). https://doi.org/10.1016/j.compfluid.2013.02.003

4. Alekseenko, E., Roux, B., Sukhinov, A., Kotarba, R., Fougere, D.: Nonlinear hydrodynamics in a mediterranean lagoon. J. Nonlinear Process. Geophys. **20**(2), 189–198 (2013). https://doi.org/10.5194/npg-20-189-2013

5. Sukhinov, A.I., Chistyakov, A.E., Alekseenko, E.V.: Numerical realization of the three-dimensional model of hydrodynamics for shallow water basins on a high-performance system. J. Math. Models Comput. Simul. **3**(5), 562–574 (2011). (in Russian)

6. Leontyev, I.O.: Coastal Dynamics: Waves, Moving Streams. Deposits Drifts, GEOS, San Moscow (2001). (in Russian)

7. Liu, X., Qi, S., Huang, Y., Chen, Y., Pengfei, D.: Predictive modeling in sediment transportation across multiple spatial scales in the Jialing River Basin of China. Int. J. Sediment Res. **30**(3), 250–255 (2015)

8. Chetverushkin, B.N.: Resolution limits of continuous media models and their mathematical formulations. J. Math. Models Comput. Simul. **5**(3), 266–279 (2013). (in Russian)

9. Sukhinov, A.I.: Precise fluid dynamics models and their application in prediction and reconstruction of extreme events in the sea of Azov. J. Izv. Taganrog. Radiotech. Univ. **3**, 228–235 (2006). (in Russian)

10. Sukhinov, A.I., Protsenko, E.A., Chistyakov, A.E., Shreter, S.A.: Comparison of numerical efficiency of explicit and implicit schemes as applied to sediment transport in coastal systems. J. Vychisl. Metody Program. Novye Vychisl. Tekhnol. **16**(3), 328–338 (2015). (in Russian)

11. Marchuk, G.I.: Numerical Solution of the Problems of Atmosphere and Ocean Dynamics. Gidrometeoizdat, Leningrad (1974). (in Russian)

12. Sukhinov, A.I., Chistyakov, A.E., Protsenko, E.A.: Mathematical modeling of sediment transport in the coastal zone of shallow reservoirs. J. Math. Models Comput. Simul. **6**(4), 351–363 (2014). (in Russian)

13. Sukhinov, A.I., Chistyakov, A.E., Protsenko, E.A.: Sediment transport mathematical modeling in a coastal zone using multiprocessor computing systems. J. Num. Methods Program. **15**(4), 610–620 (2014). (in Russian)

14. Sidoryakina, V.V., Sukhinov, A.I.: Well-posedness analysis and numerical implementation of a linearized two-dimensional bottom sediment transport problem. J. Comput. Math. Math. Phys. **57**(6), 978–994 (2017). https://doi.org/10.7868/S0044466917060138

15. Sukhinov, A.I., Sidoryakina, V.V., Sukhinov, A.A.: Sufficient conditions for convergence of positive solutions to linearized two-dimensional sediment transport problem. J. Vestnik Don State Tech. Univ. **1**(88), 5–17 (2017). (in Russian)
16. Sukhinov, A., Chistyakov, A., Sidoryakina, V.: Investigation of nonlinear 2D bottom transportation dynamics in coastal zone on optimal curvilinear boundary adaptive grids. In: MATEC Web of Conference XIII International Scientific-Technical Conference 'Dynamic of Technical Systems' (DTS 2017), Rostov-on-Don, vol. 132 (2017). https://doi.org/10.1051/matecconf/201713204003
17. Sukhinov, A.I., Sidoryakina, V.V., Sukhinov, A.A.: Sufficient convergence conditions for positive solutions of linearized two-dimensional sediment transport problem. J. Comput. Math. Inf. Tech. **1**(1), 21–35 (2017). (in Russian)
18. Sukhinov, A.I., Sidoryakina, V.V.: On the convergence of solutions of linearized on a time grid sequence problem to the solution of nonlinear problems of sediment transport. J. Math. Models. **29**(11), 19–39 (2017). (in Russian)
19. Samarskiy, A.A., Gulin, A.V.: Numerical Methods. Nauka, Moscow (1989). (in Russian)
20. Samarskiy, A.A.: Theory of Difference Schemes. Nauka, Moscow (1989). (in Russian)

Three-Dimensional Mathematical Model of Wave Propagation Towards the Shore

Alexander Sukhinov, Alexander Chistyakov, and Sophia Protsenko$^{(\boxtimes)}$

Don State Technical University, Rostov-on-Don, Russia
sukhinov@gmail.com, cheese_05@mail.ru, rab55555@rambler.ru

Abstract. To describe wave processes, we use here a system of Navier–Stokes equations containing three equations of motion in regions with dynamically varying geometry of the computational domain. The pressure correction method was used to approximate the hydrodynamic model difference schemes that describe the mathematical model of wave propagation towards the shore. This model was constructed on the basis of an integro-interpolation method using a scheme with weights. An adaptive alternating-triangular iterative method was used to solve the system of equations. The practical significance of the numerical algorithms and the complex of programs implementing them is determined by the possibility of application in the study of hydrophysical processes in coastal water systems, as well as in the construction of the velocity and pressure fields of the aquatic environment. They also make it possible to assess the hydrodynamic effect on shore protection structures and coastal structures in the presence of surface waves.

Keywords: Coastal structures · Surface gravitational waves
Mathematical model of wave propagation towards the shore
Three-dimensional wave processes · Sediment transport

1 Introduction

The study of hydrodynamic processes of coastal waters is connected with the investigation of the influence of wave processes generated either in the open sea or in the coastal zone of reservoirs. The movement of waves can exert negative effects on the behavior of the coastal zone, i.e. transformation of the bottom surface resulting from the rise of bottom sediments, abrasion (i.e. the destruction process of the banks of various water systems by waves and surf). The result of the interaction of waves with the bottom surface and the coastal slope consists of refraction, diffraction and changes in wave structure. The most significant factors are fluctuations in water surface level, wind phenomena, currents, transport of bottom materials, and deformation of the coastal slope. A characteristic feature of coastal waters is the significant influence of the bottom surface on wave processes, which makes it difficult to study tidal phenomena in coastal regions of seas and river mouths. The influence of wave processes on the coastal zone

© Springer Nature Switzerland AG 2018
L. Sokolinsky and M. Zymbler (Eds.): PCT 2018, CCIS 910, pp. 322–335, 2018.
https://doi.org/10.1007/978-3-319-99673-8_23

can be ambivalent: wave processes can have a significant effect on the accumulation and abrasion of the coastal zone of the reservoir and directly on coastal structures. We bring into focus the problem of practical application of computationally effective methods for simulation the hydrodynamic processes, making it possible to obtain a fairly accurate approximate numerical solution. At present, there is a need to construct a set of interrelated models of three-dimensional wave processes intended for modeling wave processes.

2   Statement of the Problem of Wave Hydrodynamics

The initial equations of hydrodynamics of shallow water bodies are [1–3]:
— the equation of motion (Navier–Stokes):

$$u'_t + uu'_x + vu'_y + wu'_z = -\frac{1}{\rho}p'_x + (\mu u'_x)'_x + (\mu u'_y)'_y + (\nu u'_z)'_z,$$

$$v'_t + uv'_x + vv'_y + wv'_z = -\frac{1}{\rho}p'_y + (\mu v'_x)'_x + (\mu v'_y)'_y + (\nu v'_z)'_z, \qquad (1)$$

$$w'_t + uw'_x + vw'_y + ww'_z = -\frac{1}{\rho}p'_z + (\mu w'_x)'_x + (\mu w'_y)'_y + (\nu w'_z)'_z + g;$$

— the equation of continuity in the case of variable density:

$$\rho'_t + (\rho u)'_x + (\rho v)'_y + (\rho w)'_z = 0, \qquad (2)$$

where $V = \{u, v, w\}$ are the components of the velocity vector, p is the pressure, ρ is the density, μ, ν are the horizontal and vertical components of the coefficient of turbulent exchange, g is the acceleration of gravity.

The system of Eqs. (1)–(2) is considered under the following boundary conditions:
— at the entrance

$$u(x, y, z, t) = u(t), \quad v(x, y, z, t) = v(t),$$
$$p'_n(x, y, z, t) = 0, \quad V'_n(x, y, z, t) = 0; \qquad (3)$$

— lateral border (shore and bottom)

$$\rho\mu(u')_n(x, y, z, t) = -\tau_x(t), \qquad \rho\mu(v')_n(x, y, z, t) = -\tau_y(t),$$
$$V_n(x, y, z, t) = 0, \qquad\qquad p'_n(x, y, z, t) = 0;$$

— upper limit

$$\rho\mu(u')_n(x, y, z, t) = -\tau_x(t),$$
$$\rho\mu(v')_n(x, y, z, t) = -\tau_y(t),$$
$$w(x, y, t) = -\omega - \frac{p'_t}{\rho g}, \quad p'_n(x, y, t) = 0,$$

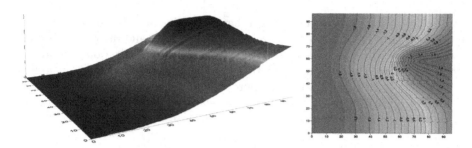

Fig. 1. The geometry of the computational domain

where ω is the evaporation rate of the liquid, τ_x, τ_y are the components of the tangential stress [4,5]. The components of the tangential stress for the free surface can be expressed as $\tau_x = \rho_a C_p (|\boldsymbol{w}|) w_x |\boldsymbol{w}|$, $\tau_y = \rho_a C_p (|\boldsymbol{w}|) w_y |\boldsymbol{w}|$, where $\boldsymbol{w}$ is the vector of the wind speed relative to the water, ρ_a is the density of the atmosphere, and $C_p(x) = \begin{cases} 0.0088, & x < 6,6 \text{ m/s}; \\ 0.0026, & x \geq 6,6 \text{ m/s}; \end{cases}$ is a dimensionless coefficient.

Figure 1 shows the geometry of the computational domain. The components of the tangential stress for the bottom, taking into account the notations introduced, can be written as $\tau_x = \rho_v C_p (|V|) u |V|$, $\tau_y = \rho_v C_p (|V|) v |V|$, where ρ_v is the density of bottom sediments.

The approximation considered below makes it possible to construct on the basis of the measured velocity pulsations the coefficient of vertical turbulent exchange, which is inhomogeneous with respect to depth [4]:

$$nu = C_s^2 \Delta^2 \frac{1}{2} \sqrt{\left(\frac{\partial \overline{U}}{\partial z} \right)^2 + \left(\frac{\partial \overline{V}}{\partial z} \right)^2}, \tag{4}$$

where $\overline{U}, \overline{V}$ are time-averaged pulsations of the horizontal velocity components, Δ is the characteristic scale of the grid, and C_s is a dimensionless empirical constant whose value is usually determined by calculating the decay process of homogeneous isotropic turbulence.

3   The Discrete Model of Hydrodynamics of Shallow Water Reservoirs

The computational domain is inscribed in a parallelepiped. A uniform grid is introduced for the numerical implementation of the discrete mathematical model of the posed hydrodynamic problem:

$$\bar{w}_h = \{ t^n = n\tau, \, x_i = ih_x, \, y_j = jh_y, \, z_k = kh_z;$$
$$n = \overline{0 \dots N_t}, \, i = \overline{0 \dots N_x}, \, j = \overline{0 \dots N_y}, \, k = \overline{0 \dots N_z};$$
$$N_t \tau = T, \, N_x h_x = l_x, \, N_y h_y = l_y, \, N_z h_z = l_z \},$$

where τ is the time step, h_x, h_y, h_z are the space steps, N_t is the number of time layers, T is the upper bound of the time coordinate, N_x, N_y, N_z are the number of nodes for the spatial coordinates, l_x, l_y, l_z are the boundaries along the parallelepiped, respectively, in the direction of the axes Ox, Oy and Oz.

To solve the hydrodynamic problem, we used the pressure correction method. In the case of a variable density, this method can be written as [7,11]:

$$\frac{\tilde{u} - u}{\tau} + u\bar{u}'_x + v\bar{u}'_y + w\bar{u}'_z = \left(\mu\bar{u}'_x\right)'_x + \left(\mu\bar{u}'_y\right)'_y + \left(\nu\bar{u}'_z\right)'_z,$$

$$\frac{\tilde{v} - v}{\tau} + u\bar{v}'_x + v\bar{v}'_y + w\bar{v}'_z = \left(\mu\bar{v}'_x\right)'_x + \left(\mu\bar{v}'_y\right)'_y + \left(\nu\bar{v}'_z\right)'_z,$$

$$\frac{\tilde{w} - w}{\tau} + u\bar{w}'_x + v\bar{w}'_y + w\bar{w}'_z = \left(\mu\bar{w}'_x\right)'_x + \left(\mu\bar{w}'_y\right)'_y + \left(\nu\bar{w}'_z\right)'_z + g, \quad (5)$$

$$p''_{xx} + p''_{yy} + p''_{zz} = \frac{\hat{\rho} - \rho}{\tau^2} + \frac{(\hat{\rho}\tilde{u})'_x}{\tau} + \frac{(\hat{\rho}\tilde{v})'_y}{\tau} + \frac{(\hat{\rho}\tilde{w})'_z}{\tau},$$

$$\frac{\hat{u} - \tilde{u}}{\tau} = -\frac{1}{\rho}\hat{p}'_x, \quad \frac{\hat{v} - \tilde{v}}{\tau} = -\frac{1}{\rho}\hat{p}'_y, \quad \frac{\hat{w} - \tilde{w}}{\tau} = -\frac{1}{\rho}\hat{p}'_z,$$

where $V = \{u, v, w\}$ are the components of the velocity vector, $\{\hat{u}, \hat{v}, \hat{w}\}$, $\{\tilde{u}, \tilde{v}, \tilde{w}\}$ are the components of the velocity vector fields in the "new" and intermediate time layers, respectively, $\bar{u} = (\tilde{u} + u)/2$, $\hat{\rho}$ and ρ is the density distribution of the aqueous medium in the new and previous time layers, respectively.

In the discrete mathematical models of hydrodynamics, we take into account the "fullness" of the control cells, which makes it possible to increase the real accuracy of the solution when the investigated region has a complex geometry, by improving the approximation of the boundary.

Denote by $o_{i,j,k}$ the "fullness" of the cell (i, j, k) [8]. The degree of "fullness" of a cell is determined by the pressure of the liquid column inside the cell. If the average pressure at the nodes belonging to the cell vertices is greater than the pressure of the liquid column inside the cell, then we say that the cell is full $(o_{i,j,k} = 1)$. In the general case, the "fullness" of a cell can be calculated by the following formula [7]:

$$o_{i,j,k} = \frac{P_{i,j,k} + P_{i-1,j,k} + P_{i,j-1,k} + P_{i-1,j-1,k}}{4\rho g h_z}, \quad (6)$$

where $P = p + \rho g z$ is the pressure.

Let us introduce the coefficients $q_0, q_1, q_2, q_3, q_4, q_5$, and q_6 to describe the "fullness" of regions located in a vicinity of the cell (control areas). The value characterizes the "fullness" of the region

$q_0 - D_0$: $\{x \in (x_{i-1}, x_{i+1}), y \in (y_{j-1}, y_{j+1}), z \in (z_{k-1}, z_{k+1})\}$,
$q_1 - D_1$: $\{x \in (x_i, x_{i+1}), y \in (y_{j-1}, y_{j+1}), z \in (z_{k-1}, z_{k+1})\}$,
$q_2 - D_2$: $\{x \in (x_{i-1}, x_i), y \in (y_{j-1}, y_{j+1}), z \in (z_{k-1}, z_{k+1})\}$,
$q_3 - D_3$: $\{x \in (x_{i-1}, x_{i+1}), y \in (y_j, y_{j+1}), z \in (z_{k-1}, z_{k+1})\}$,
$q_4 - D_4$: $\{x \in (x_{i-1}, x_{i+1}), y \in (y_{j-1}, y_j), z \in (z_{k-1}, z_{k+1})\}$,

$q_5 - D_5$: $\{x \in (x_{i-1}, x_{i+1}), y \in (y_{j-1}, y_{j+1}), z \in (z_k, z_{k+1})\}$,
$q_6 - D_6$: $\{x \in (x_{i-1}, x_{i+1}), y \in (y_{j-1}, y_{j+1}), z \in (z_{k-1}, z_k)\}$.

The filled parts of the regions D_m will be denoted by Ω_m, where $m = \overline{0 \ldots 6}$. In accordance with this, the coefficients q_m can be calculated from the formulas

$$(q_m)_{i,j,k} = \frac{S_{\Omega_m}}{S_{D_m}},$$

$$(q_0)_{i,j,k} = \frac{1}{2}\left((q_1)_{i,j,k} + (q_2)_{i,j,k}\right),$$

$$(q_1)_{i,j,k} = \frac{o_{i+1,j,k} + o_{i+1,j+1,k} + o_{i+1,j,k+1} + o_{i+1,j+1,k+1}}{4},$$

$$(q_2)_{i,j,k} = \frac{o_{i,j,k} + o_{i,j+1,k} + o_{i,j,k+1} + o_{i,j+1,k+1}}{4},$$

$$(q_3)_{i,j,k} = \frac{o_{i+1,j+1,k} + o_{i,j+1,k} + o_{i+1,j+1,k+1} + o_{i,j+1,k+1}}{4},$$

$$(q_4)_{i,j,k} = \frac{o_{i,j,k} + o_{i+1,j,k} + o_{i,j,k+1} + o_{i+1,j,k+1}}{4},$$

$$(q_5)_{i,j,k} = \frac{o_{i,j,k+1} + o_{i+1,j,k+1} + o_{i+1,j+1,k+1} + o_{i,j+1,k+1}}{4},$$

$$(q_6)_{i,j,k} = \frac{o_{i,j,k} + o_{i+1,j,k} + o_{i+1,j+1,k} + o_{i,j+1,k}}{4}.$$

In the case of boundary conditions of the third kind, $c'_n(x, t) = \alpha_n c + \beta_n$, the discrete analogs of the convective transfer operator uc'_x and the diffusion transfer operator $(\mu c'_x)_x$, obtained with the help of the integro-interpolation method taking into account the partial "fullness" of the cells, can be written as

$$uc'_x \simeq (q_1)_i\, u_{i+1/2} \frac{c_{i+1} - c_i}{2h_x} + (q_2)_i\, u_{i-1/2} \frac{c_i - c_{i-1}}{2h_x},$$

$$\left(\mu c'_x\right)_x \simeq (q_1)_i\, \mu_{i+1/2} \frac{c_{i+1} - c_i}{h_x^2} - (q_2)_i\, \mu_{i-1/2} \frac{c_i - c_{i-1}}{h_x^2} - \left|(q_1)_i - (q_2)_i\right| \mu_i \frac{\alpha_x c_i + \beta_x}{h_x}.$$

The approximations for the remaining coordinate directions can be similarly expressed. The approximation error of the mathematical model is

$$O\left(\tau + \|h\|^2\right),$$

where $\|h\| = \sqrt{h_x^2 + h_y^2 + h_z^2}$. The conservation of the flow at the discrete level of the developed hydrodynamic model has been proved, as well as the absence of nonconservative dissipative terms obtained as a result of the discretization of the system of equations. A sufficient condition for the stability and monotony of the developed model is determined on the basis of the maximum principle [6] with constraints on the spatial coordinate steps:

$$h_x < |2\mu/u|, h_y < |2\mu/v|, h_z < |2\nu/w|, \text{ or } \mathrm{Re} \leq 2N,$$

where $\mathrm{Re} = |V| \cdot l/\mu$ is the Reynolds numbers, l is the characteristic size of the region, and $N = \max\{N_x, N_y, N_z\}$.

The discrete analogs of the system of Eq. (5) are solved by an adaptive modified alternating-triangular method of variational type.

4 Method for Solving the Grid Equations

The resulting grid equations can be written in matrix form [13]:

$$Ax = f, \tag{7}$$

where A is a linear positive-definite operator $(A > 0)$. To find the solution of problem (7), we will use an implicit iterative process:

$$B\frac{x^{m+1} - x^m}{\tau_{m+1}} + Ax^m = f. \tag{8}$$

In Eq. (8), m is the iteration number, $\tau > 0$ is an iteration parameter, and B is an invertible operator, which is called the preconditioner or stabilizer. The inversion of the operator B in (8) should be substantially simpler than the direct inversion of the original operator A in (7). To construct the operator B, we proceed from the additive representation of the operator A_0 which is the symmetric part of the operator A:

$$A_0 = R_1 + R_2, \quad R_1 = R_2^*, \tag{9}$$

where $A = A_0 + A_1$, $A_0 = A_0^*$, $A_1 = -A_1^*$.

The preconditioner operator can be written as

$$B = (D + \omega R_1)D^{-1}(D + \omega R_2), \quad D = D^* > 0, \quad \omega > 0, \tag{10}$$

where D is a certain operator.

If the operators R_1 and R_2 are defined, and the methods for finding the parameters τ_{m+1}, ω and the operator D are indicated, then relations (9)–(10) define a modified alternate-triangular method (MATM) for the solution of the problem.

Finally, the algorithm of the adaptive modified alternating-triangular method of minimal corrections for calculating the grid equations with a non selfadjoint operator has the form

$$B(\omega_m)w^m = r^m, \quad r^m = Ax^m - f, \quad \tilde{\omega}_m = \sqrt{\frac{(Dw^m, w^m)}{(D^{-1}R_2w^m, R_2w^m)}},$$

$$s_m^2 = 1 - \frac{(A_0w^m, w^m)^2}{(B^{-1}A_0w^m, A_0w^m)(Bw^m, w^m)}, \quad k_m = \frac{(B^{-1}A_1w^m, A_1w^m)}{(B^{-1}A_0w^m, A_0w^m)}, \tag{11}$$

$$\theta_m = \frac{1 - \sqrt{\frac{s_m^2 k_m}{(1+k_m)}}}{1 + k_m(1 - s_m^2)}, \quad \tau_{m+1} = \theta_m\frac{(A_0w^m, w^m)}{(B^{-1}A_0w^m, A_0w^m)},$$

$$x^{m+1} = x^m - \tau_{m+1}w^m, \quad \omega_{m+1} = \tilde{\omega}_m,$$

where r^m is the discrepancy vector, w^m is the correction vector, and as operator D, we take the diagonal part of the operator A.

5 Parallel Version of the Algorithm for Solving the Grid Equations

Let us consider the parallel algorithm for calculating the correction vector [14]:

$$(D + \omega_m R_1)D^{-1}(D + \omega_m R_2)w^m = r^m,$$

where R_1 is a lower-triangular matrix, and R_2 is an upper-triangular matrix. To this end, we solve successively the systems

$$(D + \omega_m R_1)y^m = r^m, \quad (D + \omega_m R_2)w^m = Dy^m.$$

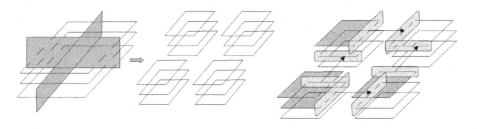

Fig. 2. Scheme of calculation of the vector y^m

Firstly, we calculate the vector y^m, starting the calculation in the lower left corner. Then the calculation of the correction vector w^m starts in the upper right corner. Figure 2 shows the calculation of the vector y^m.

Table 1 shows the results of the calculation of the acceleration and the efficiency depending on the number of processors for the parallel variant of the adaptive alternating-triangular method.

Table 1. Dependence of the acceleration and the efficiency on the number of processors.

Number of processors	Time, sec.	Acceleration	Efficiency
1	7.490639	1	1
2	4.151767	1.804	0.902
4	2.549591	2.938	0.734
8	1.450203	5.165	0.646
16	0.882420	8.489	0.531
32	0.458085	16.351	0.511
64	0.265781	28.192	0.44
128	0.171535	43.668	0.341

As Table 1 shows, the algorithm of the alternating-triangular iterative method and its parallel realization on the basis of a decomposition in two spatial directions can be effectively applied to solve hydrodynamic problems for a sufficiently large number of calculators ($p \leq 128$).

6 Measurement of Parameters of Wave Processes on the Basis of Full-Scale Observations

A full-scale experiment was conducted to measure various parameters of wave propagation in shallow water. On the basis of experimental data, we obtained the values of the spectrum of the water level-elevation function. To process the results of full-scale measurements, a trigonometric interpolation polynomial was used. The Fourier series coefficients are calculated according to the expression

$$X_k = \sum_{n=0}^{N-1} \left(x_n e^{-\frac{2\pi i k n}{N}} \right),$$

where $[x_n]$ is a given sequence of numbers with a constant discretization step of dimension N.

If $a = \mathrm{Re}\, x$, $b = \mathrm{Im}\, x$, then the trigonometric function passing through the points $\{n, x_n\}$ can be written as

$$\xi(t) = \frac{1}{N} \left[a_0 + a_{N/2} \cos(\pi t) + 2 \sum_{n=1}^{N/2-1} \left(a_n \cos\left(\frac{2\pi t n}{N}\right) - b_n \sin\left(\frac{2\pi t n}{N}\right) \right) \right]$$

Table 2 contains the depths at which the measurements were made, the wave period values, mean and maximum wave heights, the dispersion of the level-elevation function (which is depth-dependent), and the correlation coefficients for both the normal and lognormal distributions.

Figure 3 shows the result obtained with the interpolation trigonometric polynomial taking into account the fact that the spectrum of the level-elevation function lies in a certain range. The continuous red line represents the function obtained by interpolating the trigonometric polynomial, the blue crosses show the level-elevation function as obtained by means of experimental measurements.

The points in Fig. 4 correspond to the values of the wave spectrum, and the lines represent the functions distributed according to the normal and lognormal laws and having mathematical expectations and variances corresponding to actual field data.

Wave processes can be described by three quantities: mathematical expectation (wave period), dispersion and amplitude of the normal or log-normal distribution of the spectrum components. These three values were obtained by processing the results of the full-scale experiment and are used as boundary conditions for mathematical models of wave hydrodynamic processes.

Table 2. Parameters of sea waves

No	Depth, cm	Wave period, s	Average wave height, cm	Maximum value of wave height, cm	Dispersion of the level-elevation function	Correlation for the normal distribution	Correlation for the lognormal distribution
1	12.73	3.18	1.43	3.26	3.38	0.6762240	0.7281816
2	21.65	3.18	2.21	5.12	2.87	0.7197073	0.7549785
3	34.29	3.25	2.67	6.63	2.58	0.7675635	0.8080973
4	47.69	3.20	2.903	7.278	2.373	0.8043428	0.8151663
5	50.22	3.23	3.408	8.779	2.465	0.8007264	0.8223494
6	56.95	3.32	3.42	10.05	2.539	0.8252073	0.8349985
7	58.25	3.09	3.53	13.74	2.46	0.7045178	0.7501032
8	75.28	3.48	3.59	12.71	2.31	0.8046488	0.8281662
9	83.35	3.05	4.47	14.64	2.49	0.7677805	0.8044246
10	123.25	3.23	4.671	15.74	2.32	0.7871638	0.8280977

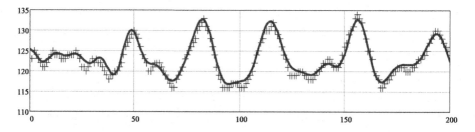

Fig. 3. Approximation of the level-elevation function (Color figure online)

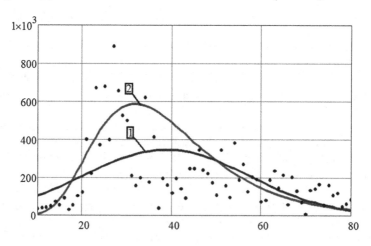

Fig. 4. The spectrum of the function of elevation level: 1 – normal distribution, 2 – lognormal distribution

7   Software Implementation of the Model of Wave Hydrodynamics

The software package consists of the following blocks: a control block (contains a cycle over a time variable; following functions are called by the block: calculation of the velocity field without taking into account the level-elevation function, calculation of the level function, calculation of the two-dimensional velocity field, verification of the presence on the water surface of the structure, and data I/O); a block for entering initial distributions for the calculation of the flow velocity and level functions (initial velocity field distributions, level-elevation functions, and initial values of cell fullness are specified); a block for constructing the grid equations of the velocity field without taking into account the level-elevation function; a block for constructing the grid equations of the field of the level-elevation function (pressure); a block for checking the presence on the surface of the aquatic environment of the structure; a block for calculating the velocity field taking into account the level-elevation function (calculation of the values of the velocity field in the next time layer); a block for calculating the grid equations by a modified adaptive alternating-triangular method of variational type; a block for output of the values of the velocity field and level-elevation functions. Figure 5 shows the scheme of the program algorithm.

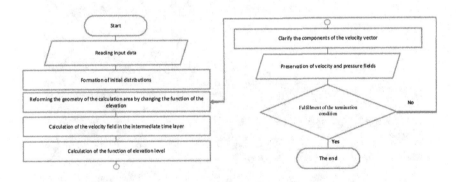

Fig. 5. Diagram of the algorithm of the program complex

8   Results of Numerical Experiments

On the basis of full-scale data, we were able to develop a three-dimensional model of wave hydrodynamic processes that describes the motion of an aquatic environment taking into account the propagation of waves towards the shore. A modern software package adapted for simulation of hydrodynamic wave processes was developed. The field of application of this package is the construction of velocity and pressure fields of aquatic environments, and the evaluation of the hydrodynamic impact exerted by surface waves on the shore. Based on the developed complex of programs, a numerical simulation of hydrodynamic wave processes in the coastal zone of a shallow water body was carried out.

The practical significance of the numerical algorithms and the complex of programs that implements them is determined by the possibility of their application to the study of hydrophysical processes in coastal water systems, to the construction of the velocity and pressure fields of aquatic environments, and the evaluation of the hydrodynamic impact of surface waves on the shoreline. The constructed program complex allows for the specification of the shape and intensity of the source of oscillations, and the geometry of the reservoir bottom. Figure 6 shows the results of numerical experiments on the simulation of the propagation of wave hydrodynamic processes when a wave recedes from the shore, taking into account the geometries of the shore of the object located in the liquid and the bottom of the reservoir.

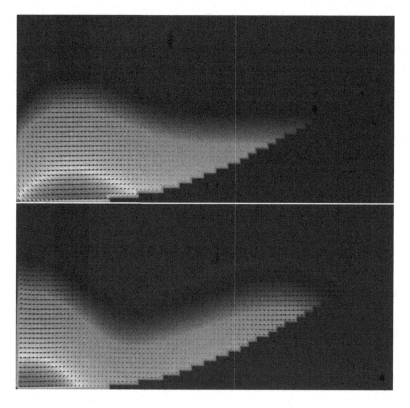

Fig. 6. The field of the velocity vector of the aquatic environment (XOZ-plane cut)

As an example of practical use of the problem-oriented program complex, we solved the problem of calculating the velocity and pressure fields. The selected modeling site is 50 by 50 m and has a depth of 2 m; the peak point rises 2 m above sea level. The disturbance source is given at some distance from the shoreline.

At the initial time, the liquid is at rest. To solve this problem, a $100 \times 100 \times 40$ grid is used, and the time step is 0.01 s.

Figure 6 shows the field of the velocity vector of the aquatic environment when the wave approaches the shore. Zones of flooding and shallowing are formed while the level-elevation function dynamically changes. Figure 7 shows that the land area was flooded by an incident wave. An accounting on flooding and dehumidification of coastal areas was carried out by recalculating the fullness of calculated cells. The proposed approach makes it possible to solve problems in domains with a complex and dynamically rearranged geometry of the boundary.

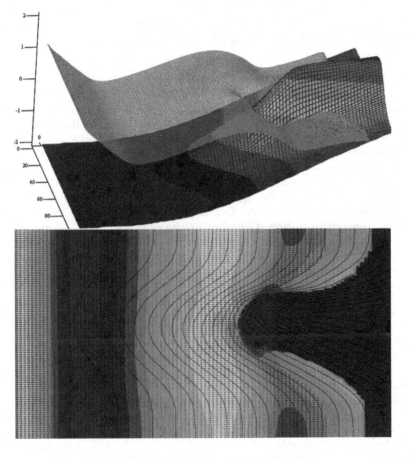

Fig. 7. Level and bottom elevation functions

It should be noted that the software package developed here has a distinctive feature, namely the wave propagation towards the shore is taken into account when modeling the propagation of the surface oscillations.

9   Conclusion

The work is devoted to the development of a model of three-dimensional wave processes aimed at simulating wave processes taking into account the wave propagation to the shore. A full-scale experiment was conducted to measure various parameters of wave propagation in shallow water. On the basis of experimental data, we obtained the values of the spectrum of the water level-elevation function. A description of the developed program complex is given. This program complex allows for the specification of the shape and intensity of the source of oscillations, and also takes into account the flooding and drainage of coastal areas. Further studies on the calculation of wave force effects on surface objects and objects of coastal infrastructure, and bottom-surface geometry are possible on the basis of the software package developed here.

References

1. Sukhinov A.I., Sukhinov A.A.: 3D model of diffusion-advection-aggregation suspensions in water basins and its parallel realization. In: Parallel Computational Fluid Dynamics, Mutidisciplinary Applications, Proceedings of Parallel CFD 2004 Conference, Las Palmas de Gran Canaria, Spain, pp. 223–230. Elsevier, Amsterdam (2005). https://doi.org/10.1016/B978-044452024-1/50029-4
2. Gushchin, V.A., Kostomarov, A.V., Matyushin, P.V., Pavlyukova, E.R.: Direct numerical simulation of the transitional separated fluid flows around a sphere and a circular cylinder. J. Wind Eng. Industr. Aerodyn. **90**(4–5), 341–358 (2002). https://doi.org/10.1016/S0167-6105(01)00196-9
3. Sukhinov A.I., Sukhinov A.A.: Reconstruction of 2001 ecological disaster in the Azov Sea on the basis of precise hydrophysics models. In: Parallel Computational Fluid Dynamics, Multidisciplinary Applications, Proceedings of Parallel CFD 2004 Conference, Las Palmas de Gran Canaria, Spain, pp. 231–238. Elsevier, Amsterdam (2005). https://doi.org/10.1016/B978-044452024-1/50030-0
4. Gushchin, V.A., Kostomarov, A.V., Matyushin, P.V.: 3D visualization of the separated fluid flows. J. Vis. **7**(2), 143–150 (2004). https://doi.org/10.1007/BF03181587
5. Alekseenko, E., Roux, B., Sukhinov, A., Kotarba, R., Fougere, D.: Coastal hydrodynamics in a windy lagoon. Comput. Fluids **77**, 24–35 (2013). https://doi.org/10.1016/j.compfluid.2013.02.003
6. Alekseenko, E., Roux, B., Sukhinov, A., Kotarba, R., Fougere, D.: Nonlinear hydrodynamics in a mediterranean lagoon. Nonlinear Process. Geophys. **20**, 189–198 (2017). https://doi.org/10.5194/npg-20-189-2013
7. Sukhinov, A.I., Chistyakov, A.E., Alekseenko, E.V.: Numerical realization of the three-dimensional model of hydrodynamics for shallow water basins on a high-performance system. Math. Models Comput. Simul. **3**(5), 562–574 (2011). https://doi.org/10.1134/S2070048211050115
8. Sukhinov, A.I., Chistyakov, A.E., Protsenko, E.A.: Mathematical modeling of sediment transport in the coastal zone of shallow reservoirs. Math. Models Comput. Simul. **6**(4), 351–363 (2014). https://doi.org/10.1134/S2070048214040097

9. Sidoryakina, V.V., Sukhinov, A.I.: Well-posedness analysis and numerical implementation of a linearized two-dimensional bottom sediment transport problem. Comput. Math. Math. Phys. **57**(6), 978–994 (2017). https://doi.org/10.7868/S0044466917060138

10. Favorskaya, A.V., Petrov, I.B.: Numerical modeling of dynamic wave effects in rock masses. Dokl. Math. **95**(3), 287–290 (2017). https://doi.org/10.1134/S1064562417030139

11. Belotserkovskii, O.M., Gushchin, V.A., Shchennikov, V.V.: Decomposition method applied to the solution of problems of viscous incompressible fluid dynamics. Comput. Math. Math. Phys. **15**, 197–207 (1975)

12. Kvasov, I.E., Leviant, V.B., Petrov, I.B.: Numerical study of wave propagation in porous media with the use of the grid-characteristic method. Comput. Math. Math. Phys. **56**(9), 1620–1630 (2016). https://doi.org/10.1134/S0965542516090116

13. Sukhinov, A.I., Chistyakov, A.E.: Adaptive modified alternating triangular iterative method for solving grid equations with a non-self-adjoint operator. Math. Models Comput. Simul. **4**(4), 398–409 (2012). https://doi.org/10.1134/S2070048212040084

14. Nikitina, A.V., et al.: Optimal control of sustainable development in the biological rehabilitation of the Azov Sea. Math. Models Comput. Simul. **9**(1), 101–107 (2017). https://doi.org/10.1134/S2070048217010112

15. Sukhinov, A., Chistyakov, A., Nikitina, A., Semenyakina, A., Korovin, I., Schaefer, G.: Modelling of oil spill spread. In: 5th International Conference on Informatics, Electronics and Vision, ICIEV 2016, 28 November 2016, pp. 1134–1139 (2016). https://doi.org/10.1109/ICIEV.2016.7760176

16. Chetverushkin, B.N., Shilnikov, E.V.: Software package for 3D viscous gas flow simulation on multiprocessor computer systems. Comput. Math. Math. Phys. **48**(2), 295–305 (2008). https://doi.org/10.1007/s11470-008-2012-4

17. Davydov, A.A., Chetverushkin, B.N., Shil'nikov, E.V.: Simulating flows of incompressible and weakly compressible fluids on multicore hybrid computer systems. Comput. Math. Math. Phys. **50**(12), 2157–2165 (2010). https://doi.org/10.1134/S096554251012016X

18. Sukhinov, A.I., Nikitina, A.V., Semenyakina, A.A., Chistyakov, A.E.: Complex of models, explicit regularized schemes of high-order of accuracy and applications for predictive modeling of after-math of emergency oil spill. In: Workshop Proceedings, vol. 1576, pp. 308–319 (2016)

Supercomputer Modeling of Hydrochemical Condition of Shallow Waters in Summer Taking into Account the Influence of the Environment

Alexander I. Sukhinov[1], Alexander E. Chistyakov[1], Alla V. Nikitina[2(✉)],
Yulia V. Belova[1], Vladimir V. Sumbaev[2], and Alena A. Semenyakina[3]

[1] Don State Technical University, Rostov-on-Don, Russia
sukhinov@gmail.com, {cheese_05,yuliapershina}@mail.ru
[2] Southern Federal University, Rostov-on-Don, Russia
nikitina.vm@gmail.com, valdec4813@mail.ru
[3] "Supercomputers and Neurocomputers Research Center" Co. Ltd.,
Taganrog, Russia
j.a.s.s.y@mail.ru

Abstract. The paper deals with the development and research of a mathematical model for hydrophysical processes which involves the use of modern information technologies and computational methods with the aim to improve the accuracy of predictive modeling of ecological condition of shallow waters during the summer. The model takes into account the following factors: movement of water flows; microturbulent diffusion; gravitational settling of pollutants; nonlinear interaction of plankton populations; nutrient, temperature and oxygen regimes; and impact of salinity. A scheme with weights is proposed for the discretization of the proposed model. This scheme significantly reduces both error and computation time. The practical significance of the paper is determined by the software implementation of the model and the determination of the limits and prospects of its practical use. Experimental software is designed on the basis of a supercomputer for mathematical modeling of possible development scenarios of shallow water ecosystems taking into account the influence of the environment. For this, we consider as an example the Sea of Azov in the summer period. The software parallel implementation involves decomposition methods for computationally intensive diffusion-convection problems taking account of the architecture and parameters of a multiprocessor computer system. The software complex contains a model for fluid dynamics which includes equations of motion in three coordinate directions.

Keywords: Mathematical model · Hydrochemical process
Field investigation · Algorithm · Supercomputer

This paper was partially supported by grant No. 17-11-01286 from the Russian Science Foundation.

L. Sokolinsky and M. Zymbler (Eds.): PCT 2018, CCIS 910, pp. 336–351, 2018.
https://doi.org/10.1007/978-3-319-99673-8_24

1   Introduction

Many studies in recent decades have dealt with prediction models for biogeo-chemical cycles as stochastic systems, for example, the paper by Straten and Keesman [1]. Numerous models were calibrated according to observational data by various researchers, such as Park [2], Bierman [3], Chen [4], Jorgensen [5], Williams [6], and others. Most models contain the Michaelis–Menten equation as a limit to the development of phytoplankton nutrients. The construction of hydrobiological models of shallow waters requires three-dimensional models of biogeochemical cycles with high resolution.

In July 2017, members of the staffs of Don State Technical University, the Southern Federal University and the Southern Scientific Center of the Russian Academy of Sciences (RAS) went on an expedition to the Sea of Azov aboard the Scientific Research Vessel (SRV) "Deneb". The main task of the expedition was to carry out a comprehensive research of the current condition and spatial-temporal changes in the hydrobiological regime of the Sea of Azov. The route of the SRV "Deneb" is shown in Fig. 1.

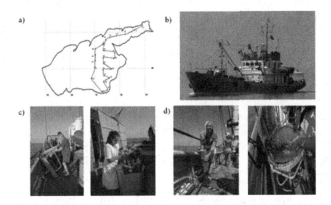

Fig. 1. Research expedition in 2017

Various systems from the "Analytical GIS" portal, developed by the Institute for Information Transmission Problems of the RAS for complex geoinformational analysis of spatial-temporal processes and phenomena, were used along with expedition data and literature sources for modeling hydrophysical processes in shallow waters (the Sea of Azov).

2   Problem Statement

The model of the biochemical transformation of biogenic nutrients (various forms of phosphorus, nitrogen and silicon) has the form [7,8,12,13,20–22]

$$\frac{\partial q_i}{\partial t} + \operatorname{div}\left(\mathbf{U}, q_i\right) = \operatorname{div}\left(\mathbf{k}_i \operatorname{grad} q_i\right) + R_i, \mathbf{k}_i = \left\{\mu_i, \mu_i, \nu_i\right\}, \ i = \overline{1, 10}, \quad (1)$$

where q_i denotes the concentration of the ith component; $\mathbf{u}$ is the velocity vector of the water flow, $\mathbf{u} = \{u, v, w\}$; $\mathbf{U} = \mathbf{u} + \mathbf{u}_{0i}$ represents the matter convective transport velocity, $\mathbf{U} = \{U, V, W\}$; $\mathbf{u}_{0i}$ stands for the velocity of the ith component of sedimentation; R_i denotes the chemical-biological source, where the index i corresponds to the type thereof: 1–3 are substance concentrations from algae *Chlorella vulgaris, Aphanizomenon flos-aquae* and *Sceletonema costatum*; 4 is PO_4; 5 is POP; 6 is DOP; 7 is NO_3; 8 is NO_2; 9 is NH_4; 10 is Si (PO_4 are phosphates; POP is suspended organic phosphorus; DOP is dissolved organic phosphorus; NH_4 is ammonium; NO_2 are nitrites; NO_3 are nitrates; Si is dissolved inorganic silicon); μ_i, ν_i are diffusion coefficients in horizontal and vertical directions.

We added an observation model for description of chemical-biological sources [12]:

$$R_i = C_i(1 - K_{R,i})q_i - K_{D,i}q_i - K_{E,i}q_i, i = \overline{1,3};$$

$$R_4 = \sum_{i=1}^{3} s_P C_i \left(K_{R,i} - 1 \right) q_i + K_{PN}q_5 + K_{DN}q_6;$$

$$R_5 = \sum_{i=1}^{3} s_P K_{D,i} q_i - K_{PD}q_5 - K_{PN}q_5; R_6 = \sum_{i=1}^{3} s_P K_{E,i} q_i + K_{PD}q_5 - K_{DN}q_6;$$

$$R_7 = \sum_{i=1}^{3} s_N C_i \left(K_{R,i} - 1 \right) \frac{f_N^{(1)}(q_7, q_8)}{f_N(q_7, q_8, q_9)} \cdot \frac{q_8}{q_7 + q_8} q_i + K_{23}q_8;$$

$$R_8 = \sum_{i=1}^{3} s_N C_i (K_{R,i} - 1) \frac{f_N^{(1)}(q_7, q_8)}{f_N(q_7, q_8, q_9)} \cdot \frac{q_7}{q_7 + q_8} q_i + K_{42}q_9 - K_{23}q_8;$$

$$R_9 = \sum_{i=1}^{3} s_N C_i \left(K_{R,i} - 1 \right) \frac{f_N^{(2)}(q_9)}{f_N(q_7, q_8, q_9)} q_i - K_{42}q_9, R_{10} = s_{Si}K_{D,3}q_3,$$

where $K_{R,i}, K_{D,i}$ are specific rates of phytoplankton respiration and mortality; $K_{E,i}$ denotes the specific rate of phytoplankton excretion; K_{PD} denotes the specific rate of autolysis; K_{PN} is the ratio of phosphatization; K_{DN} denotes the ratio of *DOP* phosphatization; K_{42} is the specific rate of ammonium oxidation to nitrites during the nitrification process; K_{23} is the specific rate of oxidation of nitrites to nitrates during nitrification; s_P, s_N, s_{Si} are normalization coefficients between N, P, and Si contents in organic matter.

The phytoplankton algae growth rate is determined by the expressions

$$C_i = K_{N,i} \min \{f_P(q_4), \ f_N(q_7, q_8, q_9)\}, \ i = \overline{1,2};$$

$$C_3 = K_{N,3} \min \{f_P(q_4), \ f_N(q_7, q_8, q_9), f_{Si}(q_{10})\},$$

where $K_{N,i}$ is the maximum specific growth rate of phytoplankton.

To describe the dependence of the nutrient content, let us use the following relations:

$$f_P(q_4) = \frac{q_4}{q_4 + K_4}; \ f_{Si}(q_{10}) = \frac{q_{10}}{q_{10} + K_{10}};$$

$$f_N(q_7, q_8, q_9) = f_N^{(1)}(q_7, q_8) + f_N^{(2)}(q_9) = \frac{(q_7 + q_8)\exp(-K_{psi}q_9)}{K_7 + (q_7 + q_8)} + \frac{q_9}{K_9 + q_9},$$

where K_4, K_7, K_9, and K_{10} stand, respectively, for the phosphate, nitrate, ammonium, and silicon half-saturation constants; K_{psi} denotes the ammonium inhibition coefficient [15].

It is necessary to add initial conditions:

$$q_i(x, y, z, 0) = q_i^0(x, y, z), \ (x, y, z) \in \overline{G}, i = \overline{1, 10}. \tag{2}$$

Let the boundary Σ of the cylindrical domain G be sectionally smooth, and suppose that $\Sigma = \Sigma_H \cup \Sigma_o \cup \sigma$, where Σ_H is the water bottom surface, Σ_o is the unperturbed surface of the aquatic medium, and σ is the lateral (cylindrical) surface. Let $\mathbf{u_n}$ be the normal component of the water flow velocity vector to the Σ surface, and let $\mathbf{n}$ be the outer normal vector to the boundary Σ. Assume that the concentrations q_i are:

on the lateral boundary: $q_i = 0$ if $\mathbf{u_n} < 0$; $\dfrac{\partial q_i}{\partial n} = 0$ if $\mathbf{u_n} \geq 0$, $i = \overline{1, 10}$;

at the bottom: $\dfrac{\partial q_i}{\partial z} = \varepsilon_{1,i}q_i$, $i = \overline{1, 3}$, $\dfrac{\partial q_i}{\partial z} = \varepsilon_{2,i}q_i$, $i = \overline{4, 10}$; $\qquad(3)$

on Σ_o: $\dfrac{\partial q_i}{\partial z} = \varphi(q_i)$, $i = \overline{1, 10}$,

where φ is a given function; $\varepsilon_{1,i}$ and $\varepsilon_{2,i}$ are nonnegative constants: $\varepsilon_{1,i}$, $i = \overline{1, 3}$, account for the descent of algae to the bottom and their deposition; $\varepsilon_{2,i}$, $i = \overline{4, 10}$ account for absorption of nutrient by bottom sediments.

3 Model Description

We assume that the process is periodic with period T_0 ($T_0 > 0$):

$$q_i(x, y, z, t) = q_i(x, y, z, t + T_0). \tag{4}$$

Let us define the functions $\mathbf{U}_n^+ = \begin{cases} \mathbf{U}_n, \ \mathbf{U}_n \geq 0; \\ 0, \ \mathbf{U}_n < 0, \end{cases}$ and $\mathbf{U}_n^- = \mathbf{U}_n - \mathbf{U}_n^+$ on the surface Σ of the domain G. Divide the interval $0 \leq t \leq T_0$ into sufficiently small time periods $t_n \leq t \leq t_{n+1}$, $n = 0, 1, .., N_t - 1$, $t_0 = 0$, $t_{N_t} = T_0$. The initial-boundary problem for the system of equations with linearized right sides has the form

$$\frac{\partial q_i}{\partial t} + \text{div}(\mathbf{U}, q_i) = \text{div}(\mathbf{k}_i \,\text{grad}\, q_i) + R_i^n(q_i); \tag{5}$$

$$R_i^n(q_i) = C_i^n(1 - K_{R,i})q_i - K_{D,i}q_i - K_{E,i}q_i, \ i = \overline{1, 3}; \tag{6}$$

$$R_4^n(q_4) = \sum_{i=1}^{3} s_P C_i (K_{R,i} - 1) q_i^n + K_{PN}q_5^n + K_{DN}q_6^n;$$

$$R_5^n(q_5) = \sum_{i=1}^{3} s_P K_{D,i} q_i^n - K_{PD} q_5 - K_{PN} q_5;$$

$$R_6^n(q_6) = \sum_{i=1}^{3} s_P K_{E,i} q_i^n + K_{PD} q_5^n - K_{DN} q_6;$$

$$R_7^n(q_7) = Q^n \cdot q_7 + K_{23} q_8^n; \quad R_8^n(q_8) = Q^n \cdot q_8 + K_{42} q_9^n - K_{23} q_8;$$

$$Q^n = \sum_{i=1}^{3} \frac{s_N C_i^n (K_{R,i} - 1) \exp(-K_{psi} q_9^n) q_i^n}{(q_7^n + q_8^n) \exp(-K_{psi} q_9^n) + q_9^n (K_7 + q_7^n + q_8^n)/(K_9 + q_9^n)};$$

$$R_9^n(q_9) = \sum_{i=1}^{3} \frac{s_N C_i^n (K_{R,i} - 1) q_i^n}{P^n \exp(-K_{psi} q_9^n)/(K_7 + (q_7^n + q_8^n)) + q_9^n} \cdot q_9 - K_{42} q_9;$$

$$P^n = (K_9 + q_9^n)(q_7^n + q_8^n) \exp(-K_{psi} q_9^n); \quad R_{10}^n(q_{10}) = s_{Si} K_{D,3} q_3^n$$

with the corresponding initial and boundary conditions. We suppose that $q_i \in C^2(G) \cap C^1(\overline{G}) \cap C^1 (0 < t \le T)$, $k_h(z), k_\nu(z), R_i(x, y, z) \in C^1(\overline{G})$. Also, we assume that the following expressions hold for each $n = \overline{0, N_t - 1}$:

$$\frac{4\mu_i}{H_x^2} + \frac{4\mu_i}{H_y^2} + \frac{4\nu_i}{H_z^2} + K_{D,i} + K_{E,i} > Tr_i, i = \overline{1,3}; \tag{7}$$

$$\frac{4\mu_i}{H_x^2} + \frac{4\mu_i}{H_y^2} + \frac{4\nu_i}{H_z^2} + K_{42} > Tr_4; \quad \frac{4\mu_i}{H_x^2} + \frac{4\mu_i}{H_y^2} + \frac{4\nu_i}{H_z^2} > Tr_5, i = \overline{4,10};$$

$$Tr_i = K_{N,i} \, min \, \{f_P^n(q_4), \, f_N^n(q_7, q_8, q_9)\}(1 - K_{R,i}), i = \overline{1,2};$$

$$Tr_3 = K_{N,3} \, min \, \{f_P^n(q_4), \, f_N^n(q_7, q_8, q_9), f_{Si}^n(q_{10})\}(1 - K_{R,3});$$

$$Tr_4 = \sum_{i=1}^{3} \frac{s_N C_{F_i}^n (K_{R,i} - 1) q_i^n}{(K_9 + q_9^n)(q_7^n + q_8^n) \exp(-K_{psi} q_9^n)/(K_7 + (q_7^n + q_8^n)) + q_9^n};$$

$$Tr_5 = \sum_{i=1}^{3} \frac{s_N C_i^n (K_{R,i} - 1) \exp(-K_{psi} q_9^n) q_i^n}{(q_7^n + q_8^n) \exp(-K_{psi} q_9^n) + q_9^n (K_7 + q_7^n + q_8^n)/(K_9 + q_9^n)}.$$

If all the conditions above are met, then the solution of the problem exists and is unique.

The field of water flow velocities calculated in [22, 23] was used as input data for the model (1)–(3).

4   Modified Alternating Triangular Method

The grid equations were obtained as a result of finite-difference approximations of problem (5)–(7) using a scheme with weights [18] and can be written in a matrix form [14, 23, 24]:

$$Ax = f, \tag{8}$$

where A is a linear positive definite operator $(A > 0)$. We applied an implicit iteration process to solve problem (8):

$$B\frac{x^{m+1} - x^m}{\tau_{m+1}} + Ax^m = f, \tag{9}$$

where m stands for the iteration number, $\tau > 0$ is an iteration parameter, B is some reversible operator called a preconditioner or stabilizer. The inversion of the operator B in (9) needs to be substantially simpler than the direct inversion of the original operator A in (8). To construct the operator B, we proceed from the additive representation of the operator A_0 as the symmetric part of the operator A:

$$A_0 = R_1 + R_2, \quad R_1 = R_2^*, \quad A = A_0 + A_1, \quad A_0 = A_0^*, \quad A_1 = -A_1^*. \tag{10}$$

The preconditioner can be written as

$$B = (D + \omega R_1)D^{-1}(D + \omega R_2), \quad D = D^* > 0, \quad \omega > 0, \tag{11}$$

where D is some operator.

Relations (9)–(11) define a modified alternating triangular method (MATM) for the solution of the problem, provided that the operators R_1, R_2 are defined and methods for determining the parameters τ_{m+1}, ω and the operator D are indicated.

The algorithm of the adaptive MATM of minimum corrections for calculating the grid equations with a non-selfadjoint operator has the form

$$r^m = Ax^m - f, \quad B(\omega_m)w^m = r^m, \quad \tilde{\omega}_m = \sqrt{\frac{(Dw^m, w^m)}{(D^{-1}R_2w^m, R_2w^m)}}, \tag{12}$$

$$s_m^2 = 1 - \frac{(A_0w^m, w^m)^2}{(B^{-1}A_0w^m, A_0w^m)(Bw^m, w^m)}, \quad k_m = \frac{(B^{-1}A_1w^m, A_1w^m)}{(B^{-1}A_0w^m, A_0w^m)},$$

$$\theta_m = \frac{1 - \sqrt{\frac{s_m^2 k_m}{(1+k_m)}}}{1 + k_m(1 - s_m^2)}, \quad \tau_{m+1} = \theta_m\frac{(A_0w^m, w^m)}{(B^{-1}A_0w^m, A_0w^m)},$$

$$x^{m+1} = x^m - \tau_{m+1}w^m, \quad \omega_{m+1} = \tilde{\omega}_m,$$

where r^m is the residual vector, w^m is the correction vector, and the diagonal part of the operator A is used as operator D [17].

The estimate of the convergence rate of the method just described can be written in the form

$$\rho \le \frac{\nu^* - 1}{\nu^* + 1}, \ \nu^* = \nu \left(\sqrt{1+k} + \sqrt{k}\right)^2, \ k = \frac{\left(B^{-1}A_1\omega^m, A_1\omega^m\right)}{\left(B^{-1}A_0\omega^m, A_0\omega^m\right)}, \tag{13}$$

where ν is the condition number of the operator C_0, defined as $C_0 = B^{-1/2}A_0B^{-1/2}$.

5   Parallel Implementation

We describe some parallel algorithms with various types of domain decomposition for solving problems (1)–(3) on a multiprocessor computer system (MCS).

Algorithm 1.

Each processor is assigned a computational domain after the initial computational domain is partitioned in two coordinate directions, as shown in Fig. 2. Adjacent domains overlap over two layers of nodes in the direction perpendicular to the plane of the partition [24].

The residual vector and its uniform norm are calculated after each processor receives information for its own part of the domain. Then each processor determines the maximum module element of the residual vector and sends its value to all remaining calculators. Now, to calculate the uniform norm of the residual vector, it is enough to find the maximum element on each processor [16].

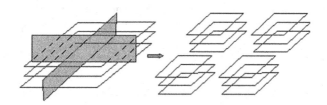

Fig. 2. Domain decomposition

The parallel algorithm for calculating the correction vector has the form

$$(D + \omega_m R_1)D^{-1}(D + \omega_m R_2)w^m = r^m,$$

where R_1 is a lower-triangular matrix, and R_2 is a upper-triangular matrix. For calculating the correction vector, we should solve the following two equations simultaneously:

$$(D + \omega_m R_1)y^m = r^m, (D + \omega_m R_2)w^m = Dy^m.$$

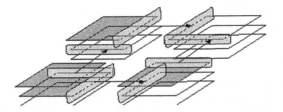

Fig. 3. Scheme of calculation of the vector y^m

Initially, the vector y^m is calculated starting in the lower left corner. Then the correction vector w^m is calculated starting in the upper right corner. The calculation scheme for the vector y^m is given in Fig. 3 (the transference of elements after calculation of two layers by the first processor is shown).

In the first step, the first processor operates with the top layer. Then the transference of overlapping elements to the adjacent processors takes place. In the next step, the first processor operates with the second layer, while its neighbors operate with the first. The transference of elements after calculation of two layers by the first processor is shown in Fig. 3. In the scheme for the calculation of the vector y^m, only the first processor does not require additional information and can independently operate with its part of the domain. Other processors wait for the results from the previous processor, while it transfers the calculated values of the grid functions at the grid nodes located in the preceding positions of this line. The process continues until all the layers are calculated. Similarly, we can solve systems of linear algebraic equations (SLAE) with an upper-triangular matrix for calculating the correction vector.

Next, scalar products (12) are calculated, and the algorithm proceeds to the next iteration layer.

We obtain the following theoretical estimates [19] for the speedup $S_{(1)}$ and the efficiency $E_{(1)}$ of parallel algorithm 1:

$$S_{(1)} = \frac{p}{1 + \left(\sqrt{p} - 1\right)\left(\frac{36}{50N_z} + \frac{4p}{50t_0}\left(t_n\left(\frac{1}{N_x} + \frac{1}{N_y}\right) + \frac{t_x\sqrt{p}}{N_xN_y}\right)\right)},$$

$$E_{(1)} = \frac{S_{(1)}}{p} = \frac{1}{1 + \left(\sqrt{p} - 1\right)\left(\frac{36}{50N_z} + \frac{4p}{50t_0}\left(t_n\left(\frac{1}{N_x} + \frac{1}{N_y}\right) + \frac{t_x\sqrt{p}}{N_xN_y}\right)\right)},$$

where p is the total number of processors; t_0 is the execution time of an arithmetic operation; t_x is the response time (latency); N_x, N_y, N_z are the numbers of nodes in the spatial directions.

We have considered the solution of the problem for a rectangular domain. In the case of a real water medium, the domain may have a complex shape. At the same time, real speedup is less than the theoretical estimate. The dependence obtained for the speedup in theoretical estimates can be used as an upper estimate for the speedup of the parallel implementation of the MATM algorithm with a domain decomposition in two spatial directions.

Let us describe now the domain decomposition in two spatial directions by the k-means algorithm.

Algorithm 2.

The k-means method is used for the geometric partition of the computational domain so as to uniformly load a MCS calculators (processors). This method is based on the minimization of the functional $Q = Q^{(3)}$ of total variance of the element scatter (nodes of the computational grid) relative to the gravity center of subdomains. Let X_i be the set of computational grid nodes contained in the ith subdomain, $i \in \{1, ..., m\}$, m is the given number of subdomains.

$$Q^{(3)} = \sum_i \frac{1}{|X_i|} \sum_{x \in X_i} d^2(x, c_i) \to \min,$$

where

$$c_i = \frac{1}{|X_i|} \sum_{x \in X_i} x$$

is the center of the subdomain X_i, and $d(x, c_i)$ is the distance between the computational node and the center of the grid subdomain in the Euclidean metric. The k-means method converges only when all subdomain are approximately equal. The result of the k-means method for model domains is given in Fig. 4 (arrows indicate exchanges between subdomains).

All points on the boundary of each subdomains are required to data exchange during the computational process. Jarvis's algorithm is used for this purpose (the task of constructing the convex hull). A list of neighboring subdomains is made up for each subdomain. An algorithm was created for data transfer between subdomains.

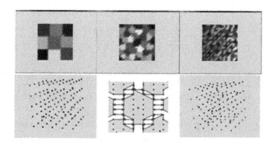

Fig. 4. Results of the k-means method for model domain decomposition into 9, 38, 150 (for a two-dimensional domain); into 6 and 10 (for a three-dimensional domain) subdomains

Theoretical estimates for the speedup and efficiency of algorithm 2 were obtained similarly to the corresponding estimates for algorithm 1:

$$S_{(2)} = \frac{p \cdot \chi}{1 + \left(\sqrt{p} - 1\right)\left(\frac{36}{50N_z} + \frac{4p}{50t_0}\left(t_n\left(\frac{1}{N_x} + \frac{1}{N_y}\right) + \frac{t_x\sqrt{p}}{N_xN_y}\right)\right)},$$

$$E_{(2)} = \frac{S_{(2)}}{p} = \frac{\chi}{1 + \left(\sqrt{p} - 1\right)\left(\frac{36}{50N_z} + \frac{4p}{50t_0}\left(t_n\left(\frac{1}{N_x} + \frac{1}{N_y}\right) + \frac{t_x\sqrt{p}}{N_xN_y}\right)\right)},$$

where χ is the ratio of the number of computational nodes to the total number of nodes (computational and fictitious).

6   Results of Experimental Studies

Parallel algorithms for the adaptive alternating-triangular method were implemented on the multiprocessor computer system (MCS) of the Southern Federal University (SFU). The peak performance of this MCS is 18.8 TFlops. MCS includes 8 computational racks. The computational field of the MCS is based on the HP BladeSystem c-class infrastructure with integrated communication modules, power supply and cooling systems. Five hundred and twelve single-type 16-core Blade servers HP ProLiant BL685c are used as computational nodes, each of which is equipped with four 4-core AMD Opteron 8356 2.3 GHz processors and 32 GB RAM. The total number of computing cores in the complex is 2048; the total amount of RAM is 4 TB.

The results of the parallel implementation of algorithms 1 and 2 for solving problem (1)–(3) are summarized in Table 1. Here $t_{(j)}$, $S_{(j)}$ and $E_{(j)}$ are, respectively, operating time, speedup and efficiency of the jth algorithm; $S_{(j)}^t$ and $E_{(j)}^t$ are the theoretical estimates of the speedup and efficiency of the j-th algorithm, $j = \{1, 2\}$.

Table 1. Comparison of speedups and efficiencies of the algorithms

p	$t_{(1)}$	$S_{(1)}^t$	$S_{(1)}$	$t_{(2)}$	$E_{(2)}^t$	$E_{(2)}$
1	7.491	1.0	1.0	6.073	1.0	1.0
2	4.152	1.654	1.804	3.121	1.181	1.946
4	2.549	3.256	2.938	1.811	2.326	3.354
8	1.450	6.318	5.165	0.997	4.513	6.093
16	0.882	11.928	8.489	0.619	8.520	9.805
32	0.458	21.482	16.352	0.317	15.344	19.147
64	0.266	35.955	28.184	0.184	25.682	33.018
128	0.172	54.618	43.668	0.117	39.013	51.933

According to Table 1, the algorithms, based on a domain decomposition in two spatial directions and on the k-means method, can be effectively used for solving hydrodynamic problems on a sufficiently large number of computational nodes.

The speedup graphs of algorithms 1 and 2 for the solution of problem (1)–(3), obtained theoretically and experimentally, are given in Fig. 5.

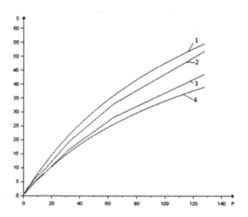

Fig. 5. Acceleration graphs of developed parallel algorithms: 1 – theoretical estimate of speedup in algorithm 1; 2 – speedup of algorithm 2, obtained experimentally; 3 – speedup of algorithm 1, obtained experimentally; 4 – theoretical estimate of speedup in algorithm 2

The estimate for comparing the efficiency values of algorithms 1 and 2, obtained experimentally, has the form

$$\delta = \sqrt{\sum_{k=1}^{n} \left(E_{(2)k} - E_{(1)k} \right)^2} \Big/ \sqrt{\sum_{k=1}^{n} E_{(2)k}^2}. \tag{14}$$

We obtained that the efficiency increases by 10 to 20% when using algorithm 2 (based on the k-means method) for solving problem (1)–(3).

7   Program Complex Description

A software complex (SC) was developed for solving problem (1)–(3). Using this SC, we have been able to calculate the fields of water flow velocities, concentrations of biogenic pollutants, phytoplankton in areas of complex shape (the Sea of Azov and Taganrog Bay).

The SC was developed for the MCS of the SFU. The SC is intended for mathematical modeling of possible development scenarios of environmental conditions in coastal systems (we considered in this regard the Azov-Black Sea basin). The

SC includes computational units allowing: to take into account factors influencing pollutant distribution in coastal systems (weather conditions, and bottom relief); to study the dependence of pollutant concentrations, the degree and size of the affected water zone on the intensity of water flows, hydrophysical parameters, climatic and meteorological factors. The features of the SC are high performance, reliability, and high accuracy of simulation results.

New computational units (modules) can be integrated into the developed SC. A new module was developed for solving the SLAE appearing as a result of the problem discretization, in which the following methods can be used: Jacobi method; minimum correction method; steepest descent method (gradient descent); Seidel method; upper relaxation method; adaptive MATM of variational type.

Sequentially condensed rectangular grids of sizes $251 \times 351 \times 15$, $502 \times 702 \times 30$, $1004 \times 1404 \times 60$, etc., were used for mathematical modeling of hydrobiological and hydrodynamic processes in a three-dimensional domain of complex shape, namely the Sea of Azov and Taganrog Bay.

The structure of the developed SC is given in Fig. 6.

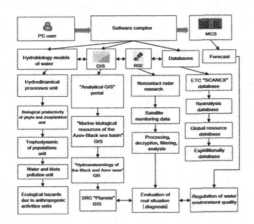

Fig. 6. The software complex

The SC includes the following units: control unit, oceanological and meteorological databases, interface systems, input-output and visualization systems. The SC has the user-friendly interface, ensuring the supplementation system of necessary information in interactive mode.

The high-level language C++ was used for the development of the SC. Message Passing Interface technology (MPI) was employed for clusters.

8   Results of Numerical Experiments

Numerical experiments were performed for modeling hydrochemical conditions of shallow waters in the summer, taking into account the influence of the environment [9]. The results are shown in Figs. 7 and 8. The influence of the Sea of Azov water flow structures on the distribution of phytoplankton concentration is shown in Fig. 7. The results of the calculation of concentrations of biogenic substances (nitrates) based on model (1)–(3) (initial distribution of water flow fields for the northern wind) are given in Fig. 8.

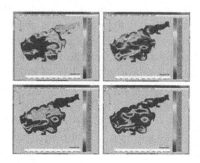
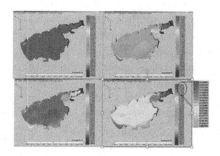

Fig. 7. Distribution of phytoplankton concentration (q_2) ($\mu_2 = 5\times10^{-11}$; $\nu_2 = 10^{-11}$)

Fig. 8. Distribution of nitrates concentration (q_7) ($\mu_7 = 5\times10^{-10}$; $\nu_7 = 10^{-10}$)

Actual data from the "Analytical GIS" portal (see Fig. 9A) and satellite data from the SRC "Planeta" [15], shown in Fig. 9B (phytoplankton spots are visible, revealing the structure of currents), were used for verification of the model (1)–(3) and validation of the adequacy of the SC.

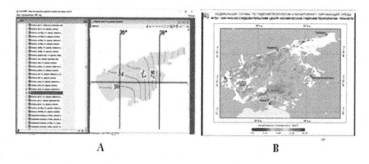

A   B

Fig. 9. A: Ecological data from the "Analytical GIS" portal; B: Satellite image of the Sea of Azov by the SRC "Planeta"

As criteria for adequacy of the developed models (1)–(3), (5)–(7), we used the error estimation of the model according to the Teil criterion [25]. Concentrations of pollution and plankton, calculated for various wind conditions, were taken into account if the relative error did not exceed 30%.

An analysis of similar SCs was carried out. The overall result is that the prediction accuracy of changes in pollutants and plankton concentrations in shallow waters increased by 10 to 20%, depending on the model problem of biological kinetics.

The SC developed by our team implements possible scenarios of ecological conditions in the Sea of Azov using numerical algorithms for model problems of biological kinetics of plankton. The results obtained by means of this SC were compared with those of similar studies concerned with the mathematical modeling of hydrobiological processes [10, 11].

9 Conclusions

Stochiometric ratios of nutrients for different phytoplankton species were studied. These ratios allow for the determination of a limiting substance. Also, observational models describing the consumption and accumulation of nutrients by phytoplankton and its growth rate were considered. We developed a three-dimensional mathematical model of transformation of phosphorus, nitrogen and silicon forms in the multi-species phytoplankton evolution problem for shallow waters. This model takes into account convective transport and diffusion transfer, absorption and release of nutrients by phytoplankton, as well as phosphorus, nitrogen and silicon cycles.

An analytical research of the continuous model developed in the study allowed us to obtain inequalities ensuring the existence and uniqueness of the problem solution. The numerical implementation of the model was carried out on a multiprocessor computer system with distributed memory. We obtained theoretical estimates for the speedup and efficiency of parallel algorithms. Experimental software was designed for mathematical modeling of possible development scenarios of shallow waters. The example of the Sea of Azov was considered in this regard. Decomposition methods of grid domains for computationally intensive diffusion-convection problems were employed for the parallel implementation, taking into account the architecture and parameters of the MCS. Two parallel algorithms were developed for data distribution among processors. The algorithm based on the k-means method yielded an increase in efficiency by 10 to 20% compared with the algorithm based on a standard partitioning of the computational domain.

Thanks to the use of the said MCS, the calculation time for the solution of the model problem decreased, while maintaining the required accuracy for modeling of hydrobiological processes in shallow waters. Note that this fact is one of primary importance in water ecology problems.

References

1. Van Straten, G., Keesman, K.J.: Uncertainty propagation and speculation in projective forecasts of environmental change: a lake eutrophication example. J. Forecast. **10**, 163–190 (1991). https://doi.org/10.1002/for.3980100110
2. Park, R.A.: A generalized model for simulating lake ecosystems. J. Simul. **23**(2), 33–50 (1974). https://doi.org/10.1177/003754977402300201

3. Bierman, V.J., Verhoff, F.H., Poulson, T.C., Tenney, M.W.: Multinutrient dynamic models of algal growth and species competition in eutrophic lakes. In: Modeling the Eutrophication Process. Ann Arbor Science, Ann Arbor (1974)
4. Chen, C.W.: Concepts and utilities of ecologic models. J. Sanit. Eng. Div. 96(5), 1085–1097 (1970)
5. Jorgensen, S.E., Mejer, H., Friis, M.: Examination of a lake model. J. Ecol. Model. 4(2–3), 253–278 (1978). https://doi.org/10.1016/0304-3800(78)90010-8
6. Williams, B.J.: Hydrobiological Modelling. University of Newcastle, Callaghan (2006)
7. Sukhinov A.I., Sukhinov A.A.: Reconstruction of 2001 ecological disaster in the azov sea on the basis of precise hydrophysics models. In: Parallel Computational Fluid Dynamics, Multidisciplinary Applications, Proceedings of Parallel CFD 2004 Conference, Las Palmas de Gran Canaria, Spain, pp. 231–238. Elsevier, Amsterdam-Berlin-London-New York-Tokyo (2005). https://doi.org/10.1016/B978-044452024-1/50030-0
8. Alekseenko, E., Roux, B., Sukhinov, A., Kotarba, R., Fougere, D.: Nonlinear hydrodynamics in a mediterranean lagoon. J. Comput. Math. Math. Phys. 57(6), 978–994 (2017). https://doi.org/10.5194/npg-20-189-2013
9. Sukhinov, A.I., Chistyakov, A.E., Alekseenko, E.V.: Numerical realization of the three-dimensional model of hydrodynamics for shallow water basins on a high-performance system. J. Math. Models Comput. Simul. 3(5), 562–574 (2011). https://doi.org/10.1134/s2070048211050115
10. Sidoryakina, V.V., Sukhinov, A.I.: Well-posedness analysis and numerical implementation of a linearized two-dimensional bottom sediment transport problem. J. Comput. Math. Math. Phys. 57(6), 978–994 (2017). https://doi.org/10.1134/s0965542517060124
11. Sukhinov A., Chistyakov A., Sidoryakina V.: Investigation of nonlinear 2D bottom transportation dynamics in coastal zone on optimal curvilinear boundary adaptive grids. In: MATEC Web of Conferences XIII International Scientific-Technical Conference "Dynamic of Technical Systems" (DTS-2017), vol. 132, pp. 13–15. Russian Federation, Rostov-on-Don (2017). https://doi.org/10.1051/matecconf/201713204003
12. Yakushev E.V., Mikhailovsky G.E.: Mathematical modeling of the influence of marine biota on the carbon dioxide ocean-atmosphere exchange in high latitudes. In: Jaehne, B., Monahan, E.C. (eds.) Air-Water Gas Transfer, Selected Papers, Third International Symposium, pp. 37–48. Heidelberg University. AEON Verlag & Studio, Hanau (1995)
13. Samarsky, A.A., Nikolaev, E.S.: Methods of Solving Grid Equations. Science, Moscow (1978)
14. Sukhinov, A.I., Chistyakov, A.E.: Adaptive modified alternating triangular iterative method for solving grid equations with non-selfadjoint operator. J. Math. Models Comput. Simul. 24(1), 3–20 (2012)
15. SRC "Planeta". http://planet.iitp.ru/english/index_eng.htm
16. Samarskiy, A.A.: Theory of Difference Schemes. Nauka, Moscow (1989)
17. Konovalov, A.N.: The method of steepest descent with adaptive alternately-triangular preamplification. J. Differ. Equat. 40(7), 953 (2004)
18. Sukhinov, A.I., Chistyakov, A.E., Shishenya, A.V.: Error Estimate of the solution of the diffusion equation on the basis of the schemes with weights. Math. Models Comput. Simul. 6(3), 324–331 (2014). https://doi.org/10.1134/s2070048214030120

19. Chetverushkin, B., et al.: Unstructured mesh processing in parallel CFD project GIMM. J. Parallel Comput. Fluid Dyn., 501–508 (2005). https://doi.org/10.1016/b978-044452206-1/50061-6
20. Sukhinov, A.I., Chistyakov, A.E., Semenyakina, A.A., Nikitina, A.V.: Parallel realization of the tasks of the transport of substances and recovery of the bottom surface on the basis of schemes of high order of accuracy. J. Comput. Methods Program.: New Comput. Technol. **16**(2), 256–267 (2015)
21. Chistyakov, A.E., Hachunts, D.S., Nikitina, A.V., Protsenko, E.A., Kuznetsova, I.Y.: Parallel Library of iterative methods of the SLAE solvers for problem of convection-diffusion-based decomposition in one spatial direction. J. Mod. Probl. Sci. Educ. **1**(1), 1786 (2015)
22. Sukhinov, A.I., Nikitina, A.V., Semenyakina, A.A., Protsenko, E.A.: Complex programs and algorithms to calculate sediment transport and multi-component suspensions on a multiprocessor computer system. J. Eng. J. Don **38**(4(38)), 52 (2015)
23. Sukhinov, A.I., Nikitina, A.V., Semenyakina, A.A., Chistyakov, A.E.: A set of models, explicit regularized schemes of high order of accuracy and programs for predictive modeling of consequences of emergency oil spill. In: Proceedings of the International Scientific Conference Parallel Computational Technologies (PCT 2016), pp. 308–319 (2016)
24. Nikitina, A.V., Semenyakina, A.A., Chistyakov, A.E.: Parallel implementation of the tasks of diffusion-convection-based schemes of high order of accuracy. J. Vestn. Comput. Inf. Technol. **7**(145), 3–8 (2016). https://doi.org/10.14489/vkit.2016.07.pp.003-008
25. Sukhinov, A.I., Nikitina, A.V., Chistyakov, A.E., Semenov, I.S., Semenyakina, A.A., Khachunts, D.S.: Mathematical modeling of eutrophication processes in shallow waters on multiprocessor computer system. In: CEUR Workshop Proceedings of 10th Annual International Scientific Conference on Parallel Computing Technologies, PCT 2016, 29 March–31 March 2016, Code 121197, vol. 1576, pp. 320–333. Russian Federation, Arkhangelsk (2016)

Author Index

Abramova, Olga A.   235
Akhatov, Iskander Sh.   235
Akimova, Elena N.   162
Amosova, Elena S.   294
Antonov, Alexander   3

Barkalov, Konstantin A.   174
Belova, Yulia V.   336
Bezrukov, Alexander   31
Biryukov, Sergey   77

Chistyakov, Alexander E.   306, 336
Chistyakov, Alexander   322

Dergunov, Denis   92
Dlinnova, Ekaterina   77
Dordopulo, Alexey   62
Doronchenko, Yuriy   62

Fedorov, Alexander   62
Frolov, Alexey   3

Gainetdinov, Azamat R.   235
Gergel, Victor P.   174
Grigorjev, Sergej K.   135
Gumerov, Nail A.   235

Il'in, Valery   186
Imomnazarov, Sherzad   266

Kireev, Sergey   266
Kireeva, Anastasiya   280
Kokarev, Mikhail   31
Kondratyuk, Nikolay   77
Konshin, Igor   3
Konyukhov, Sergey   104
Kozinov, Evgeny A.   174
Krendelev, Sergey F.   119
Kuzmina, Kseniia   251

Legalov, Alexander I.   16
Levin, Ilya   62

Marchevsky, Ilia   251
Matkovskii, Ivan V.   16
Misilov, Vladimir E.   162
Moskovsky, Alexander   104

Nikitenko, Dmitry   47
Nikitina, Alla V.   336
Novikov, Ivan Gennadievich   147

Odintsov, Igor   104

Perepechko, Yury   266
Pityuk, Yulia A.   235
Pokatovich, Gennady A.   294
Protsenko, Sophia   322

Rechkalov, Timofey   200
Ryatina, Evgeniya   251

Sabelfeld, Karl K.   280
Semenyakina, Alena A.   336
Shaykhislamov, Denis   31
Shvets, Pavel   47
Sidoryakina, Valentina V.   306
Smirnov, Grigory   77
Sokolinskaya, Irina   216
Sorokin, Konstantin   266
Stegailov, Vladimir   77, 92
Sukhinov, Alexander I.   306, 336
Sukhinov, Alexander   322
Sumbaev, Vladimir V.   336

Timofeev, Alexey   92
Toropov, Vassili V.   174
Tretyakov, Andrey I.   162
Tyutlyaeva, Ekaterina   104

Ushakova, Mariya S.   16

Varlamov, Dmitry A.   294
Vasilyev, Vladimir S.   16
Vishnevsky, Artem K.   119
Voevodin, Vadim   31, 47

Voevodin, Vladimir   3
Volokhov, Alexander V.   294
Volokhov, Vadim M.   294
Voropinov, Andrey Alexandrovich   147

Yakobovskiy, Mikhail V.   135

Zarafutdinov, Ilnur A.   235
Zhizhin, Mikhail   104
Zhumatiy, Sergey   31, 47
Zymbler, Mikhail   200
Zyubin, Alexander S.   294
Zyubina, Tatyana S.   294

Printed in the United States
By Bookmasters